ECOLOGICAL GENETICS

Ecological Genetics

E. B. FORD, F.R.S.

*Fellow of All Souls College
and Professor in the University of Oxford*

CHAPMAN AND HALL LTD
11 NEW FETTER LANE · LONDON EC4

First published 1964
by Methuen and Co Ltd
Second edition 1965
Third edition 1971
published by Chapman and Hall Ltd
© 1964, 1965, 1971, *E. B. Ford*
Printed in Great Britain by
Richard Clay (The Chaucer Press), Ltd.,
Bungay, Suffolk

SBN 412 10320 6

Distributed in the U.S.A.
by Barnes and Noble, Inc.

TO
THE MEMORY OF
SIR RONALD FISHER, F.R.S.

Contents

CONTENTS

CONTENTS

Text-figures, Maps, and Tables

1. TEXT-FIGURES

2. MAPS

3. TABLES

Preface to the First Edition

This book describes the experimental study of evolution and adaptation, carried out by means of combined field-work and laboratory genetics. That technique has been developed during the last forty years or so by my colleagues and myself, and by a small but increasing number of geneticists throughout the world. In discussing what has been achieved by these means many relevant pieces of work familiar to me have been omitted, while doubtless there are others that have escaped my attention. To those who have thus laboured without recognition here, I offer my apologies. Yet I would not include further examples were I writing again, and this for two reasons. First, my aim is not to produce a compendium in the German fashion, for I have endeavoured to develop *principles* with enough instances to illustrate them but no more. Secondly, this book is in danger of becoming too long as it is: one which is in general consulted only in libraries, not read familiarly by students. Indeed, to friends who have, rightly in a way, suggested that I ought to describe some additional piece of work or other, I have been forced to reply, 'What, then, am I to omit from the chapters already written in order to include it?'

In these circumstances, it seems appropriate to explain why the choice of examples is not a more balanced one. For considered merely as a collection of relevant 'research items', not as a means of illustrating particular concepts, it certainly includes too much zoology and too little botany. This is partly due to the fact that I am myself a zoologist and consequently I have tended to use zoological material in my investigations: moreover, other things being equal, I have favoured the inclusion of my own work, and that with which I have been associated or in close touch, because I know it best and because researches can be described most effectively by those who have carried them out or seen them in progress. Furthermore, botanists have made noteworthy advances in studying cytology and the genetics of adaptation in wild plants, but they have done relatively little in comparison with ecological geneticists in zoology to detect and analyse evolutionary change and the action of stabilizing selection in nature. Again, it may be objected that the

Lepidoptera figure too prominently as illustrative material in this book. That is partly an accident of my personal predilections and history and partly due to reasoned choice. I was led to this type of study through an interest in that group. Yet more mature judgement does indeed assign to it a high place among the forms in which the occurrence of rapid evolution can readily be detected and examined.

I am here writing for research workers, for those carrying out what it is the absurd fashion to call post-graduate studies,* and for undergraduates reading for Honours in Zoology or Botany at a University. I hope too that schoolmasters taking the more advanced forms may find something stimulating here. In consequence, I have throughout assumed a knowledge of ordinary genetics and ecology and have therefore felt myself free to use the technical terms of those subjects without explanation. These considerations have, moreover, influenced the choice of illustrations. It would, in the circumstances, be a mistake, raising the cost of this book heavily and unnecessarily, to include colour plates or to figure more than a small proportion of the species and forms discussed in detail here. Biologists, for whom this work is written, will know them, or know how to obtain access to works depicting them. I include them only when they show features not generally illustrated, together with the normal forms necessary for comparison. Consider, for instance, mimetic butterflies: these and their models can be examined in the plates available in well-known text-books. I therefore restrict the photographs of such insects to specimens demonstrating genetic variation in the effects of a major gene; of a type, that is to say, upon which selection can act to improve a mimetic resemblance within the ambit of a single switch-control, and this can be shown effectively in black-and-white.

This book was planned in 1928, and in considerable detail. At that period I believed it would be necessary for myself and others to work for a quarter of a century before it could be written. I was over-optimistic; more than thirty years were in fact needed. The subject has not been continuously pursued by me during that time. I have myself researched and written on several other topics quite unrelated to it in the interval, but it has constantly been kept in mind and pressed forward when possible. It is instructive now to examine the scheme as originally prepared. It is recognizable in that adopted here, though transformed by the developments of a generation. Thus, on the theoretical side, polymorphism had not then been defined; yet it was indicated, though by much circumlocution: while from the technical point of view, the

* Graduate studies or post-graduation studies are the proper terms.

method of estimating animal populations numerically by means of marking, release, and recapture had not been devised. These two advances have opened up great possibilities.

It may perhaps be inquired why the project of writing this book took shape in 1928. The answer is not far to seek. In 1927, R. A. Fisher, in his article on mimicry-theory, had envisaged the selective modification of the *effects* of genes in natural conditions, followed the next year by his Theory of Dominance: an extension of the same principle, not that it was generally so regarded at the time. Here was surely a basis for analysing evolution in wild populations, and as such it was recognized even at that date. In this it extended immensely the significance of the field-studies which I had begun during the 1914–18 war.

This book, then, should have been dedicated to Sir Ronald Fisher. His death during its preparation has been a great loss. I am happy to think that he approved its plan and lived long enough to read some of the chapters. He and I published our first joint paper in 1926 (*Nature*, **118**, 515), and he never failed in his help and encouragement during the forty years of our friendship. His great contributions to statistics and to the study of natural selection will be a lasting memorial to him.

My friend Mrs J. B. Clark, F.S.A., has given me constant encouragement in the writing of this book. Her interest and her skilled literary criticism have been a most material help to me.

While still an undergraduate I had begun to work with Sir Julian Huxley on genetic physiology, in the early days of that subject. I owe a great debt to him for his constant friendship and support. The advice he has so often given me in my writing and research has always proved far-seeing and constructive. He has now taken the time to read the whole of this book in detail, and his numerous comments and suggestions have been of the utmost value to me.

I wish to express my great obligation to the University of Oxford in providing my colleagues and myself with laboratories and equipment in the Department of Zoology, of which they are a part. I am deeply indebted also to the generosity of All Souls College which has in various ways lightened the task of preparing this book, especially by providing me with a room where I could write undisturbed.

For many years I worked on ecological genetics by myself, paying special attention to the experimental study of evolution. During that time I collaborated privately with a few colleagues and kept in touch with others by means of the literature. This small-scale programme was transformed by the Nuffield Foundation which, judging it worthy of support, granted the funds to establish a small group at Oxford in order to extend it. Henceforth we were able to maintain a Secretary and Technical Assistant and to meet other necessary expenses, while the Linacre Professor very kindly set aside suitable rooms for us in the Department of Zoology. It is indeed fortunate that these facilities have attracted geneticists of great originality, especially Professor P. M. Sheppard and Dr H. B. D. Kettlewell. The results obtained by the methods we have devised or developed are, in an outstanding degree, due to them; while others who have researched here for shorter periods have made valuable contributions to the subject. Yet it must be remembered that our work at Oxford could not have been carried out but for the foresight and generosity of the Nuffield Foundation. This book, then, is very largely a tribute to the policy of that organization; one which has constantly been encouraged by its Director, Mr L. Farrer-Brown. It has been fundamental to the success of these laboratories to know that his advice and balanced judgement have always been available to us.

I should like to express my great appreciation of the help given to one section of our researches by the Department of Scientific and Industrial Research. Hearing of our studies on *Maniola jurtina*, they provided additional funds to cover expenses for our field-work when analysing evolution and variation in that species. Chapters Four and Five of this book describe some of the results obtained in the course of that investigation.

A number of my friends and colleagues, in addition to Mrs J. B. Clark and Sir Julian Huxley already mentioned (p. xv), have taken up their valuable time to read the typescript of this book. Their suggestions have been in the highest degree fruitful while, in addition, they have saved me from many errors. It must be understood, however, that they are in no way responsible for the mistakes the work may contain or for the statements and expressions of opinion which I have made; with some of which, indeed, they may not agree. What I have written would have been very imperfect but for their constructive criticism. I wish, then, to express my sincere gratitude to: Dr A. J. Cain, Mr E. R. Creed, Professor C. D. Darlington, Sir Alister Hardy, Dr H. B. D. Kettlewell, The Hon. Miriam Rothschild, and Professor P. M. Sheppard.

Professor J. W. S. Pringle took the time, at a juncture particularly difficult for him, to give me valuable suggestions for the Conclusion. Professor Th. Dobzhansky did me the honour to read and comment on Chapters Six and Eleven. Mr W. H. Dowdeswell and Mr K. G. McWhirter most kindly gave me their valuable help with the account of our work on *Maniola jurtina*, as did Dr Bryan Clarke and Dr J. J.

Murray on the section dealing with the ecological genetics of snails. Dr C. A. Clarke has given me his help, for which I am most grateful, both in the account of human polymorphism and in that describing his researches, with Professor P. M. Sheppard, on the genetics of *Papilio dardanus* and other mimetic butterflies.

Dr A. J. Cain, Mr W. H. Dowdeswell, Dr H. B. D. Kettlewell, and Professor P. M. Sheppard have been so good as to provide drawings, photographs, and specimens for illustration. These are individually acknowledged in the descriptions of the plates and text-figures.

I am greatly indebted to Mrs John Davies for her patience and efficiency in taking down in shorthand, and typing, the whole of this book and in arranging and verifying the references. Miss C. Court has drawn Text-figure 11 and Map 6, while she has redrawn Text-figure 2 from photographs. Her skill has ensured their accuracy and high quality. Mr John Haywood is responsible for a number of the photographs reproduced in the Plates (see pp. xix–xx for details). I am very much obliged to him for the care which he took to produce exceptionally fine results from difficult material.

Finally, I am anxious to express my gratitude to those who have collaborated or worked in association with me in the study of ecological genetics: to Mr W. H. Dowdeswell who has done so much to further this project and with whom I have spent many months camping on uninhabited islands in the Isles of Scilly and elsewhere; to Professor P. M. Sheppard and Dr H. B. D. Kettlewell for their invaluable help and support. I admire greatly their outstanding contributions to this subject. Dr Lincoln and Mrs J. van Z. Brower and Dr L. M. Cook have taken up much time to combine with Professor P. M. Sheppard and myself in the study of *Panaxia dominula*. I feel much indebted to Mr E. R. Creed and Mr K. G. McWhirter who, in association with Mr W. H. Dowdeswell and myself, have done so much practical and theoretical work on the variation and evolution of *Maniola jurtina*.

The Genetics Laboratories,   E. B. FORD
Department of Zoology,
The University Museum,
Oxford.   28 December 1962.

Preface to the Second Edition

In the Second Edition of this book I have made a few adjustments which have become necessary as it is now several years since some of the chapters were written. I have also corrected a number of minor mistakes such as are liable to occur in a text of 150,000 words. However, none of them, fortunately, was serious.

E. B. F.

9 December 1964.

Preface to the Third Edition

The first two editions of this book appeared successively in 1964 and 1965 so that few additions or adjustments were needed in the second of them. A great deal of further work on ecological genetics has now been carried out. That is natural in a rapidly expanding science which is arousing wide interest. Furthermore, the analysis of protein variation by means of electrophoresis is too recent to have impinged upon the subject five years ago, but today it is providing highly important information upon cryptic variation. This technique has extended, but not transformed, the conclusions reached before its introduction and a discussion of the results so obtained is now included in this book. Thus it has been necessary to add many new passages and to make numerous adjustments in the present edition. I am indebted to the publishers for allowing it to appear in an extended and up-to-date form.

We see the impact of ecological genetics on a number of subjects in addition to the experimental study of evolution with which it was primarily concerned: for instance, in medicine, nature conservation and in the fight to reduce atmospheric pollution. The two latter aims have within the last year become of transcendent interest in the United States of America, a reaction likely to accelerate the efforts upon similar lines that are being made in other countries.

The concepts of ecological genetics have much to contribute to such work in which we inevitably detect the operation of powerful selection, so fundamental to this subject. I have rejected the plan of adding a chapter devoted to conservation and methods for detecting at an early stage, and for mitigating, the effects of industrialisation, as it would be too technical. Those interested in these matters will, however, find information particularly relevant to them in the sections of this book dealing with industrial melanism in the Lepidoptera, with mine-tips polluted by salts of heavy metals and with the evolutionary effects of pest control: and for these they are referred to the detailed entries in the Index.

It will be understood from the plan of this book laid down in the original Preface that no attempt is made to discuss the results of recent research bearing on the general propositions of genetics (such as dominance-modification, the concept of the cistron and many others) a knowledge of which is assumed here. For it is necessary to keep strictly to the purpose of the work; which is to review the principles of ecological genetics, with such examples and references as may be necessary to illustrate them but no more. Moreover, it is evidently quite outside the scope of what is intended, to treat the subject historically.

I am deeply indebted to those of my colleagues who have been so good as to read and give their expert opinion upon the many new passages required to bring this edition up-to-date. They are of course not responsible for such errors and omissions as may be found in them. I would especially mention the outstanding help that I have received from The Hon. Miriam Rothschild, Professor P. M. Sheppard, Professor K. G. McWhirter and Dr E. R. Creed, also, on special aspects of the subject, from Professor L. P. Brower, Professor C. A. Clarke, Dr L. M. Cook, Professor Th. Dobzhansky, Mr W. H. Dowdeswell, Dr H. B. D. Kettlewell and Mr D. R. Lees. Others have also been to considerable trouble to read through certain passages dealing with work on which they are experts. They are mentioned with my grateful thanks, at the appropriate places in this book.

As already explained (pp. xv-xvi), the Nuffield Foundation for long gave financial support to the study of ecological genetics at Oxford until that subject was sufficiently established for our laboratory expenses to be fully borne by the University. The annual cost of our field work on the butterfly *Maniola jurtina* is still defrayed by a sum generously placed at our disposal by the Nuffield Foundation. For this we are most grateful. It enables us to continue and expand our researches on the experimental analysis of evolution in that insect.

I am greatly indebted to those who have supplied photographs or material and given permission for them to be used in the plates: The Director, Department of Entomology, Natural History Museum, London, 4, 7, 9; Professor Th. Dobzhansky, 17; Mr W. H. Dowdeswell, 2(1) and (2), 6(2); Mr John Haywood, 1, 3(1) and (2), 4, 5(1) and (2),

6(1), 7–11, 13–15, 16(2); Dr H. B. D. Kettlewell, 4, 7, 13–16; the late Sir Edward Poulton, 8; Professor P. M. Sheppard, 10–12, Mrs J. B. Clark and Mr D. L. Blackwell kindly assisted me in obtaining the primrose flowers shown on Plate 3(1).

Department of Zoology,   E. B. F.
Oxford.   15 July 1970.

CHAPTER ONE

Ecological Genetics

It is a surprising fact that evolution, the fundamental concept of biology, has rarely been studied in wild populations by the fundamental techniques of science, those of observation and experiment. Consequently the process has seldom been detected and analysed in action. However, I have for many years attempted to remedy that omission by a method which has in fact proved effective: one which combines field-work and laboratory genetics. It is now being practised by an increasing number of biologists, some of them in collaboration with myself. Among researches of this type conducted independently outside Britain, I would especially mention those of Th. Dobzhansky (Chapter Eleven), which are distinguished by their originality and penetration.

The field-work needed in these investigations is of several kinds. It involves detailed observation, usually conducted on successive generations, having strict regard to the ecology of the habitats. Also it often requires long-continued estimates of the frequency of genes or of characters controlled on a polygenic or a multifactorial basis, also of population-size.

Such field-work makes use of experiment also. This is conducted in a variety of ways: for instance, by establishing artificial colonies. These are founded by a given number of individuals with a known frequency of the gene or genetically controlled character to be studied. Their fate is followed in successive broods using controls of an ecological type the validity of which will be discussed in relation to specific instances (see pp. 73, 141–4). Here it suffices to say that, employing appropriate statistical analyses, comparisons are made between two or more experimental populations maintained in habitats which differ in certain definite ways. Alternatively, the characteristics of one or more of them may be contrasted with those of the original stock from which they were derived.

It will be appreciated that reference is here made to the founding of new colonies in nature, not to the more familiar technique of establishing them in the laboratory in population-cages or by other means. This too has notable advantages, whether samples of a wild community, or those with specially adjusted gene-frequencies be set up. It has its dangers also. These derive from the highly abnormal conditions of the environment, which may not at all represent those to which the species is subject in the field. That need be no handicap provided the fact be fully recognized, but neglect of it may directly lead to false conclusions.

Other types of field-experiment involve the marking of specimens, in a way invisible to their enemies, so that movement from one habitat to another can be detected (pp. 185, 298). This method also makes it possible to compare the extent to which distinct phenotypes survive in different conditions. There are situations also in which elimination by predation could only be detected and correctly apportioned when the opportunities for doing so are artificially increased. Thus it may be necessary to augment some wild populations by large numbers of bred specimens carrying certain genes and their alleles in known proportions (pp. 185-6, 298).

An early instance of a somewhat similar kind is provided by the work of Gerould (1921) which demonstrates, I think for the first time, the selective elimination of a mutant gene by predators hunting by sight. He had been studying a recessive mutant of the butterfly *Colias philodice* in which the larvae, the pupae, and the eyes of the imago are blue instead of the normal grass-green shade; apparently because the gene concerned prevents the digestion, or the utilization of yellow-green but not of blue-green chlorophyll. No doubt a chemical block of that kind can occur in many insects pigmented with chlorophyll derivatives. Thus S. Beaufoy has obtained rare blue larvae of *Pararge aegeria* (Lepidoptera, Satyridae).

Gerould found that when segregating broods of these *C. philodice* larvae were reared in the open, the conspicuous blue specimens were killed by sparrows while the normal green ones escaped. These latter match their food-plant very closely and the result demonstrates the value of their cryptic coloration and one at least of the handicaps to which the mutant is subject.

The forms studied in the field must be bred in the laboratory in order to elucidate their genetics. But we can go further than this. By appropriate tests it is sometimes possible to analyse the evolution of their genetic control: such as selective changes in the effects of the genes, including their dominance-modification (pp. 144-5, 308), or the evolution of linkage and the creation of super-genes within the switch-mechanism responsible for a polymorphism (pp. 179-80).

2

The term 'ecological genetics', which describes the technique of combined field and laboratory work outlined here, has recently come into general use. I have, however, for many years employed it in lectures and scientific discussions, in which it has proved self-explanatory. Ecology, which denotes the inter-relation of organisms with one another and with the environment in which they live, may be regarded as scientific natural history. Consequently, ecological genetics deals with the adjustments and adaptations of wild populations to their environment. It is thus, as already indicated, essentially evolutionary in outlook. Indeed it supplies the means, and the only direct means, of investigating the actual process of evolution taking place at the present time.

Yet ecological genetics is not synonymous with evolutionary genetics or with population genetics. Evolutionary genetics includes laboratory work and the development of mathematical theories which do not employ ecology. On the other hand, some aspects of ecological genetics do not bear directly upon the process of evolution, though they may do so indirectly. These, for example, involve genetic studies of adaptations to differing environments, and of the distinctions both between races and between related species (Chapter Fifteen). They may tell us also what is meant by the statement that 'the same' form occupies distinct areas (p. 149). Nor do these latter types of inquiry form a part of population genetics: a subject which, on the other hand, is often conducted on laboratory cultures or developed purely mathematically.

It must, however, be remembered that even when we study the genetics and ecology of local races and of distinct species whose characteristics remain constant, we are in fact faced with important evolutionary phenomena. For evolution, in the technical meaning of the word (as opposed to its unfortunate etymological sense), does not necessarily result in change, any more than do the powerful selection-pressures which may control it. The stability of an organism when exposed to the inevitable variations of any natural environment in reality demonstrates decisively the working of evolution and of selection, which must be actively forcing the population to maintain an average type.

All evolutionary research other than that based upon the technique of ecological genetics, though it may be of the highest value, is relatively indirect and inferential. This applies to comparative anatomy, embryology, and physiology, to the construction of phylogenetic trees, to evidence from palaeontology and geographical distribution, and indeed to that derived from all the more conventional sources. A study of domestic animals and cultivated plants, as well as work on the effects

3

of mutagenic agencies and on selection in laboratory populations, can tell us much concerning the mechanism of evolution but not concerning its mode of operation in natural communities.

The methods of chromosome analysis developed with striking success by C. D. Darlington do, however, provide remarkably direct information on the past course, and sometimes on the future possibilities, at least of plant evolution (see, especially, Darlington 1965a). On the basis of chromosome numbers and structural changes, he has made it clear (*l.c.*, p. 164) that

> wild plants, the staple and ancient economic plants, and the ornamental plants have been undergoing evolutionary changes over three ranges of time, roughly speaking over millions, thousands and hundreds of years.

He adds,

> if the three compartments were indeed water-tight their comparisons might be highly speculative. But the chromosome approach has in fact joined the three together. It has done so for the unexpected reason that different plants in each field are seen to evolve at entirely different speeds on the chromosome scale with respect to the external or classical scale. The chromosomes change as though controlled by a three-speed gear-box. Which, indeed, is true, since number, structure and gene content vary independently with different orders of effects. In some plants, therefore, the chomosomes change visibly in direct relation with the external symptoms of evolution. In others they move much faster; and in others again they hardly move at all.

Darlington points out that this is partly due to the fact that the stability of the chromosome numbers is correlated with the length of the reproductive cycle. The basic values frequently vary within species, and usually within genera, in annuals; they vary with the tribe in perennials, while their stability often transcends families in long-lived shrubs and trees. For chromosome changes, when initiated, easily upset fertility and since an error in the sexual mechanism does not become effective until maturity, it will, as Darlington (*l.c.*, p. 103) says, take its toll 'a few weeks later in a weed but only many years later, after enormous biological capital has been locked up, in a tree'. Thus he deduces a rule when assessing evidence on evolution. On the one hand, that in those forms, usually short-lived plants, in which the chromosomes are the less stable, their numbers are of special importance below the taxonomic level of genus; on the other, that in long-lived perennials in which the chromosomes are the more stable, their numbers are of special evolutionary significance above the level of genus.

4

Furthermore, chromosome analysis may throw light upon the evolution still open to a plant and the restrictions which will affect its course in the future. For example, as Darlington (*l.c.*, p. 137) says,

> when Karpechenko found that there were three chromosome series in *Brassica*, 8, 9, and 10, the past history and future possibilities of the confused groups of cabbages and turnips, swedes and kales became clear. The reason for this is that different basic numbers and different stages in the polyploid series imply inter-sterility and therefore determine successive steps in evolution.

It is at present doubtful how far such chromosomal evidence is likely to be available in animals; for they seldom reproduce vegetatively, so as to carry them over the difficult initial stages of the spread of polyploids and polysomics. In consequence, such forms are far rarer in them than in plants, though not so rare as generally supposed. On the other hand, our knowledge of animal cytology, and especially comparative cytology, is at present extremely poor compared with that of plants. Only when we possess, for at least some parts of the Animal Kingdom, a survey fully comparable with the *Chromosome Atlas of Flowering Plants* (Darlington and Wylie, 1955), towards which that of Makino (1951) makes a valuable approach, will it be possible to assess the importance of chromosome numbers in the study of animal evolution: we can be certain, however, that it will be considerable.

In addition, the analysis of chromosome reconstructions could sometimes be decisive in elucidating the history of local races, especially when contiguous. There is no doubt that small inversions and duplications are frequent both in animals and plants. If the material were suitable, an examination of their meiosis might reveal, in the first-generation hybrids, unpaired loops absent from either parental form. Such an observation would strongly favour the suggestion that the populations in question had evolved in isolation, undergoing independent chromosome adjustments meanwhile, even if they had later extended their ranges and met (see pp. 335–7).

Comparative chromosome studies may thus throw direct light upon the past history and future trends of evolution. Ecological genetics demonstrates its operation, and helps us to analyse its working, at the present time. Thus the two subjects are complementary.

In order to examine the genetic adjustments of organisms to their habitats, and to investigate evolution in natural conditions, new concepts and new techniques were required: that they have been supplied is, to an outstanding degree, a tribute to the genius of Sir Ronald Fisher.

It is principally to him that we owe the development of modern statistical methods applicable to biological problems in the field as well as in the laboratory. He has, moreover, been responsible for other advances fundamental to ecological genetics. Among them I would especially mention his proposition that the *effects* of the genes can be modified by selection so as to produce evolution in wild populations (Fisher, 1927), and his successful explanation of dominance (1928*a*); also his proof that one member of a pair of alleles is rarely of neutral survival value compared with the other and can only spread very slowly when it is so (Fisher, 1930*b*), a demonstration which constitutes an essential step in the analysis of polymorphism (pp. 96–7). These and other deductions of great originality are developed in his *Genetical Theory of Natural Selection* (1930*a*), an outstanding achievement of evolutionary biology.

Many of the techniques used in ecological genetics will be mentioned in the course of this book, but one of them is so important that it must be referred to at the outset: that is to say, the quantitative analysis of animal populations by the marking and recapture of specimens. This has several distinct uses. First, there is the obvious one of identifying individuals previously examined and liberated. The marking can be employed not only to detect and estimate migration but also the extent to which the members of a community distribute themselves within it and scatter from its periphery. Moreover, when ascertaining the frequency of particular forms, and sometimes therefore of the genes controlling them, it is necessary to mark the specimens comprising successive samples, unless indeed they are to be killed.

Secondly, numerical estimates of animal populations can be obtained by the marking, release, and recapture technique, provided appropriate statistical devices and safeguards be used. This possibility was put forward by Lincoln (1930) on the basis of results obtained by ringing birds; also, as a purely theoretical suggestion, by Ford and Ford (1930). The two papers were in press at the same time, though that of Lincoln appeared first.

As will be explained (p. 9), the Lepidoptera provide exceptionally suitable material for the study of ecological genetics in some of its aspects, but technical difficulties were at first encountered in marking these insects in a suitable way. Each specimen must show not only on how many occasions but on what dates it has previously been caught. As a result of preliminary work on Cara, an island off the coast of Argyll on the west of Scotland, in 1937 and the year following on Tean in the Isles of Scilly (pp. 62–3), it was found that a number of dots of cellulose paint could be placed on the wings of even such a small butterfly as the Common Blue, *Polyommatus icarus* (Dowdeswell, Fisher, and Ford, 1940). These seal the scales on to the wing membrane and are permanent and waterproof. Most of the quick-drying waterproof paints now available in metal tubes with controlled supply for writing also prove very suitable for this purpose. The colour and position of each dot is used to indicate the date of capture. Methods of calculating population-size from the data so obtained are described by Fisher and Ford (1947), Leslie and Chitty (1953), and Ford (1953a). When carried out in the field upon a population of moths, the numbers of which were known, the result obtained by marking release and recapture proved to be remarkably accurate (Cook, Brower, and Croze, 1967): the estimated and true values were, respectively, 1,074 and 1,093. It is to be noticed that the technique involves obtaining an estimate of the average death-rate. This is an important piece of information in itself (pp. 328–31). It may make it possible to decide whether or not the individuals constituting different populations have a similar expectation of life, or whether distinct forms of the same species are being differentially eliminated. We are here in a very real sense detecting the operation of natural selection.

In addition to the Lepidoptera, numerical estimates by these marking methods have been successfully obtained in tsetse-flies (Jackson, 1936), birds (Blackwell and Dowdeswell, 1951), Snails (Cain and Sheppard, 1954; Cain and Currey, 1968a), and other forms. Special difficulties are, however, encountered with small mammals, since these may learn that traps are a source of food so that their returns to them may not be made at random. Moreover, the vole *Microtus agrestis*, and doubtless other species, seem attached to quite small areas of ground, ten to fifteen square yards, so that marked individuals may not mix freely with the general population (Elton, 1942, p. 206; Chitty, 1937–8).

A notable refinement of the capture, release, and recapture technique of population analysis has been developed by Kettlewell (1952). It consists in the use of radio-active tracers for marking. The imagines resulting from the treated larvae of insects can be recognized in this way, a purpose for which neither paint nor other superficial markers are applicable owing to the ecdyses and to pupation. For species with a relatively long life-history (such as Lepidoptera), Kettlewell used sulphur-35, emitting low-energy beta-rays, as this has a half-life of 87·1 days (Cook and Kettlewell, 1960). For short-lived species (such as locusts) phosphorus-32, with a half-life of 14·3 days, is appropriate (Kettlewell, 1955a).

7

A given number of larvae is removed from the population, fed on plants which have been allowed to take up the sulphur or phosphorus isotope, and returned to the colony. The population-size of the imagines is obtained by the ordinary marking, release, and recapture methods while the individuals are also scored with a Geiger counter to determine how many of them are derived from the radio-active larvae. The total larval population at the time when the treatment was carried out is then evaluated as:

$$\frac{\text{radioactive larvae released} \times \text{total imagines caught}}{\text{radioactive imagines caught}}$$

Consequently it is possible to estimate the subsequent mortality taking place in natural conditions, assuming that the treated and untreated larvae survive equally well. Information on this point can of course be obtained by a controlled experiment in the laboratory: in the instances so far studied (*Panaxia*, pp. 131–2, and Locusts) there was no evidence of any difference in death-rate between the two groups in spite of a heavy over-all pupal mortality.

When first envisaging the possibilities of ecological genetics in the mid nineteen-twenties, it was clear that if the study of evolution in wild populations were to become feasible, it would be necessary to identify situations in which the process takes place fast enough for each project to be handled within a reasonably short space of time. It has been found to do so in three situations (Ford, 1958): (1) when marked numerical fluctuations affect isolated communities; (2) when polygenic characters are studied either (*a*) in populations inhabiting ecologically distinct and isolated areas or (*b*) even in the absence of isolation if subject to very powerful selection-pressures; (3) in all types of genetic polymorphism. The definition of the latter condition and the appreciation of its special evolutionary properties have been essential steps in the development of ecological genetics. At first it seemed that a fourth situation promoting rapid evolution must become available when a species spreads into new territories. Yet the examples of that event which it was possible to study forty years ago did not involve adjustments of an obvious kind. I deduced that such changes must certainly take place in these populations but that they were largely physiological. Today we know this interpretation to be correct, and methods for examining such cryptic adaptations are now being developed (pp. 173–7).

The choice of material suitable for investigation is of the utmost importance. Certain requirements can at the outset be laid down as a guide in this matter. The species selected must in general be such that their adaptations are not physiological only but can be assessed by characters easily recognized and measured, though the techniques of electrophoresis are now overcoming this restriction. Something of their ecology should, if possible, already be known. They should occur, at least in their particular habitats, in reasonably high densities; while for much of the work, those which form colonies are desirable. They should be easily collected in many types of weather and be sufficiently robust to withstand handling, while the animal material should be susceptible of marking by one of the recognized techniques. In particular, the generations should be frequent enough in nature for a number of them to be passed through in the time which can be allotted to the investigation, while variation should be of a kind that can readily be measured and subjected to statistical analysis. The forms chosen should be easily paired and reared in the laboratory.

It is obvious enough that it will rarely be feasible to select species which combine all these qualities; they must possess as many of them, and of the types especially desirable in the proposed research, as possible. Even with the help of these criteria, the choice of outstandingly good material well adapted to the purpose in hand and allowing the work to expand in various directions seems rather an art than a science, in the sense that it usually proves difficult to rationalize. Actually, the more rewarding decisions are generally reached by a thorough, though perhaps unconscious, grasp of the essentials of the problem to be solved, estimated in the light of much knowledge and experience.

Among many other forms, the Lepidoptera have up to now been extensively used in ecological genetics. This is partly, but by no means wholly, accidental. The earlier work on the subject had in fact largely grown from observations on this group. However, their wing-patterns do provide exceptional opportunities for detecting phenotypic variation; and it will be noticed that, in general, they possess a large number of the desirable qualities listed above. It is true that they seldom produce more than two generations per year in temperate climates, though the number can sometimes be increased in the laboratory.

It is certainly wise to experiment upon those animals and plants which one has learned how to handle and whose ecology one knows, but it is essential not to become the slave of one's material. We must ever be ready to employ a different species, belonging perhaps to a wholly distinct group, if circumstances so dictate. One of the reasons why Winge attained outstanding success in the analysis of sex-determination was that, being primarily an ichthyologist, he wisely worked upon fish so long as they served his purpose. Yet he was prepared at need to turn from them to flowering plants when these were likely to supply answers to the questions before him.

There has in the past been a tendency to study evolution in material drawn solely from a particular group: birds or insects, Papilionaceae or Fungi. We are passing beyond that stage in biology now. It imposes an entirely unnecessary restriction upon the data and types of evidence available.

It is indeed surprising that the evolution of natural populations has seldom been examined by means of observation and experiment. The reasons for this are of course various. Few geneticists are also ecologists. There has been a tendency for mathematicians to sit at home and deduce how evolution ought to work rather than base their analyses upon the results of observation in the field, whether conducted by themselves or by others: however, several names, especially that of R. A. Fisher, spring to the mind as honourable exceptions to this stricture. To such unsupported theorizing I have ever been opposed. An example of its dangers is provided by the recent recognition that advantageous qualities are frequently favoured or balanced in particular environments by far greater selection-pressures than had hitherto been envisaged; a discovery which could only be made by direct observation, yet it is one which profoundly influences evolutionary concepts.

Another factor which has militated against the wide use of such field-work derives from choice of material. Apart from domestic animals and cultivated plants, the proportion of genetic research which has been conducted on *Drosophila melanogaster* and on *Neurospora* must be very large. The advantages of these forms for studying the physical basis of heredity, and the vast contributions arising from their use, need no stressing. Yet, as I have pointed out elsewhere (Ford, 1960a), it is hardly possible to speak of the ecology of *Neurospora*, while *Drosophila melanogaster* seems to provide remarkably poor ecological material: for little is known of its larval, and almost nothing of its imaginal, ecology.

Even so, certain types of laboratory work bearing upon evolution in wild populations could well be carried out on *D. melanogaster*, though this has hardly been attempted: for example, selection experiments demonstrating the evolution of linkage, though they have in fact been successful when tried (e.g. Parsons, 1958). The brilliant results of T. H. Morgan (1929) showing genetic modification of the effects of the gene for 'eyeless' (together with the selection experiments of Castle and Pincus on hooded rats, 1928), clearly pointed the way to experimental studies on dominance-modification. The tests which have so far been undertaken on that subject have proved convincing and show that results can be quickly and easily obtained, but they have not been conducted on *Drosophila*. Nor has work on a related topic: the gradual evolution in wild populations of adaptive polymorphic characters under the switch-control of single genes or super-genes; a subject to be discussed, in particular, in Chapter Thirteen and on pp. 308–9.

Studies on *Drosophila melanogaster*, though highly diverse and completly fundamental, have become traditionalized along certain lines: those laid down by a few original minds and followed by great numbers of geneticists. If ecological genetics is to be successfully developed, we need new thinking, new methods, new material.

CHAPTER TWO

Numerical Changes
in Animal Populations

As already explained (p. 8), the programme of ecological genetics conducted for forty-five years, outlined in the previous chapter, has exploited those situations which so greatly increase the speed of the evolutionary processes that they can conveniently be observed and studied.
The occurrence of marked numerical fluctuations in isolated populations
was the first of these to be used in that way.

Recurrent changes in density affect plant and animal communities to
an extent, and in a manner, that is by no means generally appreciated.
Haldane (1953), in his excellent survey of this subject, points out that
organisms rarely become extinct and never increase without limit, so
that there must in general be regulating forces which favour numerical
increase in an area when a population is small and numerical decrease
when it is large. Having mentioned that such controlling agents have
been called 'density-dependent factors' he remarks that 'negative density-dependent factors' would be a better name, to distinguish them
from those of opposite tendency promoting numerical instability. One
would have thought that stabilizing and unstabilizing density-dependent
factors would be preferable terms. These include, on the one hand,
situations in which the death-rate rises relatively with expanding numbers (owing, for instance, to greater opportunities for infection) and on
the other, the difficulties in finding a mate, or achieving fertilization,
when a species becomes rare over a large area. Haldane also draws
attention to the conclusion reached by Nicholson and Bailey (1935)
that the survival of the fittest produces numerical increase only when the
fitness concerns a negative density-dependent factor. We shall encounter
one situation when it does not do so in considering the improvement of
a mimetic resemblance (pp. 241, 272–9), and of course there are many
more.

On the other hand, when fluctuations in numbers do in fact occur, they
depend fundamentally upon the fact that the environment of plants and

animals is never constant. Consequently such numerical changes are not necessarily in the same phase from one to another isolated population of a species nor, in general, are the cycles of abundance and rarity of any regular periodicity, unless seasonal. If conditions favour an organism it may become commoner; it will do so because some aspects of selection are then relaxed. This will result also in greater variability, since genetic variation is in equilibrium between mutation and recombination on the one hand and selection on the other: it may also produce greater environmental variation, but with this we are not here concerned. At such times, therefore, forms survive which would be destroyed in more rigorous conditions. Consequently, numerical increase with diversity prepares the way for numerical decline with relative uniformity, and the reverse. For as soon as a less advantageous situation supervenes, the numbers, and the variability, will be reduced once more by the elimination of the more unsuitable types. Indeed, apart from polymorphism (Chapter Six), there is then a powerful trend towards uniformity, as selection for the optimum form will be particularly strong. Little departure from it will be possible until changing conditions once more prove favourable and allow another numerical expansion.

It is to be noticed that the actual period of increase is the one associated with marked diversity. For once a large though approximately constant population is achieved, selection again reduces the average product of each pair to one pair only; though, of course, unfavourable variants will not be so strictly eliminated as during the phase of numerical decline.

Such fluctuations are of great importance. They do not merely ensure that departures from the form that is typical in each locality are permitted as the numbers rise, checked when they are constant, and opposed even more strongly as they fall: they allow evolution to take place far more rapidly than it could otherwise do. At the period of extreme variability, opportunities of an exceptional kind may be provided for genes to stand their trial in fresh genetic combinations, with some of which they may interact in new and occasionally advantageous ways. The great range of recombinations realized in two or three generations at such a time would take an immense period to achieve in a numerically stable community. Any advantages which thus accrue will be at a premium when stricter selection is restored and checks the increase; still more so subsequently as the numbers again become smaller. It is to be noticed also that those mechanisms will be favoured which tend to bring and hold together any groups of genes which interact with one another for the good of the individual (pp. 110–16).

Numerical fluctuations in a colony of *Melitaea aurinia*

Apart from the concept last mentioned, which has been added more recently, I had reached and stated these conclusions on abundance and variability forty years ago (Ford and Ford, 1930), and have adjusted them but little on reformulating them on a number of subsequent occasions. Thus they remain much as originally conceived, partly theoretically and partly as the result of a long series of observations on a colony of the Marsh Fritillary butterfly, *Melitaea* (*Euphydryas*) *aurinia* (for colour photographs of living imagines see Ford, 1962, Plate 20). It is a depressing fact that no other equally comprehensive observations of such phenomena have yet been undertaken. Consequently these original results must in the first place be described once again to provide an illustration of the principles here discussed.

It is unfortunate that they were not quantified by the method of marking, release, and recapture; but at the period in question this had not yet been devised, and no way was known of making a numerical estimate of an insect population. That defect is, however, not too serious in this instance, since the difference between the minimum and the maximum density was so spectacular as to leave no doubt of the magnitude of the change involved. In recent years several other colonies of the Marsh Fritillary at the upper or lower limits of extreme fluctuations in numbers have been reported, and it is important to notice that these are not synchronous from one to another. One of them is at present being analysed by modern techniques in the hope that it will be possible to obtain a complete account of a cycle in its abundance and variability, with the appropriate measurements; but it is likely to be a number of years yet before this work can be completed.

The colony of the Marsh Fritillary butterfly which we originally analysed provided an instance, the first to be recorded (Ford and Ford, *l.c.*), in which the effects of fluctuation in numbers were related to variability and evolution. We studied it by direct observation for nineteen years consecutively. Records of its condition during the previous thirty-six years had been left by collectors, and preserved specimens taken during that time were available for comparison with one another and with those captured subsequently by ourselves. Thus information on this population was available for a total period of fifty-five years, and during that time an extreme numerical fluctuation took place.

The colony, which was in Cumberland, occupied the low-lying parts of eight fields sloping down to woods which the butterflies never penetrated, even for a few yards. The larvae normally feed upon the Devil's-bit Scabious, *Scabiosa pratensis*, but occasionally take to honeysuckle, *Lonicera periclymenum*, in adjacent hedges if their main food-supply runs short; and on this plant, and upon the imported *Symphoricarpus rivularis*, they can also be reared. When young, they hibernate in webs upon the Scabious (see Ford, 1962, Plate 3), and they can then withstand long submergence when the ground becomes flooded. The imagines are on the wing only in the latter part of May and the first half of June, for there is but one generation in the year. They are rather slow-flying and seldom wander even for a hundred yards outside their habitat. At a short but varying distance from the woods (10 to 200 yards or so), the fields in question slope slightly up to drier ground and here even a stray specimen was hardly ever seen. The colony was effectively separated from any other in the neighbourhood by three miles of dense woods and heather moors, and by agricultural land, extending for much greater distances.

The records began in 1881, when the insect was quite abundant. It continued to be so and indeed to increase, so that by 1894 it had become excessively common and the imagines were said to occur 'in clouds'. Subsequent to 1897 the numbers began to decline, and from 1906 to 1912 they were quite small. From 1912 to 1920 the butterflies were rare. It was during this period, in 1917, that my father, H. D. Ford, and I began our personal observations of the colony. At that time, as well as for the succeeding two seasons, we could find only two or three specimens by careful search throughout the day, where once they had flown in thousands. From 1920 to 1924, however, a great increase in numbers took place, so that by the latter year the butterflies were present in vast quantities once more, and several could often be caught with a single stroke of the net. From 1925 onwards the population was approximately stabilized at this high density and so it continued, though perhaps tending to become rather less until the observations ceased in 1935.

The amount of variability was small during the first period of abundance and throughout the phase of declining numbers and rarity. Indeed, a constant form occurred at that time (see Ford, 1962, Plate 39, Figures 5 and 6) and departures from it were quite infrequent. It should here be remarked that this estimate of low variability is reliable, being stressed

15

by contemporary entomologists and confirmed by the constancy of the specimens which they caught and preserved in collections. For butterfly collectors prize and search for varieties, the occurrence of which is therefore much exaggerated in the specimens which they accumulate. Thus even apart from any records, a low frequency of variation in a series of preserved specimens strongly indicates its genuine rarity, though a high one cannot be taken to demonstrate the reverse.

Yet when the numbers were rapidly increasing from 1920 to 1924 an extraordinary outburst of variability took place. Hardly two specimens were alike, while marked departures from the normal form of the species in colour-pattern, size, and shape, were common (see Ford, 1962, Plate 39, Figures 7–9). A considerable proportion of these were deformed in various ways (Figure 7); the amount of deformity being closely correlated with the degree of variation, so that the more extreme departures from normality were clumsy upon the wing or even unable to fly. When, after 1924, the rapid increase had ceased, these undesirable types practically disappeared and the population settled down once more to an approximately constant form which, however, was recognizably distinct from that which had prevailed during the first period of abundance (*ibid.*, Figures 10–12). Thus an opportunity for evolution had occurred, and the insect had evidently made use of it. It will be remarked how closely the observed changes in numbers and variability followed the course, and produced the effects, that I have suggested on general grounds. It should especially be noticed that we are not here dealing with the segregation of recessives due to inbreeding in a colony reduced to very few specimens, for the great variability did not in fact appear at that stage in the cycle but when the population was rapidly expanding.

There is no reason to think that the slight difference in colour and marking which distinguished the old stable form from the new is in itself of any adaptive significance. We can, however, be sure on general grounds that it is associated with various changes of a less obvious nature which are so. Indeed, even in this instance the strong probability that the genes responsible for size and colour-pattern are also associated with the control of physiological processes in the body is indicated by the close correlation between the more extreme varieties and deformities of various kinds.

Moreover, we have good reasons, combined with direct evidence, for thinking that the action of at least the major genes is always multiple. Among the hundreds of them which have by now been studied in

Drosophila melanogaster, not one seems to have been encountered which is without effect upon the viability, length of life, or fertility of the fly. That is to say, such genes have an important influence upon the working of the body, though the characters used by geneticists to identify their presence may be of the most trivial kind: a slight change in eye-colour or bristle-number which may indeed be of negligible survival-value. Such evidence accumulates whenever it is looked for, and many examples will be found scattered through this book (see for instance pp. 163–4). The truth of this assertion might, however, be doubted when applied to polygenic conditions. Yet it has recently been shown that the polygenes which control spotting in the butterfly *Maniola jurtina* have a profound selective effect (pp. 53–5).

It is still occasionally emphasized that many of the characters distinguishing races, sub-species, and species appear to be selectively neutral, with the implication that the evolution of such forms has been to a considerable extent non-adaptive. Thus Waddington (1957, p. 85), who by no means stresses that view, says:

> Many, if not most, nearly-related species differ in characters the selective value of which is not obvious. It is, of course, always possible that minor differences in, say the distribution of bristles on the body of a fly, or in the proportions of some appendage of the body, have some unsuspected importance in the animal's life and that differences in these respects have been brought about strictly by natural selection. . . Nevertheless, it is by no means obvious that one is justified in relying upon such hypothetical selective advantages to explain all cases.

The point at issue, which these remarks miss, is not whether the bristles on a fly or the proportions of one of its appendages influence survival but whether the genes controlling them do so, and the high probability of this is indicated by the facts just mentioned. It is indicated also by the behaviour of *Drosophila* cultures when maintained in the laboratory. For it is well known that apparently unimportant characters such as bristle-number do not vary at random in the stock bottles, as they would do if the genes which determine them were selectively neutral.

The winter population of many species, especially of short-lived organisms passing through several broods in the year, may be reduced far below the total which summer conditions can support. Selection is therefore relaxed when the more favourable climate of spring allows the numbers to rise again. Consequently at that time forms survive which would be eliminated in a more rigorous climate. In these circumstances the long-term changes recorded in *Melitaea aurinia* may in some degree be reproduced in miniature as a seasonal cycle. Of this, *Drosophila pseudoobscura* provides an instance (p. 226). In the Mount San Jacinto area, Southern California, one karyotype (the CH inversion) becomes commoner from April to June after which its frequency declines markedly. Laboratory experiments conducted by Birch (1955) have shown that this phase is capable of relative increase only in an expanding population: an example illustrating in a somewhat different way the propositions laid down at the beginning of this chapter.

Numerical fluctuations in *Panaxia dominula*

Marked numerical fluctuations in isolated populations do not invariably have the effects so far described. Their influence may be partly or wholly obliterated by powerful selection acting in other ways or favouring some particular phenotype.

No population of animals in the world has been so fully quantified as that of the moth *Panaxia dominula* at Cothill in Berkshire. The details of this work and its evolutionary significance in regard to polymorphism are discussed in Chapter Seven. One aspect of it is, however, relevant here. This species has a single generation in the year and the number of insects emerging each season has been estimated by the technique of marking, release, and recapture (pp. 134–7) from 1941 to 1970 inclusive. During that time the annual total has fluctuated greatly, ranging approximately from 14,000 to 18,000 in 1957 to 216 in 1962 (see Tables 1 and 9). Moreover it may differ considerably from one generation to the next: for instance, from 10,000 to 12,000 in 1954, compared with 1,500 to 2,500 the year following. Considering indeed that a single female normally lays about 200 fertile eggs, the opportunities for a spectacular increase in numbers are obvious enough.

Similar estimates, beginning in 1949, have been made by Sheppard (1953, 1956) at a second locality for *P. dominula*, Sheepstead Hurst. Though this is only a mile and a quarter from Cothill there is evidence to show that the two colonies are fully isolated from one another (p. 143). The approximate annual totals of flying insects at both places are compared for seven years in Table 1 (those for Cothill being extracted from Table 9, p. 136). It will be apparent that the fluctuations in numbers are largely independent in the two habitats, so that the population-sizes are probably controlled to a considerable extent by local changes in the environment. Of course they may in addition be affected from time to time by more widely acting conditions, generally, no doubt, climatic.

For instance, both the Sheepstead Hurst colony and one at Faringdon, thirteen miles away, became almost extinct in 1959 when that at Cothill, though remaining quite large, was also considerably reduced compared with the year before.

The great increase at Sheepstead Hurst in 1953 is very striking. Sheppard (1956) says that this

> may well be due to the fact that the vegetation was badly burnt by fire in March of that year. Many larvae undoubtedly perished, but in some areas plenty remained, and these were heavily predated by cuckoos. However, the cover which protects the small mammals was destroyed, so that *P. dominula* larvae were probably much less heavily predated than usual. If this is a true interpretation of the facts, small mammals are an important factor controlling population-size in this moth.

It will be noticed that two years afterwards, the Sheepstead Hurst colony had dropped back approximately to its previous numbers, suggesting that these represent something like the average which the area can normally support.

TABLE I

Year	Cothill	Sheepstead Hurst
1949	1,400–2,000	5,000–6,000
1950	3,500–4,700	3,600–4,800
1951	1,500–3,000	4,000–6,000
1952	5,000–7,000	6,000–8,000
1953	5,000–11,000	20,000–26,000
1954	10,000–12,000	8,500–13,500
1955	1,500–2,500	3,000–5,000

Population-sizes of the moth *Panaxia dominula*
in two isolated colonies one mile apart.

A variety *medionigra*, affecting the colour-pattern and controlled on a unifactorial basis, though absent from Sheepstead Hurst, is not uncommon in the Cothill locality. Its gene-frequencies were calculated throughout the whole period during which the numbers of that colony were studied and indeed for two years previously. It will be seen from Table 9 that they also fluctuate, but in a way that bears no simple direct relationship to the changes in population-size. The latter doubtless influence the spread of the variety, but they must be one only of a number of components, equally or more important in this matter, which do so. Indeed, though these alterations in the frequency of the *medionigra* gene occur in a state of nature, it has been proved that they are not due merely to chance but to the working of selection (p. 137). In this instance then, observed numerical fluctuations are not in any simple way correlated with a particular observed genetic change.

Numerical fluctuations in a *Maniola jurtina* population

Any character, whether controlled on a unifactorial or polygenic basis, may survive a cycle of great reduction in numbers and return to its former proportions when the population increases again, provided that very powerful selective forces are operating to maintain it at its previous level. To produce such a result, their magnitude would indeed be far beyond that envisaged twenty or twenty-five years ago, but quite in keeping with that now being recognized as usual in wild populations. An instance of this kind is provided by our own work on the Meadow Brown butterfly, *Maniola jurtina*, in the Isles of Scilly, to be described more fully in Chapter Four (and see Plate 1).

This insect is generally common in grassy areas in Britain but, though quite powerful on the wing, even a hundred yards of unsuitable terrain proves almost a complete barrier to it. It occurs over most of Tresco, Isles of Scilly, an island of 962 acres, except where grass does not grow. There is, however, a small enclave, the Farm Area, covering only two hundred by seventy yards, which supports a distinct colony of the butterfly. Here it is isolated from the main population by the sea to the west, a lake to the east, farm buildings, estate workshops, and a village to the north, and to the south and south-east by country which it cannot colonize, consisting of woods, bracken, bramble, and rocky ground with gorse and short turf. The locality thus enclosed is ecologically exceptional. A number of huts had been built there during the 1914–18 war. Their concrete foundations remain and around them has grown up vegetation consisting of garden escapes and wild plants of a type most unusual for these islands: *Carpobrotus* (*Mesembryanthemum*) which is established at the edge of the shore and in sandy patches elsewhere, *Hyoscyamus*, *Verbascum*, *Glaucium*, *Arctium*, *Achillea*, *Digitalis*, thistles, long grass of various kinds, brambles, bracken, and nettles.

As a measure of variability and a means of detecting evolutionary changes in *Maniola jurtina*, we use a character which can be studied quantitatively; that is to say, the number of spots on the underside of the hind-wing. These can vary from 0 to 5. Though perhaps of no biological significance in themselves, the polygenes which control them are of importance for the survival of the individual (pp. 53–6). The frequencies of these spots differ in the two sexes and we shall in the present context be concerned only with those of the females. Their distribution in that sex proved to be of a most exceptional kind in the Tresco Farm Area, having a single clear-cut mode at 2. Nevertheless, it was effectively constant in type in 1954, when these observations began, 1955 and 1956 (for which period its homogeneity is measured by $\chi^2_{(6)} =$ 5·428, with P approximately $= 0·5$). In the Main Area of the island, on the other hand, the number of individuals with 0, 1, and 2 spots was nearly equal, a condition which had also remained constant over the three years in question ($\chi^2_{(6)} = 4·064$, with P $> 0·5$) and indeed since the insect was first studied there in 1950. The two distributions differ significantly from one another ($\chi^2_{(3)} = 13·813$, for which P $< 0·01$).

From 1954 to 1956 about a hundred to a hundred and fifty insects must have been flying daily in the Farm Area during the main period of emergence, which lasts from mid-July to early September: an estimate based upon detailed observations and the marking, release, and recapture of specimens. The day-to-day population of the large Main Area of the island is unknown but must have amounted to some thousands. In 1957, however, there was a general reduction in vegetation following a quite exceptional period of long-continued dry weather in the late winter, spring, and early summer. The Farm Area was particularly affected since its soil is extremely sandy; moreover it offers no choice of habitat. Here, therefore, *Maniola jurtina* was nearly exterminated; so much so that at the height of the season, and in perfect conditions, some hours of collecting by two people produced only four specimens. This probably represented something like the total flying population at the time, since the area is so small that it can be examined in detail. The following year (1958) the vegetation there had somewhat recovered and the number of butterflies had risen again. Two collectors obtained thirty-six insects in two days and established that about fifty to one hundred might be flying there daily.

The twenty females, which the 1958 captures in the Farm Area included, made it clear that the colony had retained its quite exceptional characteristics, with a large mode at 2 spots. For the comparison with the 1954–6 period before the numerical collapse of 1957, $\chi^2_{(1)} = 0·605$, with P $> 0·3$ (using Yates' correction). That is to say, the selection-pressure was so powerful that the colony was forced back to this particular spot-distribution even when repopulated from a very low density. It will be noticed that an alternative explanation, that of migration into the locality, is excluded; for the spotting of the butterflies in the adjoining Main Area of the island was throughout of a quite different type (Dowdeswell, Ford, and McWhirter, 1960).

Thus it is clear that fluctuations in numbers may be utilized by a population greatly to increase the speed with which it can adjust itself to changes in its environment. It is clear also that the effects of such fluctuations may be obliterated when selection for a particular stabilized type is sufficiently powerful and constant and the environment remains unchanged.

Some effects of climate and food on animal numbers

In addition to those fluctuations in numbers so far discussed which the interaction of organisms with their environment automatically generates, it is obvious that periodic changes in climate may profoundly influence the distribution, density, and evolution of plant and animal populations. The Ice Ages provided very long-term examples of them, but these are the province of pleistocene geology and palaeontology, not of ecological genetics. Nor are we here concerned with the major climatic trends of prehistoric and historic times, as exemplified by the general deterioration in weather which appears to have affected England in the ninth century A.D. compared with the early part of the Roman occupation of the country.

Similar but less extreme climatic trends also occur, and these may occupy much shorter periods, some scores of years. They are mainly, but not exclusively, to be detected by their ecological repercussions, for they may greatly affect the numbers and the range of many species. It is deplorable that such events have hardly ever been studied by the techniques of ecological genetics, so that no attempt has been made to detect and analyse their influence upon the frequency and adjustment of genetic characters in the populations concerned. For this reason, only a brief reference will be made here to such changes in the density and distribution of animals. They are mentioned indeed merely to direct attention to certain situations which are likely to promote rapid evolution.

The range of many species of Lepidoptera in Great Britain contracted markedly during the latter half of the nineteenth century (Ford, 1962). Thus the Comma butterfly, *Polygonia c-album* (Nymphalidae), formerly widespread in southern England, became restricted to Herefordshire and the adjoining counties. So did the Wood White, *Leptidea sinapis* (Pieridae), with a few additional localities in Northamptonshire,

Devonshire, and elsewhere; while the White Admiral, *Limenitis camilla* (Nymphalidae), was no longer to be seen outside Hampshire, especially the New Forest, and a limited number of other favoured places. Indeed a long list of such instances could be compiled not only among the Lepidoptera but in other groups also.

Most of the species involved have made a spectacular recovery this century, from about 1915 onwards. *Polygonia c-album* has spread throughout southern England and is now common in many regions where it was unknown fifty years ago. *Leptidea sinapis* has always been a local insect and so it remains today, but it has returned to numerous woodlands where it had been reported in the early part of the last century; though more probably it had survived in some of them at so low a density as to escape notice. But some distant localities, such as that in Westmorland where it certainly became extinct, were too isolated for the species to recolonize and in these it has never re-appeared. *Limenitis camilla* has become generally common once more in the forests of southern England. A further instance of the same kind is provided by *Pararge aegeria* (Satyridae). Before 1943 it had not been recorded within ten miles of Oxford this century, except for a single specimen in 1922, though the region is particularly well documented. It is now widespread and numerous there. A general reversal of this whole trend has recently set in.

The spread of Lepidoptera during the last fifty years or so has taken place at a time when the English countryside has been changing adversely for them. Woods have been cut down wholesale, especially during the two wars, and urbanization and industrialism have made vast inroads throughout the period. Some general climatic change producing greater warmth is certainly suggested by these facts, and is indeed supported by the present world-wide retreat of glaciers.

Fluctuations in numbers have been studied extensively in some groups of Vertebrates, and their possible evolutionary effects have aroused considerable interest. Unfortunately this has stimulated theoretical speculation more often than experiments and observation in the field.

Short-term numerical changes are less marked in birds than in insects. Thus detailed investigations of the Heron, *Ardea cinerea*, in England have shown that this species becomes scarcer after each hard winter, the occupied nests being reduced to two-thirds or half, followed by a rapid return to the former density. Lack (1954), summarizing the available evidence, is of opinion that even a tenfold fluctuation is rare in birds. This is to be expected in species of large size and small egg-production, in which a clutch of a dozen is uncommon, compared with the relatively minute Lepidoptera in which it is usual for a female to lay two hundred to three hundred eggs; for slight differences in ecological adjustments can evidently have more profound effects when the normal elimination-rate is so much greater. Some moths indeed remain fairly constant in numbers for many years but are subject to occasional huge increases. Their population may then become ten thousand times larger than at its lowest level: a situation encountered by Schwerdtfeger (1941, 1950) in his studies of insect pests in the German coniferous forests.

One other aspect of fluctuations in numbers must at least be mentioned: the fact that in a few species they occur in regular cycles. These are repeated rather accurately but are not synchronous throughout an extensive range.

This subject has received much attention, which it certainly deserves, though it is not of general application. The existence of such regular periodicities has been proved only in three groups of northern rodents, together with their predators, and in certain gallinaceous birds, secondarily affected.

Thus a cycle in numbers of approximately four years has been demonstrated in Lemmings on the Arctic tundra and, consequently, in the Arctic Fox and Snowy Owl which depend upon them for food (Elton, 1942). A similar cycle, also of about four years, occurs in the Voles (*Microtus*) of the open forests and grasslands between the tundra and the main Arctic Conifer belt (Elton, *ibid.*). These are hunted by the Red Fox, Marten, Great Grey Shrike, and a few other species which, accordingly, are also involved. The Varying Hare has, on the other hand, a 'ten-year' (nine to eleven year) cycle in abundance and rarity, reflected also in its predators; especially the Lynx, American Goshawk, and Horned Owl (Chitty, 1948). It has been claimed that another North American rodent, the Muskrat, has a 'ten-year' cycle also (Elton and Nicholson, 1942). Both these cycles, that of four years and that of ten, in fact fall slightly short of the stated periods, which are used merely as an approximation to specify them, but the averages differ both with species and locality.

The effect on the gallinaceous birds, the Ruffed Grouse and others, is probably a secondary one. Some of the predators turn to them as alternative prey when the rodents upon which they normally depend become scarce.

Since the regular periodicity of the predators and game birds in this association seems to depend upon that of the rodents, it is in these latter that its cause must be sought. It was formerly, but it seems incorrectly, ascribed to a cyclical climatic change. A correlation between the variations in rodent density and the phases of the sun-spot cycle, though long advocated, was disproved by MacLulich (1937). There appears to be no valid ground for the suggestion that rodent numbers are associated with an ozone cycle in the atmosphere (Lack, 1954, p. 212), and no other widely acting climatic periodicity of the required length has been found. Even if it were, it would be difficult to correlate it with the rodent fluctuations, which are not in phase with one another in different parts of the animals' range. Moreover, they are not due to predation in any simple sense. For it is generally agreed that though the predators increase greatly as their prey becomes more abundant, they are nevertheless far too rare to influence the immense hordes of rodents present at the maxima so as to produce the sudden and spectacular decline in numbers which then ensues.

Lack (l.c.) reasonably suggests that these regular cycles in abundance and rarity are dependent upon food-supply, perhaps secondarily associated with disease when the animals are weakened by starvation. It looks as if the rodents increase almost in a geometric ratio, the last stage of which will, of course, involve an immense excess compared with the penultimate one and produce a population which the country cannot support, while predation merely slows down the potential multiplication to a certain extent. Such a situation must inevitably generate a rhythm in abundance and rarity provided it is little affected by other influences: which, however, in normal circumstances it will be. But as Lack remarks, the fact that

> such regularity is confined to a few rodents (and dependent species) suggests that in these animals the ecological conditions, including the food-chains, are unusually simple and that climatic factors (which are normally irregular in their incidence) have a much less disturbing effect than elsewhere.

As expected, the cycles are longer when the reproductive rate is slower; that is to say, ten years in the Varying Hare compared with four years in the lemming or vole.

One aspect of population-size has been discussed by Wynne-Edwards (1962), who considers that many animal communities are kept by 'social interactions' at a lower density than their food-supply would impose, so preventing the over-utilization of resources. He holds that gregarious behaviour enables certain species to adjust their reproduction in this sense. Because ordinary selection would work against such a tendency, Wynne-Edwards invokes 'group selection' to achieve it: the populations which fail to check their increase before eating out their food-supply will largely perish through starvation, being replaced by those which had evolved this type of limitation.

That concept, which is marginally related to ecological genetics, has evoked considerable discussion, a great deal of which has centred upon the reproductive activities of seals and birds; for these were employed considerably by Wynne-Edwards to support his views. A few words may therefore be said about them.

McLaren (1967) who has analysed group selection in seals, pointed out that these animals may return year after year to crowded colonies where their chances of successful reproduction are reduced or annihilated: for there the females will have little choice of breeding sites and will, owing to competition, experience shortage of food, causing poor early nutrition for the young which, moreover, are frequently crushed to death in such places. This state of affairs occurs where other apparently suitable sites in the neighbourhood are unoccupied. As McLaren remarks, we may question whether this situation is achieved against the short term needs of the population, or whether it is the result of ordinary selection. He shows that the published data indicate that the most social seals have the highest potential rates of increase: a fact difficult to reconcile with group selection but easily explained on conventional lines. Females of the most social species have but small choice of breeding sites and therefore exercise little 'intelligence', while their young mature rapidly. On the other hand, the solitary breeders benefit from accumulated knowledge of the environment, while their young, being subject to less immediate competition, mature more slowly. Indeed species with the lowest potential rate of increase breed biennially at most because they give their young more extended care. The fatal crushing and other deleterious aspects of crowded breeding grounds can be explained as excess of sexual selection and there is no need to postulate group-selection in seals; nor does there seem to be in birds.

Lack (1965) finds that normal natural selection will provide an explanation for variation in clutch-size, age at first breeding, and territorial behaviour, for which group-selection was postulated. He shows that the clutch-size of birds that feed their young in the nest has been adjusted to correspond with the brood-size from which most young survive: a view tested experimentally. For example, the same quantity of food collected per day can be used to raise a smaller brood more quickly or a larger one more slowly. When the risk of predators destroying the nestlings is great, selection for rapid growth may be advantageous to the species even though it generates smaller broods.

The spread of species in new territories

The numerical changes so far discussed have taken place within the existing range of the species. But opportunities for a great increase in numbers occasionally arise in another way: when some plant or animal spreads in a new territory or a new country owing to artificial introduction or to some exceptional act of migration. That situation is one which must be associated with evolution. For an expanding population, liable as it is to promote evolutionary changes, then finds itself in an entirely new environment; one which must require far greater readjustments to ensure successful survival than are needed to meet the ecological variations normally taking place within the original area of distribution.

In many instances, however, species have colonized new lands without undergoing recognizable modifications. One thinks of the importation of the rabbit into Australia and New Zealand, the spread in both these regions and in North America of the Small White butterfly, *Pieris rapae*, also of the American Monarch butterfly, *Danaus plexippus*, across the Pacific Islands to Australia, while the Canadian Pond Weed, *Anacharis canadensis*, speedily became a serious pest on its introduction into Europe. In these and a great many more instances there was an extremely rapid increase in numbers, necessarily allowing the survival of forms which would be eliminated in a stable population and providing exceptional opportunities for testing the effects of genes in new combinations. The possibility for evolutionary adjustment consequent upon stricter selection arises when the rapid expansion ceases, as perhaps it has done in the introduced Canadian Pond Weed, or when the population is actually reduced again. This has already occurred in the rabbit in New Zealand; also in the Gypsy moth, *Lymantria dispar*, in the United States where that Palaearctic species had for a time proved highly destructive in some areas.

Extensions of range that are less striking, yet surely valuable sources of material for investigation, are provided by certain moths, heretofore of great rarity in England and apparently only casual immigrants there. These have in recent years been found breeding in some of the woods in southern Kent. *Catocala fraxini*, the largest of the European Plusiidae, and *Phoberia lunaris* are examples of them. Competent entomologists seem satisfied that they have not been present at a low density in the past and overlooked. Moreover, *Catocala fraxini* has extended its range northwards in Europe and become common in Denmark. Thus its presence in Britain during the same period is likely to be due to successful recent colonization from the Continent.

A spectacular extension of range, accompanied doubtless by a great increase in the population, has been reported in several species of birds in spite of their normally more constant population-sizes. That of the Fulmar Petrel, *Fulmarus glacialis*, is the best documented, owing to the work of James Fisher (1952). In the British Isles, it bred only on St Kilda, one hundred miles west of Scotland, until the last quarter of the nineteenth century. Subsequently, its nesting-sites have spread not only round Scotland and the Scottish Isles but round Ireland also; and in England and Wales they are now to be found in suitable places along the whole west coast and down much of the east coast. Fisher seems to have demonstrated successfully that this extraordinary occurrence is due to a change in the amount and distribution of the Fulmar's food. These birds, fundamentally plankton-feeders, will also eat offal, which they formerly obtained from the carcasses of whales caught in the northern fisheries. When that trade declined, ice-carrying trawlers came into service. Vast quantities of refuse derived from cleaning the fish are thrown overboard from these vessels and provide abundant meals for the Fulmar in more southerly waters.

Some of these instances, as well as others of a similar kind, occurred to the mind when in the mid nineteen-twenties I was considering situations in which rapid evolution could be detected and studied (p. 8). Even then, it seemed inevitable that they must involve evolutionary adjustments, but that in some forms, such as those just mentioned, these were cryptic, being largely physiological and chemical.

Opportunities for examining that aspect of evolution have now become available owing to the development of electrophoresis. Mutation is due to the loss, gain, or substitution of one or several nucleotides in a deoxyribonucleic acid chain. This results in a change in one or more amino-acids in the protein which the mutant controls. A proportion of such changes, estimated by Lewontin and Hubby (1966), though on doubtful evidence, as about a half, affect the electric charge of these substances, causing them to move faster or slower in an electric field. Their migration can be made to take place in a starch-gel, and is then detectable by means of the different enzyme-staining methods now available. By this technique all three genotypes at a locus may be identified (Hubby and Lewontin, *l.c.*). Analyses along these lines have already been carried out in a number of animals (Mammals, including Man, and Insects) and in plants (pp. 104–6, 173–7, 341–2).

We have here a means of disclosing cryptic variation affecting the proteins, so that many purely physiological adaptations can now be identified. It should therefore become possible to detect a considerable proportion of adjustments, the existence of which was long ago predicted in general terms in species colonizing new territories such as those already mentioned.

The butterfly *Thymelicus lineola* (Hesperiidae) provides an interesting addition to these. It is one which has undergone a great extension of range without alteration in its visible characteristics. This is a European species, with a distribution extending to south-east England in a manner suggesting a Holocene colonist. It was introduced into America at London, Ontario, apparently shortly before 1910. It has spread thence into the neighbouring parts of Canada and the U.S.A. and has appeared also in New Brunswick, New England, and lately in New Jersey. In some of these places it has become so common as to constitute a minor agricultural pest. A recently discovered colony in British Columbia may perhaps represent a separate introduction from Europe. Johnson and Burns (1966) have already been able to detect great protein (esterase) variation in a butterfly, *Colias eurytheme* (Pieridae), when examined by means of starch-gel electrophoresis; for instance, 77 per cent of the individuals caught at Austin, Texas, were heterozygous for one esterase zone (EST E), while 83 per cent were so among those from San Antonio. Their results suggest that *T. lineola* may well prove a rewarding insect to study by this means (Burns, 1966). Probably the technique will indicate that it has undergone characteristic adjustments, indicated by esterase variation, in the different regions into which it has penetrated and that, in spite of its unchanged appearance, the race now established in America has come to differ, no doubt adaptively, from the parent one in Europe. Such a method of analysis should be widely applied to other instances of the same kind.

Elton (1958) has produced an excellent survey of the introduction of animals and plants into new territories, though he does not discuss their subsequent evolution. However, his work furnishes a useful basis from which such inquiries might start.

There have, indeed, been many occasions on which the spread of a species into a new region has been accompanied by visible changes and adjustments. It is not my purpose to compile a list of these, since the aim of this book is to develop *principles*, with instances sufficient to illustrate them but not more (p. xi). I shall therefore indicate a few examples only.

A famous historical one may first be mentioned, since it might still be studied usefully by the techniques of ecological genetics. The original home of the Black Rat, *Rattus rattus*, seems to have been India. Thence it has spread, through human agency, to other parts of the world where it has given rise to distinct forms, partly by means of interspecific crosses (Matthews, 1952). It reached Britain in the Middle Ages imported from Asia Minor, whence it also made its way along South Europe and North Africa, becoming modified into the Tree Rat, *R.r. frugivorus* (white-bellied brown), and the Alexandrine Rat, *R.r. alexandrinus* (reddish-brown with dusky belly). The colouring of the *frugivorus* form is uni-factorial and dominant to that of *alexandrinus*, while both are recessive to the blackish, dusky-bellied type, *R. rattus rattus*, which seems to have been the one found in Britain before it was replaced by *Rattus norvegicus* in the eighteenth century. The Black Rat now only maintains itself here at ports where it can constantly be recruited from abroad, but this prevents the formation of a population properly adapted to local conditions (pp. 331–2). The three races differ in other characters in addition to colouring, and these are presumably polygenic.

Matthews (*l.c.*, pp. 170–1) states that not only are there brown forms of the Black Rat but there are also black forms of the Brown, *R. norvegicus*. He points out that, unlike the situation in *R. rattus*, they occur sporadically throughout the range of *R. norvegicus* and says: 'They evidently have no advantage over the others, either from their coloration or from any linked physiological characters, for if they had, natural selection would be expected to have produced one of three alternatives from them. If they had a general advantage they would have replaced the normal type everywhere; or if a regional advantage, they would have produced a cline; or finally if only a local advantage and some of them are geographically isolated, they would have produced a subspecies.' Matthews seems to have forgotten balanced polymorphism (Chapter Six).

The most successful of bird colonists is the House Sparrow, *Passer domesticus*. It was introduced into North America from England and Germany in 1852 and has now become one of the commonest birds in that continent, adjusting itself by the formation of local races to the great diversity of environments which it encounters there. Its adaptations have been studied intensively by Johnston and Selander (1964), their work leading to important conclusions. They find that marked differences affecting size, proportion, and especially colour have evolved as the Sparrows have colonized regions distinct in their ecology. The birds have become particularly dark along the north Pacific coast, especially around Vancouver where they only arrived in 1900, and pale in the arid south-west from southern California to Texas; yet the species had not penetrated into the Death Valley before 1904. It is recognizably dark and distinct round Mexico City, where it was not established until 1933. The situation in Hawaii is especially remarkable. The Sparrow was only introduced there in 1870 or 1871, brought from New Zealand where English specimens had been released during the period 1866–8. Yet the Hawaiian form is so striking that I doubted if it were *Passer domesticus* when I first saw it in Honolulu. Its colour is a rufous brown, with an absence of dark markings; also, the bill proves to be longer than elsewhere.

Johnston and Selander point out that the characteristics of the various local races are recognizable not merely by differences in the means of the respective populations but that they may affect all the individuals comprising each of them. The speed at which these forms have evolved demands special notice: only 111 generations in Honolulu (to the date when Johnston and Selander wrote), while in many instances the process has occurred wholly within the present century: in thirty years round Mexico City. Yet the minimum time generally envisaged for the evolution of geographical races in birds is five thousand years (Moreau, 1930). The Sparrow indicates something different from that. The situations it has encountered certainly promote rapid evolution; but its reactions to them exemplify the effects of the powerful selection demonstrated so widely in natural populations by the techniques of ecological genetics.

The Jersey Tiger moth, *Euplagia quadripunctaria* (Hypsidae), extended its range to England last century. The increase in numbers which followed that event seems to have enabled it to adjust rapidly and visibly to a novel environment at the edge of its range. Some degree of mystery surrounds that occurrence and the events leading to it. The species is related to the Scarlet Tiger, *Panaxia dominula* (Chapter 7 and Plates 4 and 5). It is, however, rather larger, about 64 mm across the expanded fore-wings. These are white, heavily marked with brownish-black bars and the thorax is blackish with white lines. The abdomen is normally scarlet, as are the hind-wings, and these bear three black spots. The species has one generation in the year and the larvae feed upon low-growing plants; *Taraxacum, Lamium album, Senecio vulgaris* and others.

The imago will often fly by day, when it is conspicuous, but it seems to be more active at night and may then be attracted by a light. The hind-wings are concealed when the insect is at rest and the broken pattern of the fore-wings then renders it, on the whole, difficult to detect. The colour of the hind-wings is warning and the insect is distasteful (Walker, 1966).

The extraordinary abundance of *Euplagia quadripunctaria* in a valley in Rhodes has become famous. Here, and in a few other and smaller colonies in the Dodecanese but not elsewhere, the imagines undergo a genuine aestivation. This occurring in part only of the range is almost unrecorded in the imagines of the Lepidoptera, but V. Scali (paper in press) has recently detected it in females of *Maniola jurtina* in central and southern Italy.

E. quadripunctaria is quite common in southern Europe and is found in various places northwards to the Channel Islands. Half a dozen specimens or so had been captured widely scattered in England and a few even in Wales, from 1859 onwards. However, it was in 1871 that the species was first reported in the south Devon area which has become its British home. That is to say, the strip of coast from about Bolt Head to the estuary of the River Exe, a distance of approximately thirty-five miles. It extends inland for a maximum of ten miles, for it reaches the verge of Dartmoor in the west. In this area it still occurs and it is especially frequent near Newton Abbot, Dawlish, and Starcross. Though negative evidence is normally difficult to obtain, there does seem to be good reason for thinking that this insect only established itself here in the second half of the nineteenth century. In 1869* Edward Newman published a large treatise on the British 'Macro Moths', including all the species that had been recorded here, even those which had been taken only once or twice. Yet he omits the Jersey Tiger. It is incredible that he should have done so were it established in its present locality in south Devon. For not only is his book extremely thorough, but this insect is large, striking, and is seen flying and resting exposed by day; while the area it now inhabits was well known to entomologists in Newman's time, as he himself points out in another connection (see his page 444).

* The year of publication is not entered on the title page of this book; but the Preface, which speaks of the work as complete, is dated 1 June 1869.

In Devon, the hind-wings may be of the normal scarlet, clear yellow or of an intermediate shade. The latter varies from pinkish-yellow to terracotta but does not seem to overlap the other two classes. It is generally reported that the three colour-forms are about equally common in England, but though the statement is not based on adequate random samples, it excludes the non-scarlet phases from being merely rare segregants.

The genetics of this colour-variation have not been ascertained. The frequencies, in so far as they are known, do not suggest that the intermediate group represents the heterozygotes: it may be that a single pair of alleles decides whether the hindwings are yellow or potentially scarlet, while another pair converts the latter to the pinkish shades which may vary environmentally or by modifiers; but this is uncertain. The Hon. Miriam Rothschild has bred the species easily, though Walker (*l.c.*) remarks that the larvae are notoriously difficult to rear in Britain. A few small families from known females fertilized in nature by unknown males have been preserved in collections. It would be well worth while to study this situation in detail, as we have here a species which has developed a colour polymorphism on extending its range to south-west England. For though the yellow and pinkish forms seem to be found occasionally in northern France they appear to be rare except in the population which has established itself in south Devon.

CHAPTER THREE

Genetic Drift
and the Founder Principle

It has been explained in the last chapter that organisms automatically generate their own cycles of abundance and rarity and that the changes in selection-pressure with which these are associated may greatly increase the speed of evolution. There are, however, additional ways in which the reduction of a population to a very low level, whether permanently or recurrently, may bring about certain types of evolutionary change. That is to say, through the operation of Random Genetic Drift, or the Founder Principle, and by other means related to one or both of these concepts. This, therefore, is a convenient place in which to outline them. Once that is done, it will be possible to assess their importance (which, in respect of Drift I believe to have been greatly exaggerated) as the relevant evidence, derived as it is from numerous distinct sources, becomes available here and there in subsequent chapters of this book.

Random Genetic Drift

Random Genetic Drift is the name given to the effect of the sampling errors which can influence the genetic constitution of a population. It has been widely held that these have a profound influence on evolution, a view for which Sewall Wright (e.g. 1931, 1948) is primarily responsible. He does not of course regard them as alternative to selection nor, in spite of his assertions to the contrary (e.g. Wright, 1948), do those who consider that he much over-stresses their evolutionary significance. The question in dispute relates to the relative importance of chance fluctuations as compared with the differing survival values of alleles and gene-combinations in varying circumstances.

The principle of random genetic drift is easily understood. It involves the properties of any group made up of two (or more) distinct types in definite proportions; and these are directly applicable to the distribution of a pair of alleles, A and a in an animal or plant population. Consider the situation in which the two kinds of genes are present in equality in a given generation (n), so that A has a gene-frequency of 50 per cent. Suppose that it has no advantage or disadvantage compared with a either in the gametes or any stage in the life history of the organism, then A and a should be represented in a $1:1$ ratio in the next (n + 1) generation also. This, however, would be approximate only. It will not always be exact owing to chance differences in fertilization and survival. For instance, should the numbers in an isolated colony be fairly large, say 500 individuals, a might very easily come to occupy two loci too many and A two loci too few in generation n + 1; that is to say, assuming the organism to be a diploid, 498 A and 502 a. Thus in this instance a now exceeds expectation by 0·4 per cent. If on the other hand, the population were small, say 10 individuals, A and a might by mere sampling errors depart from equality to the same absolute extent and respectively occupy two loci too many and two loci too few; that is, with a in excess, 8 A and 12 a. Here a exceeds expectation by 20 per cent.

When the small population breeds again, it produces only two-thirds as many gametes carrying A as a. Sometimes their random assortment will tend towards restoring the original equality between the types of alleles; with equal frequency it will have the opposite effect and increase the disparity between them. Suppose it does so and by the same absolute amount as before; the frequencies in the next (n + 2) generation will be 6 $A:14$ a. Two fifths of the A genes have now been lost and a is more than twice as numerous as A. It is easy to see that such a process could sometimes continue until one type had disappeared. It could then be restored only by mutation or the immigration of specimens carrying it. Yet in the large population a chance assortment involving a further loss of 2 A and gain of 2 a genes in the n + 2 generation will result only in a shift of gene-frequency from original equality to 496 $A: 504$ a. Thus the effects of chance are more important in small than in large communities.

It is of course one of the elementary principles of genetics that if two autosomal alleles of equal survival value are carried in the gametes of a diploid in a ratio of $p:q$; then the genotypes of the next and subsequent generations will be stabilized in the proportion $p^2:2pq:q^2$. That statement, though true for very large populations, is untrue for small ones owing to the operation of chance; that is to say, of random genetic drift.

Consider, in addition, the action of selection. Suppose that A has a 1 per cent advantage over a in the sense that, on the average, for every 99 a genes surviving in a breeding population, 101 A genes will have survived in it. We may relate this to the example of genetic drift already outlined. Random survival changed the proportion of alleles in the large population, of 500, from equality in generation n to 498 A:502 a in generation n + 1. But selection of the degree just specified will convert this to 503 A:497 a. That is to say, such selection is here more effective than chance. A similar absolute excess of a genes had in the small population shifted the allelic ratio from equality to 8 A:12 a in the same (n + 1) generation. The 1 per cent selective advantage possessed by A will modify this to 8·1 A:11·9 a. Thus, in the large population selection has overpowered drift, but in the small one, drift has overpowered selection, having increased a in opposition to it. Clearly then, a small selective advantage will tend to obliterate the effects of chance when acting upon large numbers, but may be unavailing in a population of limited size. Consequently, the power of selection to shift gene-ratios depends not only upon its intensity but upon the size of the population on which it acts. Evidently, however, a strong selection-pressure can carry all before it, even within a very restricted group. Thus in the instance cited, a selective advantage exceeding 20 per cent would outweigh the opposing drift even in the population of ten individuals.

Up to about 1940, it was assumed that the advantage possessed by genes spreading in natural populations rarely exceeded 1 per cent. As it is now known that it quite commonly exceeds 25 per cent and is frequently far more (Chapter Four onwards), it will be realized how restricted is the field in which random genetic drift is of importance.

Wright (1948) has studied the evolutionary aspects of this subject mathematically. He concluded that in a very small isolated population in which, as we have seen, drift may be relatively effective compared with selection, nearly all loci will become homozygous, even though the result will usually be disadvantageous. Thus the organism could not be fully adapted to its environment. Moreover, because of its low genetic variability it could not adjust itself to changing conditions even when these exert powerful selection-pressures, and must become extinct. On the other hand, the evolution of a very large interbreeding community would be almost entirely selective; a situation which, under constant conditions, would again reduce genetic variability by the elimination of inappropriate alleles and gene-combinations. In these circumstances, the species would be well adapted to its environment, but further improvement or readjustment would be very slow since this would largely be dependent upon fresh mutation.

Wright, therefore, maintained that the most favourable situation for rapid evolution is that provided by an abundant species sub-divided into many small groups each predominantly inbreeding but subject to a limited amount of out-crossing owing to occasional migration between them. In these circumstances, both drift and selection would operate and, since their action would rarely have the same results, they would prevent one another from producing a high degree of homozygosity, and the gene-frequencies would fluctuate. Such partly isolated communities would therefore possess a high degree of genetic variability: new advantageous combinations would arise and evolution could take place much more rapidly than in the other two situations.

It will be worth while to make a few brief general comments on these views at once, while reserving their more detailed discussion for later chapters. In the first place, it certainly is probable that isolated populations which permanently remain very small, perhaps comprising some dozens of individuals only, do in fact tend to become extinct. For fluctuations in numbers may in such circumstances reduce them to vanishing point, or to so low a level that they cannot recover. Yet, their evolution must be more adaptive than Wright contemplated since they must be considerably swayed by the forces of selection, remembering that these are far greater than had until recently been realized.

On the other hand, it is completely unrealistic to consider the behaviour of a large interbreeding community under constant conditions since that situation, far from being realized, is not even approached in nature. The great effects of environmental instability are, moreover, reinforced by the continued repetition of qualitative and quantitative ecological changes which may be only a remote consequence of them: for an organism has not the same adaptive requirements when abundant as when rare, or when the plant and animal forms which impinge upon it are so. It is partly for this reason that Wright's well-known landscape simile, depicting adaptive peaks, and valleys of less rigorous adaptation, is dangerously misleading. He remarks (Wright, 1932, p. 362) that

> the environment, living and non-living, of any species is actually in continual change. In terms of our diagrams [that is, of contour maps] this means that certain of the high places are gradually becoming depressed and certain of the low places are gradually becoming higher.

But this in fact invalidates the whole of this far-fetched comparison, in which, if it is to be made at all, a terrestrial landscape should be replaced by the surface of the sea, where the position of a wave-crest ('a peak') at one phase becomes that of a trough ('valley') at another. The concept then, of relatively easy passage from hill to hill during reduced selection, and of isolation on a summit during increased selection, becomes meaningless. Moreover, when Wright considers that selection tends to produce genetic uniformity in a large population, he entirely ignores the polymorphism concept (Chapter Six); this ensures that many major genes, other than the deleterious type constantly eliminated and replaced solely by mutation (or immigration from environments where they are favoured), attain balanced frequencies with their alleles and do not pass to homozygosity. Furthermore, Mather and Harrison (1949a and b) have produced good evidence to show that polygenes are associated into balanced linkage groups, which makes it questionable whether a gene having very small effects can be appreciably influenced by random genetic drift.

It will indeed be shown that the subdivision of a large population into relatively small groups promotes its rapid evolution, but not for the reason which Wright holds nor in the conditions which he envisages. As my colleagues and I (Dowdeswell and Ford, 1953) have long maintained and demonstrated, the efficacy of such subdivision consists in this: that when populations occupy a series of restricted habitats they can adapt themselves independently to the local environment in each of them, while when spread over a larger area they can be adjusted only to the average of the diverse conditions which obtain there. This, however, requires that the adaptations should not constantly be broken down by a trickle of immigrants from one small colony to another. It is for this reason that the respective populations may differ from each other on a series of small islands but resemble one another on a series of large ones (Chapter Four). Moreover, it is precisely why, as will later be shown, immigrants can prevent the evolution of local races (pp. 331-2); also why selection favours isolating mechanisms, restricting the wandering habits of animals (p. 339) and promoting homostyly or polyploidy in plants (pp. 340-1) at the edge of their range, where survival may only be attained by accurate adjustments to the exacting local conditions.

In general, it is necessary to point out that when considering isolated small populations a distinction should be made between those which are concentrated, perhaps at a high density, in a restricted area, on the one hand, and those sparsely scattered over an extensive trace of country on the other. The two situations lead to very different results even when the total number of individuals is the same. The one engenders a high degree of adaptation to strictly defined local conditions, the other tends to produce an average type suited to a diversified habitat: the dangers arising at the lowest density reached during fluctuations in numbers are also of different kinds. They are minimized by the varied ecological conditions presented by a large area but augmented there by the risk that very rare individuals may fail to find a mate. These effects are reversed in a restricted locality.

Wright attaches much importance to random genetic drift in small but strictly inbreeding populations and to its co-operation with selection in those which are partially isolated; consequently he is anxious to demonstrate that evolution, at least up to the sub-species and species level, is largely non-adaptive. Here his remarks appear contradictory. For on the one hand he states (1940) that 'it is probable that most of the mutations which are important in evolution have much smaller selection coefficients than it is practicable to demonstrate in the laboratory' while, on the other, he attempts to support his views by means of observed instances. These, as indeed his own remarks suggest, are quite worthless as negative evidence, since we certainly could not detect a 1 per cent (or indeed a larger) selective difference between the variations concerned. In fact, his choice of examples has proved singularly unfortunate. In his 1940 article he picks upon the chromosome inversions of *Drosophila pseudoobscura*, remarking that Sturtevant and Dobzhansky (1936) have shown that these 'behave as approximately neutral Mendelian units'. In subsequent years Dobzhansky himself (1947a) has demonstrated that such apparent neutrality was wholly deceptive, and that these inversions are of powerful selective importance (see also Dobzhansky and Levene, 1948). Wright also cites the human blood-groups as 'neutral as far as known'. Actually their adaptive importance could be, and in fact was, deduced in advance of the now well-known discovery that they are associated with specific diseases and considerable pre-natal differences in survival in a way which is at the present time opening up new fields of medical inquiry (Chapter Six). In his 1932 paper he also quotes the evidence, now recognized as worthless (pp. 200–3), of non-adaptive evolution in the land snails of Hawaii (Gulick, 1905) and of Tahiti (Crampton, 1925), as well as other instances that modern ecological work would similarly reject. We have in fact strong reasons for regarding the spread of non-adaptive characters as a rare and temporary event (Chapter Six). It has always seemed to the critics of Sewall

39

Wright's theories that random genetic drift can be effective in comparison with selection in very small populations of up to a hundred or two, but these are generally faced with extinction. The recent discoveries suggesting that selection-pressures in nature are far more powerful than had been supposed tend, of course, to reduce that upper limit still further. They indicate that results such as those obtained with *Maniola jurtina* in the Farm Area on Tresco (pp. 20–2), demonstrating the overmastering effect of selection in populations reduced to a total of perhaps a hundred or less, are probably quite normal.

On this subject, Wright (1948, p. 289) makes a statement which might almost suggest mere mathematical speculation unrelated to ecological facts. It is best to quote it in full:

> It has sometimes been held that non-adaptive differentiation due to sampling can only occur in populations of a few hundred but given enough time and sufficiently neutral segregating genes, this can occur in completely isolated populations whenever $4 Nv$ is less than 1, v being the mutation rate from the allele most likely to be fixed under the conditions.* Thus with mutation rates of 10^{-6}, there can be random fixation up to a population size of 250,000.

Students of ecological genetics will realize that organisms are not given 'enough time' nor 'sufficiently neutral genes' (see pp. 96–7) to allow the effects of chance to overpower selection in colonies of up to a quarter of a million, or anything like it.

The theory of random genetic drift put forward by Wright, and so far discussed, relates to those populations which remain small for long periods of time. It takes no account of the constantly recurring fluctuations in numbers which are such an important factor of plant and animal communities nor, in consequence, of the special opportunities for selection which these provide. They do so as a result of *changes* in density, irrespective of absolute size which, even at the minimum, may be large. It is another issue that such fluctuations may from time to time reduce a considerable population sufficiently for the effects of chance temporarily to overwhelm the working of selection.

The latter possibility is considered by Waddington (1957, p. 86) under the name of 'intermittent drift'. He carefully points out that my colleagues and I (e.g. Dowdeswell and Ford, 1953) have long distinguished this from what he usefully describes as 'permanent drift', in which the numbers remain for long periods small enough for random changes in gene-frequency to influence evolution.

* $N =$ the effective population size.

It is perfectly clear that opportunities for the working of intermittent drift must occur when populations become very small, yet even when this takes place cyclically its effect must be negligible owing to subsequent events. For, as we have seen, the restoration of stricter selection, after maximum density has been attained, eliminates all but the more advantageous genes and gene-combinations. It is worth noticing that the colony of *Melitaea aurinia* described on pp. 14–17 was remarkably invariable and remained constant from generation to generation when its numbers were at their minimum, perhaps comprising no more than a hundred imagines during a season, though it was then that such drift should be operating. On the contrary, the great variability which did in fact supervene occurred during the period of rapid numerical increase and continued while this lasted, even though the numbers had, towards the end of that time, become very great. The fact that after the maximum density had been attained the population was very invariable, and remained constant in appearance year after year, indicated that the characters concerned were controlled by selection. Thus it should be stressed that such an outburst of variability due to relaxed selection-pressure, as indicated by a rapidly expanding population irrespective of its absolute size, is a consequence not of intermittent drift but of the equilibrium by which, as already indicated on pages 11–12, genetic variability is normally maintained. Waddington (*l.c.*), however, used intermittent drift in an attempt to explain the fact that female spotting in the butterfly *Maniola jurtina* has remained similar on three large islands in the Scilly archipelago but differs characteristically from one to another on a number of small ones. The numerous and cogent reasons for rejecting that suggestion will be discussed in Chapter Four.

It has already been pointed out that the effect of selection depends not only upon its intensity, but upon the size of the population in which it acts. There is another situation, as well as that provided by a very small group, in which chance survival is of importance in evolution: that is to say, when an advantageous genetic variation is still a great rarity. It may be due to a gene which has become useful owing to a change in the environment since it previously arose by mutation or, similarly, to a reconstruction within a chromosome, or to the occurrence of a polysomic or a polyploid. It must at first be largely a matter of luck whether or not any such variant survives, though of course it will be biased in favour of doing so to an extent determined by selection. Consider, for instance, the situation in which a mutant gene M^B possesses a 1 per cent advantage over its allele M^A in some environment and, having just arisen, exists only as a heterozygote and in but a single individual in a numerically stable population. The probability of its transmission to the succeeding generation will then be $1 \cdot 01 : 0 \cdot 99$, for and against. The chances of its survival, at first so largely fortuitous, increase should it happen to establish itself in more than one of the progeny. They increase also if the advantage of the gene becomes greater. This may happen as a result of a change in the environment or if it is incorporated into a gene-complex which enhances its beneficial effects, a result which selection will achieve should the gene survive long enough.

The influence of numerical fluctuations upon a very rare advantageous variation is not simple. As the numbers increase, some aspects at least of selection become less rigorous. If the characters produced by the gene (or other genetic change) are thus affected, its advantage wanes, but it has a greater opportunity of chance survival owing to the fact that each pair is on the average now leaving more than two offspring. The reverse of this is true as the population becomes smaller, but selective advantages then tend to be increased. The opportunities presented to any gene of experiencing new gene-complexes during changes in population-size have been discussed on pp. 13–14. Naturally a highly advantageous mutant stands a fair prospect even of initial preservation; the whole situation being comparable with the spread of a relatively common allele in a very small population.

The founder principle

Those fluctuations in numbers which reduce an isolated population to a very low level can influence its evolution in ways not so far discussed. For in such circumstances, the arrival of rare immigrants will have a far greater effect than at any other time, since they will not then be submerged by a mass of residents but may form a significant proportion of the whole.

In normal conditions, such stragglers from another community can have little effect since they will be surrounded by individuals with a gene-complex balanced to fit their own local environment. On the other hand, they will themselves bring with them one more or less adapted to the locality whence they came. Any offspring they may now produce therefore tend to be eliminated, as the hybrid gene-complex is very unlikely to be properly adjusted. The situation is indeed very similar to the well-known one in which two large populations extend their range until they meet when, for similar reasons, they often remain distinct instead of merging into one another (pp. 334–7). However, a rare arrival can now and then make genetic contributions to the large population in which it may become immersed. It does so by introgression (pp. 332–3); that is to say, the hybrids it produces may comprise individuals approaching one or the other parental type and a range of intermediates between them. Those most like the adjusted population in which they find themselves may breed successfully enough to introduce into it a few new useful genes and super-genes.

When, however, an immigrant reaches a population which chances to be at a minimum, and very small, density, the circumstances are different. Too great a proportion of the next generation may then be ill-adapted hybrids, leading to the extinction of the already depleted group. If, however, at this time it may be nearing the stage when the numbers are about to expand rapidly again, the ensuing favourable conditions may allow the hybrids to survive. They would then contribute materially to the greatly increased variability upon which stricter selection can act at the next phase in the cycle.

This situation is evidently an approach to the one in which wanderers found a new colony in an isolated locality where the species has not previously existed, or where it has become extinct. That event is the subject of a penetrating analysis by Mayr (1954). Many species have a more or less continuous range beyond which they occur here and there as small isolated populations. Mayr points out that each of the latter is generally characterized by clearly marked qualitative features, often producing distinct races, in sharp contrast with the situation in the main area of distribution. Here there is far less phenotypic diversity, and the forms inhabiting different parts of it are principally to be recognized quantitatively and are connected by clines.

Mayr interprets these facts, the truth of which is evident, in the following way. The genetic components of any population in an interbreeding and widespread species are continuously recruited by genes from elsewhere, both far and near. In such circumstances, those of them will be at an advantage which interact in a fairly uniform and favourable way in a wide range of genetic settings. On the other hand, the selective values of the genes may change markedly in a newly founded and peripherally isolated population (though, if effective, the isolation need not, of course, be peripheral). They do so according to Mayr partly because the number of possible genetic interactions of each allele is much diminished, while one of the effects of the great reduction in size (a few immigrants instead of a large population) will be a strong increase in homozygosity. Consequently, those genes which react favourably as homozygotes will be at a premium rather than those which fit harmoniously into a wide range of gene-complexes. Moreover, as Mayr (*l.c.*, p. 166) remarks: 'the better integrated such a gene-complex is, the smaller the chance that a novel mutation will lead to improvement.' Thus in the conditions experienced by the founders, there are opportunities for the use of mutants that have been eliminated heretofore. This situation, then, is one which produces a rapid change in genefrequency at many loci: to a 'genetic revolution', as it were. Mayr therefore maintains that the major evolutionary novelties have often arisen in such circumstances and have only spread later when they have been perfected. He indeed uses this conclusion to explain the well-known fact that entirely new groups seem often to have appeared rather suddenly and without intermediates in the fossil record; for remains from the restricted populations, where they might have had special opportunities to evolve, can seldom be found. It must be stressed, however, that such an interpretation of palaeontology in the light of microevolution is speculative only.

There can be little doubt that Mayr is correct in maintaining that the evolution of a newly-formed colony may proceed along very different lines from that of the large population whence came the founders of it; also that the type of genetic revolution which he envisages may be one of the causes of this. Certainly, however, as he indeed indicates, it is not the only one; a matter to be considered further in Chapter Fifteen in relation to species on the edge of their range. Moreover, as explained on pp. 72–3, when populations are cut off in restricted areas they can be adjusted by selection to the special features of their environment in a way that is not possible when they occupy a large and usually, therefore, a more diversified territory. This consideration applies to all isolated populations of small extent whether derived from founders or produced by the subdivision of a more extensive habitat due to the creation of barriers in it. These may arise through cultivation, urbanization, or through a variety of natural causes.

It is perhaps true that the areas inhabited by isolated interbreeding communities can be classified into three types in regard to their effect upon evolution. These are related to their ecology and the powers of dispersal possessed by the animals and plants occupying them.

(1) Those in which the separate localities differ widely from one another, though each has approximately similar characteristics throughout, and is small enough for the individual members of the population

to reach all parts of it. This may mean a hundred square yards or less for snails, or five hundred square miles or more for certain birds. These conditions favour the formation of distinct local races.

(2) Regions including a number of different ecological habitats sufficiently closely intermingled over a considerable area for every individual to experience most of them. Here there is a tendency for each population to develop features suited to the average of its environment. Since averages are likely to approximate to one another, several such communities, occupying more diverse and on the whole larger areas than those of type 1, will tend to resemble each other rather closely.

(3) Regions broken up into a number of dissimilar environments changing in a graded fashion each of which is substantially larger than the area normally exploited by the individuals of a species. Such conditions promote the formation of clines within each isolated population.

It is, of course, in type 1 that the founder principle may operate, either because occasional wanderers reach such habitats for the first time or because they recolonize them after the species has become extinct there, perhaps as a result of an extreme fluctuation in numbers. It will be realized, however, that the special characteristics of the populations in such places are not necessarily the outcome of reduction to a few individuals or of growth from a few immigrants. That conclusion can be deduced from general principles: proofs of it in particular instances will be considered in Chapter Five.

CHAPTER FOUR

Polygenic Characters Evolving in Isolation

Polygenic characters can be rapidly adjusted to the special ecology of isolated habitats. This is due to the large amount of continuous variation under genetic control to which they are subject. It is due also to the small effect produced by each one of the genic substitutions involved, which can do little to disturb the balanced gene-complex. Indeed such adjustments are achieved principally by a shift of the mean within the ordinary range of variability to which the population is already adapted. In addition, this may result in a small amount of new variation, by extending in one direction the extreme frequency-distribution of the character concerned. But the whole process takes place gradually, allowing opportunities for the progressive repairing of any harmful effects which may accrue. It contrasts strikingly with the segregation of the major genes, which generally gives rise to discontinuous variation owing to the relatively large effects for which each is responsible. These tend to be disadvantageous because, like all mutants on their first appearance, they are unrelated to the needs of the organism and therefore disturb the balance of the gene-complex which, owing to their magnitude, they do to a marked extent. In consequence, some genetic adjustment is needed when the organism makes use of a new major gene.

It seems, then, that a study of polygenic characters is likely to reveal instances of rapid evolution provided the effects of selection, or other agencies responsible for it, can be accumulated: that is to say, it is often desirable to carry out our observations on isolated populations, free from contamination by individuals differently adapted.

Undoubtedly Mayr (1947) is in general right in regarding such 'allopatric' evolution (involving geographical isolation, that is to say) as almost essential for the formation of races and, in consequence, for the speciation to which that process occasionally leads. In doing so he is considering bisexually reproducing animals only; for it is clear, and fully

46

allowed by him, that quite different evolutionary mechanisms are possible with other mating-types as, for instance, in the Protozoa. They are widely possible also in plants owing to the immensely more frequent establishment in them of both polyploids and polysomics, the success of which, especially in their earlier stages, depends so considerably upon vegetative reproduction: such changes in the chromosome outfit can obviously produce new races and species sympatrically (within a single community). Apart from them, my colleagues and I had ourselves held geographical isolation, past or present, to be necessary for race-formation and for speciation; except perhaps in certain groups such as fish. We have, however, now been confronted with an instance to the contrary in our own work (pp. 80–95): one in which selection is powerful enough to break a single fully-interbreeding population into two races occupying contiguous geographical areas with no boundary between them, so providing an instance of 'sympatric' evolution. This possibility is not envisaged by Mayr in his discussion (*l.c.*). That is not surprising since, at the time he wrote, it had not been realized how powerful are the selection-pressures which normally maintain and adjust the characters of wild populations.

Selection for a balanced gene-complex in *Panaxia dominula*

Before dealing with sympatric evolution (Chapter Five), which must certainly be exceptional, it is necessary to consider the normal allopatric type in relation to polygenic variation. Sheppard (1954) has drawn attention to a remarkable occurrence which demonstrates with exceptional clarity certain aspects of that process. H. B. D. Kettlewell bred the Scarlet Tiger moth, *Panaxia dominula* (pp. 128–46 and see Plate 4, Numbers 1 and 2), for ten generations, of which there is one in a year, selecting for a reduction in the dark markings of the imago. This led to a progressive increase in the size of the white and yellowish-white spots until in later generations they coalesced and became much more extensive, while the black on the hind pair was greatly reduced. The colour-pattern so produced was very variable, but even the less extreme individuals were so abnormal that it might have been difficult by mere inspection to identify the species to which they belonged (Plate 4, Numbers 3–7). In the course of this work Kettlewell introduced four aberrations into his stocks: *albomarginata*, *crocea*, *brunescens*, and *juncta*. Their genetics are not fully understood: probably, but not certainly, they are all polygenic, as without doubt was the variation to which they

47

were each subject, and this apart from any environmental component which may also have contributed to it. However, these particular forms were not responsible for some of the more marked alterations in pattern which took place (the reduction in spotting on the hind-wings, for instance) which were certainly controlled by polygenes.

As he was leaving England for a considerable time, Kettlewell decided to preserve this selected strain by founding an artificial population with it. For that purpose he planted the larval food in the grounds of the Tring Museum, Hertfordshire. Here the stocks would not be contaminated, since this is a colony-forming species absent from the neighbourhood. Indeed it rarely wanders even a few hundred yards from its habitat, so that the Cothill and Sheepstead Hurst colonies in Berkshire are completely isolated from one another, though but one and a quarter miles apart.

The broods of larvae were released at Tring by Kettlewell in 1948. They produced very few insects the following year. By 1951 the imagines were fairly common and it was noticed by the Museum Staff that they differed profoundly from the type which had been liberated and resembled quite closely the ordinary English form. The next year thousands were flying, and in 1953 the larvae were so abundant that most of them starved to death and the colony became nearly extinct. Some of these starving larvae were, however, collected and reared, and they produced moths which had almost but not quite returned to the normal appearance of *P. dominula* (Plate 4, Numbers 8 and 9).

Very powerful *natural* selection must have been operating in the Tring colony to reverse in so few generations the new pattern built up in the laboratory stocks. Yet, as pointed out by Sheppard, this could not have acted directly upon the wing pigmentation. The moth is highly distasteful (Lane, 1956) and obtains protection in that way; indeed, though cryptic when at rest, it exhibits a warning coloration if disturbed (p. 130), and this had been enhanced by the changes which had taken place in Kettlewell's experimental broods. Moreover, Sheppard (1952a) has shown that the chief stimuli operating in the courtship of this species are olfactory rather than visual. Thus the selective forces acting upon the insects after their release doubtless favoured a balanced gene-complex which had been destroyed by the artificial selection to which the broods had previously been subjected. Indeed Mather and Harrison (1949a and b) have demonstrated a similar reversion in laboratory stocks of *Drosophila melanogaster* after artificial selection had been discontinued. However, Kettlewell's results seem to represent the only

48

instance of the kind which has been studied in the field. They are therefore of outstanding importance. It is noteworthy that selective evolution on the scale encountered in the Tring colony of *Panaxia dominula* would completely override the effects of random genetic drift even in a very small population.

Selection for spotting in *Maniola jurtina*

By far the most extensive work upon the evolution of polygenic characters in isolation has, however, been carried out upon the Satyrine butterfly *Maniola jurtina*, known in Britain as the Meadow Brown. It has a single generation in the year, but the growth-rate of the larvae is very variable so that the period of emergence is protracted and, though the imagines have an average life of approximately thirteen days only, they are to be seen on the wing for about two months; from the third week in June onwards in normal seasons in the south of England. This is one of the commonest and most widespread of butterflies in Britain, occurring up to the north coast of Scotland, throughout Ireland and on most of the Islands, including the Outer Hebrides and the Orkneys, but not the Shetlands. In general, its distribution covers the western half of the Palaearctic region except for the extreme north. The larval food plant consists of grasses: a number of species are eaten, especially *Poa annua*. The eggs hatch in a fortnight to a month and feeding, which takes place principally at night, begins in the late summer. It continues sporadically throughout the winter, the larvae becoming quiescent in cold weather. After this partial hibernation they start to feed regularly in early May, when they are about 12 mm long and are generally in their fourth instar. They attain a length of about 25 mm when their growth is complete, and the pupal period lasts from three weeks to a month.

This is essentially a butterfly of meadows, hay fields, and roadsides. It is to be found also on rough ground and in the grassy rides and clearings of woods. It tolerates equally well the conditions of acid marshes and those found on the strongly alkaline soils of chalk downs, so long as the grass there is not too short. It is quite a powerful insect on the wing and the males in particular can be difficult to catch. Among its advantages as a species for experimental study should be mentioned the fact that it will fly, or can easily be put up, in dull weather; even in the rain, provided the temperature be not too low, especially if there has been a little sunshine since the previous night.

We required a character that could be studied quantitatively in *Maniola jurtina* in order to compare populations and to detect changes in them. For that purpose we chose the number of spots on the underside of the hind-wings. These are placed sub-marginally, in a curving row (Plate 1). They may be absent, or present in any number up to 5 (indeed, three specimens with 6 spots have been found among the many thousands examined).

McWhirter (1969) has established the heritability of spotting in *Maniola jurtina* as 0·14 in the males and 0·63 ± 0·14 in the females. For this purpose he obtained the butterflies in the Isles of Scilly and bred their progeny at an approximate temperature of 15 °C. He points out, however, that in the special circumstances of this work, the usual method of linear regression employed for his calculations is subject to certain objections. This arises from the small size of the broods, due to the difficulty of rearing *M. jurtina*, the great difference in heritability between the two sexes and the fact that the spot-frequency of the parents had resulted from their genotype interacting with the natural environment, while the progeny were bred in the laboratory. These defects are overcome by calculating the variance-ratio of spot-numbers within and between the broods. This provides another, and in this instance more accurate, assessment of heritability. On the material available this does not differ significantly in the males, but amounts to 0·83 in the females.

Professor McWhirter informs me (personal communication) that the heritability of spotting in these butterflies increases as the temperature rises. At about 22 °C it amounts to 0·78 ± 0·16 in the females and reaches 0·47 ± 0·2 in the males. It is of course frequently found that the genetic control of variability differs in the two sexes (partial sex-controlled inheritance) up to the well-known condition of total sex-control, when the effect of an autosomal gene is obliterated in one sex and always manifest in the other, as in the whitish females of *Colias* (pp. 161–4).

The two sexes, as well as the various races, are characterized by different spot-distributions. The males are nearly always unimodal at 2 spots (Plate 1, Number 3). There are a very few populations with the mode at 1 or at 3, so that other types are possible. The females, on the other hand, are much more distinct from one region to another: they can be unimodal at 0 or at 2, bimodal in varying degrees at these values, or they may take quite other forms of spotting such as unimodal at 3 or equal frequencies at 0, 1, and 2. These are the primary or 'first-order' differences. Within them, certain 'second-order' arrangements are recognizable. Thus in the male spot-distributions, even though generally unimodal at 2, there is in some districts an excess of the lower or of the higher spotting (0 and 1 or else 3 and more spots); and these conditions respectively produce lower and higher spot-averages per individual (obtained by adding together the number of spots in a given sample of one sex, and dividing by the number of individuals used). Similarly in the unimodal females the excess at 0 spots may be very marked, amounting to 60 per cent or more of the total, or it may be less extreme, including fewer than 60 per cent of the specimens (the 'Old English' and

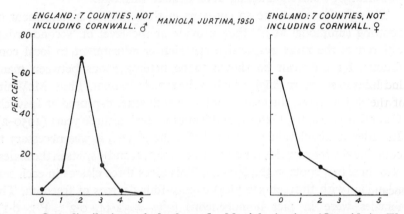

FIG 1. *Spot-distributions of the butterfly* Maniola jurtina (*Satyridae*). *The 'Southern English' type, characteristic of the populations from the North Sea to the Devon–Cornwall border. Males on the left; females on the right.* Heredity (*1952*), *6, 104.*

the 'New English' stabilizations respectively, p. 56). Corresponding differences, often very consistent, are to be recognized within the other female spotting-types. There is, moreover, a close correlation between the relative amounts of spotting in the two sexes. The lowest averages are to be found in Ireland (males about 1·3, females about 0·2), through the Isle of Man, Scotland, and southern England, to the high values of Cornwall and Scilly (males about 2·2 and females about 1·1): evidently a cline in spot-frequencies, though one that is subject to local irregularities (Dowdeswell and Ford, 1953).

Throughout a large area in southern England, from west Devon to the North Sea and northwards to the Midlands, it has been found that the 'first-order' spot-distributions are of a uniform type: with the males, as usual, unimodal at 2 spots and the females unimodal at 0, Figure 1 (Dowdeswell and Ford, 1952; Beaufoy et al., 1970). Even in the latter sex, which is capable of assuming such diverse values elsewhere, this form of spotting is maintained in spite of great environmental differences. It is the same in the Atlantic climate of west Devon as in the semi-Continental one of east Suffolk, with average rainfalls of 40 to 50 and 20 to 25 inches per annum respectively. It is the same, too, in the exposed conditions of a dry chalk down, with its alkaline soil, as in woods, acid meadow-land, and marshes. The ecological distinctions between regions and habitats such as these are immense. They indicate that the spotting of the *Maniola jurtina* populations found in them must be stabilized powerfully by selection acting upon genetic variation.

Though the males are so constant in their general arrangement of spotting (unimodal at 2), they provide at the level of 'second-order' differences the more responsive criterion of adjustment to local conditions, for they can be shown to be heterogeneous between many localities in southern England when the females cannot. Thus, McWhirter (1957) analysed thirteen samples of both sexes captured at Ipswich, Canterbury, Rugby, Oxford, and Taunton during three years (1950–2). In order to avoid values too small for the χ^2 test, it was necessary to combine the frequencies for 0 and 1 and for 3, 4, and 5 spots in the males; also for 2 to 5 spots in the females. This gives three classes in each sex, allowing (with thirteen samples) twenty-four degrees of freedom. The females were in fact homogeneous ($\chi^2_{(24)} = 32.44$, $0.2 > P > 0.1$); the males, on the other hand, were decisively heterogeneous ($\chi^2_{(24)} = 52.68$, $P < 0.001$); due, so it seemed, to a tendency for their spot-values to decrease towards the west. If so, the samples of that sex taken respectively in the eastern and western parts of the area should each be relatively homogeneous, and so it proved. The localities in the east are represented by Ipswich and Canterbury, for which $\chi^2_{(6)} = 6.91$, $0.5 > P > 0.3$; while for those more to the west (Rugby, Oxford, and Taunton) $\chi^2_{(4)} = 4.63$, $0.5 > P > 0.3$. When they are subdivided, the male spot-values are therefore no longer heterogeneous. We have here evidence of a subsidiary east to west cline within southern England. It is, however, detectable only in the males. Its consistency in succeeding years, and from one to another different type of habitat within the areas both of higher and of lower spotting, indicates that it also is maintained on a genetic rather than on an environmental basis.

It seems unlikely that the spots themselves are of selective value, but such stability as that mentioned indicates that the genes controlling them are powerfully influenced by selection. Dowdeswell (1961, 1962) has now provided experimental support for this conclusion.

At Middleton East, a locality near Winchester, is an area of seventeen acres of chalk down which supports a *Maniola jurtina* population of five to ten thousand imagines, estimated by the technique of marking, release, and recapture. Here the butterflies are isolated by dense encircling forest which acts as a complete barrier to them. Within the habitat thus enclosed Dowdeswell sampled them in five years (1955–9). The males, as always in southern England, are unimodal at 2 spots, but they have shown a slight steady increase in spotting common perhaps to all British populations which, on the large samples available, has made them heterogeneous when judged over the whole period (P < 0·01). The females are, as expected, unimodal at 0 spots and they have remained consistently stable, except for a slight increase in spotting in 1958. Their homogeneity over the five years falls within the range 0·2 > P > 0·1.

From 1957 to 1959 sampling was carried out at different times during the season. This demonstrated complete homogeneity through the whole period of flight for both sexes in 1957 and 1958 and for the males in 1959 also. The only exception was a large female collection taken late (16 August) in the latter year. This differed to an extent bordering on significance ($\chi^2_{(2)} = 5·66$; 0·1 > P > 0·05) when compared with all the other females grouped together. However, the effect was not great enough to produce heterogeneity within the female samples for 1959 (P = 0·2). Thus among the insects flying at Middleton East, female spotting, and the first-order distinction of male spotting, has provided a fixed basis for comparison.

Dowdeswell (*l.c.*) collected the larvae of *Maniola jurtina* in this colony during 1957, 1958, and 1959, and subsequently, by sweeping the grasses at night with a net. Those principally eaten there are *Helictotrichon pubescens* and *Brachypodium sylvaticum*. When obtained after hibernation in this way, he has been able to rear larvae in the laboratory with relatively few deaths (10 per cent in the larvae and 6 per cent in the pupae). This must contrast strikingly with the great elimination taking place in nature, remembering that each butterfly lays about two hundred and fifty eggs. Moreover, it is clear that heavy destruction is still in progress in all the later stages, having regard to the numbers of larvae that can be obtained by sweeping at night: a method so ineffective compared with catching the imagines. The only larvae subject to a high mortality when reared in captivity were those obtained in late June 1959, at the end of their season (pp. 56–7).

When the butterflies reared artificially with so little elimination are compared with those taken in the field and therefore subject to a high death-rate in their early stages, no difference in spotting can be detected in the males (for this comparison, the lowest yearly value of P, that for 1959, exceeds 0·5). It is far otherwise with the females.

In that sex, the insects caught upon the wing during each of the four available years (1957–60) had much lower spot-values than those reared in the laboratory. Moreover, the latter, unlike the wild population, were

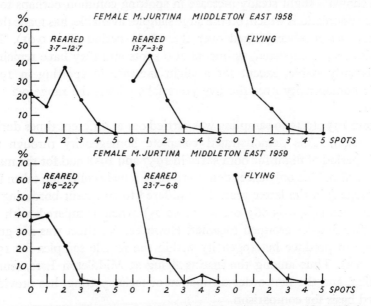

FIG. 2. *Spot-distributions of the butterfly* Maniola jurtina (*Satyridae*), *1958 above, 1959 below. Specimens reared from young wild larvae, and emerging early and late in the season respectively, compared with wild imagines from the same locality. Note that the spot-frequencies of the specimens reared throughout 1958 and early 1959 are homogeneous and differ significantly from those of the wild imagines, which are also homogeneous in the two years. The spotting of those reared late in 1959 is, however, homogeneous with that of the wild butterflies* (data from Dowdeswell, 1961, 1962).

subject to a significant intra-seasonal variation in 1957 and 1958, for the earlier butterflies to emerge had higher spot-values than the ones which appeared subsequently; although even they were much more spotted than those found flying, except in 1959. This difference indicates heavy elimination of the more spotted types in natural conditions. Considering the early emergence in the years with fullest data (Dowdeswell, 1961), the selection in nature against females with two or more spots contrasted with the unspotted type can be calculated (Woolf, 1954). It amounts to 69 per cent (compared with bred material), with 95 per cent confidence limits at 87 and 26 per cent, in 1959 and to 74 per cent, with similar confidence limits at 88 and 41 per cent, in 1960.

Considering the enormous selection-pressure against the more spotted types, it is strange that these persist in the population. One would have expected them to have disappeared; indeed the limited data from Ireland suggest that they have almost done so in that country. Perhaps, as K. G. McWhirter suggests, the survival of higher spotting in the females is due to its evident advantage in the males, so generally uni-modal at 2 spots. One would, however, have supposed that the English females would have been buffered against the higher values by a more complete evolution of sex-controlled variation.

The females bred in 1959 differed from those of previous years. During the first part of the emergence (18 June to 22 July) they had, as before, much larger spot-values than the wild specimens ($0.01 > P > 0.001$). However, those emerging subsequently (23 July to 26 August) were very distinct from them, for their spotting actually resembled that of the insects caught upon the wing ($0.3 > P > 0.2$) (Figure 2).

It did so for a reason which confirmed rather than obscured the effect of selection upon spotting. In 1959 a high proportion of the larvae obtained by sweeping late in the season were large individuals which had persisted in-stead of pupating previously as heretofore. Approximately 77 per cent of these proved to be parasitized by the Hymenopteron *Apanteles tetricus* which, apparently, is always fatal in its effects. It was the remaining 23 per cent or so of healthy larvae that produced the last part of the emergence. That is to say, the laboratory stock was spotted like the butterflies flying on the downs when, and only when, it was derived from material subjected to heavy elimination; in this instance by parasites, which seem to have been unduly common in 1959: perhaps the quite ab-normally warm and dry summer of that year had favoured them. It certainly looks as if those larvae destined to give rise to the more spotted females were in some way unduly susceptible to the attacks of *Apanteles*. A less extreme manifestation of that effect would explain the intra-seasonal female variation of *Maniola* noticed in the wild populations of the two previous years (1957 and 1958), for the larvae pupating later in the season would represent a group which had been the more subject to infection. Consequently they might be expected to produce imagines with lower spot-averages than those pupating earlier.

We have here a demonstration that the genes which control spotting produce other effects which are of importance for the survival of the individual. It is remarkable that parasitism by this Hymenopteran destroys to an unequal degree various genetic types of its larval hosts. Conceivably these differ in their habits, scents or reaction against the parasite, but we have no evidence on the point.

It is an important fact that, in certain circumstances, wild populations of *Maniola jurtina* approximate to the spotting of the reared specimens. When they do so they are also subject to an intra-seasonal change of a similar, though less extreme, kind; for then the individuals caught upon the wing have a high spot-average at the beginning of the emergence, after which it drops rather suddenly to a lower value. Such heterogeneity occurs when the species has become exceptionally common, as it did in many places in 1956, and has consequently been subjected to diminished selection. It occurs also when a population is shifting from one to another type of spotting.

Up to and including 1955, the *Maniola jurtina* populations in southern England east of Cornwall were of the 'Old English' type (p. 51), the females having a large single mode at 0 spots, generally comprising 60 per cent or more of the specimens. In the extreme south-west of this region, Tiverton (Devon) and Taunton (Somerset), they were 'New English' (p. 51), with a higher spot-average; though still unimodal at 0, less than 60 per cent of the specimens belonged to that class.

In 1956 a wave of high spotting affected the majority, but not all, of the populations throughout the southern English stabilization-area. This produced a shift from the 'Old' to the 'New English' type, and even occasionally resulted in a small secondary mode at 2 spots (the 'Pseudo-Cornish' form). The males also were affected: though remaining consistently unimodal at 2 spots, specimens with more than 2 became commoner and those with less than 2 became rarer. These changes were probably connected with the weather or with the degree of parasitization in 1956, for the species was everywhere exceptionally abundant that year. In 1957, however after an abnormally mild winter and prolonged spring drought, conditions always unfavourable to *M. jurtina*, it was much rarer and the wave of high spotting receded. By 1958 the southern English populations had in general returned towards the 'Old English' type, except in Devon and Somerset where they remained 'New English'. It will be noticed that these events are in accord with the facts just indicated: that there is normally a selective elimination of the higher spotted female types which, however, is mitigated in favourable conditions when the selection-pressure is less severe (pp. 12–13).

The colonies thus affected showed the intra-seasonal heterogeneity with significantly high spot-values at the beginning, compared with the rest, of the emergence. Indeed it seems likely that such intra-seasonal shifts are not only associated with changes in stabilization but that they may be one of the means by which these are brought about. The populations have always been homogeneous throughout their emergence when of the 'Old English' type. This applies not only to the *Maniola jurtina* communities in southern England east of Cornwall before and after 1956 to 1957, but also to those during that period which, like that at Middleton East, were never involved in the change to the 'New English' form (Beaufoy *et. al.*, 1970).

It will be remembered that, considering Britain as a whole, *Maniola jurtina* is nearly always unimodal at 2 spots in the males, while spotting in the females can take very diverse distributions. That is to say, male spotting is in its 'first-order' effects the more buffered against adjustments of the gene-complex. This is in accord with its relative lack of response to selective elimination as indicated by the results obtained in rearing larvae.

The heavily significant distinction between the Middleton East specimens reared and those caught by Dowdeswell (1961) has here been attributed to the great difference in selection to which these were exposed. Two alternative explanations of it can, however, be put forward, and, having regard to the important implications involved, it is necessary to consider whether or not there is substance in them.

In the first place, there was a possibility that the observed difference in spotting between the laboratory stocks and the butterflies taken on the wing might be due, at least in part, to non-random sampling. For there is an obvious tendency in sweeping to secure a heavy excess of those larvae which feed high up on the grasses. It might be that these tend to produce the specimens with the larger number of spots. To test this possibility, Dowdeswell examined the breeding cages at night and separated the larvae feeding near the bases of the leaves from those feeding higher up. Within the limits of the experiment, the results tended to show that the individuals are in fact of fairly consistent behaviour in this matter; also that there is no association between position when feeding and the imaginal spot numbers (for this comparison, males $P > 0.3$; females $P = 0.7$). Incidentally, it will be noticed that even had such a correlation been established, it would nevertheless have demonstrated that the genes for spotting have effects upon the larval habits on which selection could certainly operate.

The extreme stability of female *jurtina* in the widest range of ecological diversity to be found in Britain (pp. 51–2) itself indicates that the conditions of the laboratory compared with the field are unlikely to influence spotting through its non-genetic component (p. 50). However, if it be maintained that it is the artificial environment which produces the abnormally high spot-values, it may reasonably be asked why it does not always produce them. For the bred specimens are spotted in a manner identical with that of the natural population when they are derived from late larvae which had previously been subjected to heavy elimination by parasites. Yet when larvae equally late but much less parasitized are reared artificially in precisely the same way, they produce imagines with a far higher spot-average than normal. The control thus appears to be a valid one. Moreover, when southern English populations east of Cornwall have changed their spot-values or expanded their numbers, they have approached the spotting of the broods bred in the laboratory.

We may therefore accept that the genes controlling the spots on the underside of the hing-wings in *Maniola jurtina* are important for the life of the individual. Also that larvae and pupae destined to produce females with the high spot-values are strongly selected against in natural conditions and are the more liable to be attacked by the parasite *Apanteles tetricus*.

Maniola jurtina in the Isles of Scilly

The various races of *Maniola jurtina* in the Isles of Scilly illustrate clearly a number of important aspects of ecological genetics. This archipelago is situated thirty miles west-south-west of Land's End, Cornwall, and it yearly provides thousands of visitors from North America with their first sight of the Old World. It consists of a large number of islands and rocks, the more north-easterly of which are arranged roughly around a central roadstead about two and a half miles by two miles in extent. This is very shallow, little of it exceeds two fathoms, and it represents the low-lying central plain of an area of vanished land, the higher ground around the edge of which constitutes the surviving islands (Map 1).

As upon the mainland, the spotting of *Maniola jurtina* in Scilly is usually unimodal at 2 spots in the males, though not invariably so (p. 69). In the females on the other hand it may differ greatly and consistently from one island to another, and it is in that sex that it is here discussed except where stated to the contrary.

Our studies have been confined to the north-eastern part of Scilly, just described, and here the islands fall into two distinct size-groups;

large islands of 682 acres or more and small ones of 40 acres or less, a difference in area of at least seventeen times. The results obtained from 1946 to 1959 inclusive, involving fourteen generations of the butterfly, have established a clear distinction between the populations inhabiting the two types (Dowdeswell and Ford, 1953; Dowdeswell, Ford, and McWhirter, 1957). We found that the female spot-distributions are

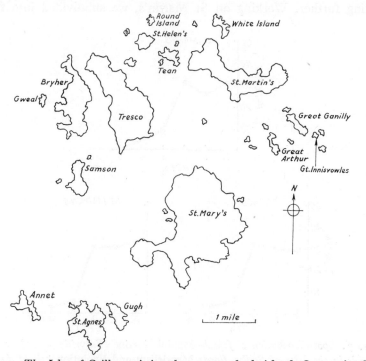

MAP I. *The Isles of Scilly, omitting the most southerly islands.* Symposia of the Society for Experimental Biology, 7, 263.

constant year after year upon each island, except for three special instances to be described later (pp. 65–9). Yet they were similar on each of the three large islands where we have worked, St Mary's, Tresco, and St Martin's, with approximately equal frequencies at 0, 1, and 2 spots (Figure 3), while they differ markedly from one to another of five small islands on which they have been examined (Figure 4). These are Tean, St Helen's, White Island (see p. 67), Arthur, and Great Ganilly, and it will be noticed that diverse as are the spotting types of the populations inhabiting them, they have not reproduced the 'large island' stabilization, assuming that this really exists as a distinct entity at all. Alternatively, it might be suggested that the approximate uniformity at 0, 1, and 2 spots characteristic of it is mainly the result of combining a number of closely associated but distinct populations with different spot-frequencies: for example, unimodal at 0 and 2 respectively. Yet if so, it seems most improbable that the samples would be so balanced that the modes should by chance cancel out to produce the 'flat-topped' effect on all three islands, and in each year. Nevertheless the point deserved testing further. Working on St Martin's, we subdivided into three

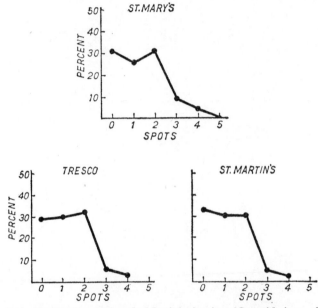

FIG. 3. *Spot-distribution of female* Maniola jurtina (*Satyridae*) *on three large Islands in the Isles of Scilly. Ford:* Mendelism and Evolution *7th ed. (1960).*

parts the area whence the butterflies giving the large island spotting had been obtained (Dowdeswell, Ford, and McWhirter, 1960). These subdivisions were each rather diversified, but they could be distinguished by their topography and cultivation and had in the past provided something like equal contributions to the total. On sampling them separately, it was found that all the three female populations inhabiting them were homogeneous ($\chi^2_{(4)} = 3.68$, $0.5 > P > 0.3$). That is to say, there was no evidence that the 'large island' spot-distribution was produced by combining several effectively distinct communities each with different spotting-types.

Various suggestions have been made to account, partly or wholly, for the similarity of female spotting in *Maniola jurtina* on the large islands and for its diversity from one to another of the small ones. They have, of course, also to take into account its constancy in each of these habitats year after year, except for certain instances to be mentioned subsequently

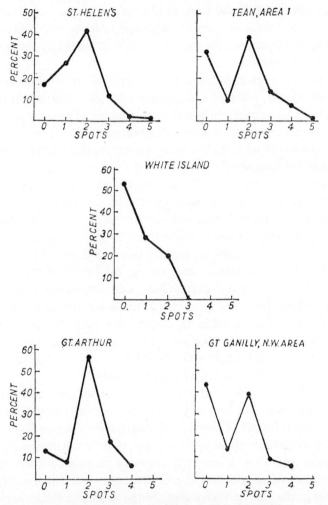

FIG. 4. *Spot-distributions of female* Maniola jurtina (Satyridae) *on five small Islands in the Isles of Scilly; that on White Island is prior to 1958. Ford:* Mendelism and Evolution *7th ed.* (*1960*).

(pp. 65–9). Those explanations which have proved untenable may first be discussed, followed by the one which can now be fully accepted.

In the first place, it might be supposed that the butterflies resemble one another on the large islands because of migration between them. This is an impossible view. We have found that even a hundred yards of unsuitable territory is an almost complete barrier to this insect (p. 65), and there is at least a mile of sea between the islands in question. Moreover, in two out of the three channels separating them, those from St Mary's to St Martin's and from St Martin's to Tresco, small islands are interposed, and these support *M. jurtina* populations with quite different female spot-frequencies. Thus the concept of similarity due to migration must be rejected.

Secondly, the dissimilarity of female spotting from one to another of the small islands, taken in conjunction with its similarity on the large ones, was in some quarters hailed as an example of random genetic drift of the permanent type, though its stabilization from year to year does not accord easily with such a view. However, the numbers of the insects involved dispose effectively of such a suggestion. There were up to 1953 three populations on Tean (pp. 64–5). These have been estimated by the technique of marking, release, and recapture as totalling approximately 15,000, 3,000, and 500 imagines per season (p. 65), nor were they affected by any considerable yearly fluctuations. St Helen's seems to harbour as many or slightly more, perhaps 15,000 to 20,000. Both these are small islands and, among the others of that size-group which we studied, the butterfly has been consistently rare only on Arthur; where the yearly total may perhaps amount to less than 1,000 imagines. These facts, taken in conjunction with the great selection-pressures to which it has been shown that *M. jurtina* is subject in Scilly (pp. 21, 66, 71), are decisive in eliminating permanent genetic drift as the cause of the phenomena we are discussing.

Thirdly, they might be attributed to fluctuation in numbers in either of its two aspects. We have not ourselves observed any marked change in population-density of the type considered in Chapter Two affecting the five small islands under consideration, save for the increasing rarity of *M. jurtina* on Tean from 1958 onwards associated with the great deterioration of its habitats there. Doubtless such fluctuations have occurred in the past and have contributed to rapid evolutionary adjustment to local conditions. These, as pointed out on pp. 16, 40–1, are correlated with the numerical *changes* rather than with absolute population-size, though doubtless small communities are the more prone to generate them.

A second aspect of fluctuation in numbers which requires attention here is that described by Waddington as 'intermittent drift' (pp. 40–1). This as already mentioned, represents a situation in which a population fluctuating numerically at intervals reaches, and for sufficiently long maintains, so low a density that random drift can materially affect its evolution. Waddington himself (1957) uses this concept to explain the characteristic difference in the female spotting of *M. jurtina* on the large and small islands of Scilly. He suggests that on the various small islands the butterfly has in the past been reduced to a few founders or has actually become extinct. The gene-frequency of the survivors, or of occasional rare immigrants recolonizing the habitat, could then affect the genetic constitution of the local races derived from them. This would tend to produce different spotting-types on each small island. He apparently postulates that the large islands, on the other hand, might be expected to carry a sufficient population to retain considerable numbers even at the minimum period of such fluctuations. Presumably, therefore, an original Scillonian type could survive on them, so accounting for the similarity of spotting upon each.

This is the same type of explanation which Dobzhansky and Pavlovsky (1957) have subjected to a brilliant experimental study. They used the F_2 hybrids of *Drosophila pseudoobscura* from crosses between flies originating respectively from Texas and California and differing by an inversion in the third chromosome. With this material they established twenty experimental populations in the laboratory: half of these were founded by a few flies (ten males and ten females) and half by many (four thousand, with approximate equality of the sexes). Though the heterozygous inversions have an adaptive advantage over both homozygotes, they found that after eighteen months the frequency of the Texan chromosome-type had declined from 50 per cent to between 20 and 35 per cent in the populations descended from many individuals and to 16 to 47 per cent in those descended from few. This difference is significant while, in addition, the heterogeneity in the latter group is the greater. Thus it was suggested that when the number of founders is small, their genes are of crucial importance in determining the genetic constitution of the colony descended from them. In such circumstances, that is to say, genetic drift is more important than selection. In the conditions of the experiment this is perhaps not surprising, since all the populations were maintained in an environment which was kept as uniform as possible.

Dobzhansky and Pavlovsky, however, used the facts obtained from these important investigations to interpret the distinctions between the *Maniola jurtina* populations on the large and the small islands in Scilly, suggesting that the special features of the latter may be due to the origin of each from a few founders only. But the effects of selection in nature and in population cages maintained in constant conditions in the laboratory are clearly open to very different interpretations.

These explanations are not consistent with the facts, and a single series of our observations is, in reality, sufficient to dispose of them (pp. 70–2). However, it will be useful to develop three further lines of evidence which do so in addition, since these illustrate very well the techniques of ecological genetics and provide information upon other subjects.

In the first place, we have in three localities been fortunate enough to observe the evolution of a new form of spotting differing from the old as much as one small-island type differs from another. Two of these were strictly related to striking ecological changes and in order to describe them it will be necessary to give a brief account of Tean, where we originally demonstrated the value of *Maniola jurtina* for studying natural populations.

This is one of the largest of the small islands, being about 1,000 yards long and having a maximum width of 400 yards; its area is 35 acres. Up to 1953 it was divisible into five distinct regions comprising two very different ecological types (Map 2). Areas 1, 3, and 5 consisted of rough rocky or hilly ground, rising in the main or central part (area 3) to 120 feet. They supported a growth of gorse, bracken, bramble, and long grass and provided suitable habitats for *M. jurtina*. These are separated from one another by two windswept necks of land, respectively 200 yards long (area 2) and 150 yards (area 4). They consisted of short

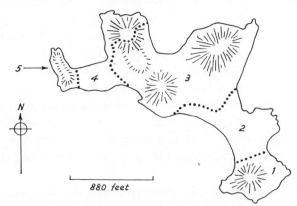

MAP 2. *The island of Tean, Isles of Scilly. The collecting areas are numbered, and their boundaries are shown by dotted lines.* Symposia of the Society for Experimental Biology, *7*, 263.

lawn-like turf growing upon a very sandy soil. The insect was unable to colonize them and its three populations were thus isolated from one another. The junction between areas 4 and 3 is shown in Plate 2 (1). One could see the butterflies setting out in numbers over the two 'lawns' from either end and, finding them continuously inhospitable, turning back as over the sea, after about 10 yards; in the middle none was to be seen. Indeed here as elsewhere we find that 100 yards or less of unsuitable terrain proves almost a complete barrier to them, though they will constantly traverse as great a distance in a few minutes within one of their colonies.

The populations on Tean were, like the localities they inhabited, of very unequal size. As already mentioned, they consisted of about 3,000 (area 1), 15,000 (area 3), and 500 imagines (area 5). Nevertheless, from 1946 to 1950 inclusive, the female spot-frequencies were similar in all three areas each year, being bimodal with a lesser mode at 0 and a greater at 2 (Figure 4). In 1951 this similarity was strictly preserved in areas 1 and 3, but a slight though significant change had occurred in the small area 5 ($\chi^2_{(2)} = 8 \cdot 02$, P = 0·02). Here the magnitudes of the two modes were reversed. It is likely that this was an adjustment to changing conditions in that restricted locality, made possible by the isolation provided by area 4.

About this time the ecology of Tean was subject to a profound modification due to the removal in the autumn of 1950 of a small herd of cattle which had long been maintained there. Their absence had made little difference to the vegetation in the following summer, that of 1951, but by 1953 (we were working on the mainland in 1952) the grass had grown quite long upon the two 'lawns' and *M. jurtina* was to be found all over them, obviously gaining adequate shelter from the wind. Thus the whole population of the island was continuous that season. In areas 1 and 3 it had remained as it had always been, judged by its spotting, and it was not surprising that the colony in area 5, since it was no longer isolated, had returned to the same type there.

Area 3 had heretofore been kept fairly open owing to the trampling of the cattle. In their absence the gorse, bramble, and bracken grew into an impenetrable jungle choking the long grass and so making it uninhabitable to *M. jurtina*. This process was slower than the one affecting the turf in areas 2 and 4, and it had not become effective until 1954. By that year Tean had been transformed. Two populations of the butterfly had each extended into an area which formerly they could not inhabit: that in area 1 had colonized area 2 and that in area 5 had colonized area 4. These were now held apart by area 3 which had become a barrier.

The female spotting of the colony in areas 1 and 2 was still unaltered and bimodal in 1954 (for the comparison with the previous year $\chi^2_{(3)} = 3\cdot74$, P approximately $= 0\cdot3$). The other colony, that in areas 5 and 4, conversely had changed completely, becoming unimodal at 2 spots, Figure 5 (the difference from 1953 is measured by $\chi^2_{(3)} = 8\cdot8$, P $< 0\cdot05$). This situation then stabilized, and persisted in subsequent years, so that its statistical significance became very great. It is note-worthy that to produce the result actually observed, there must have been selection of 64 per cent against non-spotted individuals in 1954, with 95 per cent confidence limits at 83 and 27 per cent.

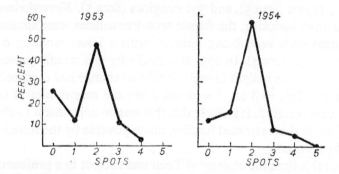

FIG. 5. *Spot-distributions of female* Maniola jurtina (Satyridae) *in the western section of the island of Tean, Isles of Scilly, showing the change associated with altered ecology between 1953 and 1954.* Ford: Mendelism and Evolution 7th ed. (1960).

It may well be asked why an extension of range had so marked an effect in one community but none in the other. The answer seems to be supplied by the ecology of the areas concerned. The population in area 1 had spread into area 2 without adjusting its spot-values because the vegetation of the two regions had become very similar. Not only had area 2 ceased to be lawn-like, but bracken had encroached across most of it, providing an environment comparable with area 1. On the other hand, the population in area 5 had spread into a very different territory, one in which the butterfly could maintain itself due solely to the growth of long grass. The sheltering bracken, so characteristic of areas 1, 2, and 5, and formerly of 3, is absent from area 4, the ecological conditions of which are thus unlike those occurring anywhere else on the island. It seems clear that the readjustment from one to another stabilized spot-distribution was produced by the altered ecology resulting from the removal of the cattle, since the two events coincided (Dowdeswell and Ford, 1955; Dowdeswell, Ford, and McWhirter, 1957).

An equally striking instance in which an alteration in the spotting of *Maniola jurtina* was associated with a clear-cut ecological change, occurred on White Island, Isles of Scilly. It will be seen from the map on page 59 that this lies off the north-west end of St Martins. The two are connected at low tide by about 225 metres of sand and seaweed covered rocks.

TABLE 2

						Totals	
Spots	1951	1955	1956	1957	1958	1951, 1955	1956–8
0	—	—	1	—	—	—	1
1	2	1	1	—	—	3	1
2	33	20	31	3	11	53	45
3	27	10	48	8	41	37	97
4	13	1	38	9	21	14	68
5	1	2	5	4	11	3	20
	76	34	124	24	84	110	232

Male spot-frequencies of *Maniola jurtina* on Great Ganilly

White Island consists of two rocky land masses each rising to about 40 metres above sea-level. They have differing micro-climates since one, the southern, runs north and south with a spine of high ground down its eastern side while the other, the northern, bent at right angles to it and more exposed to the ocean, is approximately triangular with a central hill. These parts are connected by a low isthmus, raised but little above high tide and separating the higher ground by 50 metres.

The *M. jurtina* population on White Island was first examined in 1953, and was assessed by random collecting to an approximately equal extent on both areas. The spotting proved to be homogeneous throughout the period 1953–7, the females having a single high mode at 0. Their total spot-frequency from 0–5 over these years was: 209, 108, 76, 1, 0, 1 = 329, with a spot-average of 0·68. During a great gale in the winter of 1957–8 the sea washed over the isthmus connecting the two parts of the island, obliterating the plants and leaving a belt of sand and stones from shore to shore and about 30 metres wide. As this evidently constituted a partial barrier to the butterfly, we have subsequently collected the two halves separately.

Starting in the summer of 1958, that following the inroad of the sea, the two populations became strikingly distinct from one another ($\chi^2_{(3)} = 69\cdot39$). The respective totals are each homogeneous over the period 1958–68 and amount to:

	Spots					Spot
	0	1	2	3+	Total	Average
North	317	164	87	10	578	0·64
South	147	148	140	25	460	1·09

The northern females combined for these years had not diverged significantly from those collected over the whole island from 1953–7 ($\chi^2_{(3)} = 5\cdot56$), but the latter differed very significantly from the southern females subsequent to 1957 ($\chi^2_{(3)} = 50\cdot96$).

The butterfly was rare on White Island in 1969 and the weather conditions unsatisfactory. The female capture in that year and in 1970 amounted to:*

	Spots						Spot
	0	1	2	3	4	Total	Average
North 1969	15	9	4	1	—	29	0·69
North 1970	17	13	9	1	2	42	1·00
South 1969	15	6	4	—	—	25	0·56
South 1970	16	15	7	1	2	41	0·98

In spite of the smaller number of spotless specimens in 1970 giving a higher spot average, the results suggest that the insects on the northern part of the island have tended to remain at, and those on the southern to return to, the condition prior to 1958. The samples from both areas (north and south) and both years (1969 and 1970) are homogeneous ($\chi^2_{(6)} = 3\cdot9$). The southern ones obtained in these two seasons, which can therefore be combined, differ from those found there from 1958 to 1968 ($\chi^2_{(2)} = 6\cdot2$) but agree with the captures in that section of the island from 1953 to 1957 ($\chi^2_{(2)} = 0\cdot9$). That reversal is presumably related to the fact that the barrier created in the winter storm of 1957–8 has now largely been obliterated by the growth of plants across it.

In general, the male samples have been small since that sex, being

* I am much indebted to Mr K. G. McWhirter and Mr P. T. Handford for allowing me to quote their results for these two seasons before they have published them.

more active than the females, is difficult to catch in the exposed conditions of White Island. It also appears on the wing earlier. It has, however, tended to corroborate the situation found in the females.

Thus the separation into two parts of a previously continuous population caused a complete change in one of them while the other has throughout maintained its original stabilization. This difference is doubtless related to the distinct micro-climates which characterized the areas which they inhabit. For the full data on the situation on White Island and the calculations involved in analysing it, see Creed et al. (1964) and Creed et al. (in press).

Another small island on which we have observed a change in spot-frequency is Great Ganilly (Map 1). This runs from north-west to south-east and is approximately 900 yards long. Like several others in Scilly, it is composed of two hills connected by a low sandy neck, in this instance about 50 yards in length. *Maniola jurtina* flies freely over this peninsula, as it does on the higher ground on either side of it.

In the first two years that we sampled the island, 1951 and 1955 (Table 2), the males were unimodal at two spots, which is the condition found almost everywhere. From 1956 onwards, they have been unimodal at 3 or 4 spots. The female spot-distribution differ on the two parts of the island (pp. 91–2), and there is a tendency for those of the males to do so too; as indicated from 1961 onwards when they have been collected separately in these areas. The male samples from that date are respectively homogeneous and the totals are as follows:

				Spots, 1961–8 (males)				Spot
	0	1	2	3	4	5	Total	Average
North	—	1	37	48	63	7	156	3·24
South	—	—	34	83	76	18	211	3·37

The difference is not quite significant ($\chi^2_{(2)} = 4·9$). Even if we restrict the comparison to those with the lower spot-frequency, the group which flew north of the isthmus, and therefore differ least from the situation unimodal at 2 spots, the frequencies differ heavily from the spotting of the 1951 and 1955 population ($\chi^2_{(2)} = 29·9$). The change from low to high spotting occurred between 1955 and 1956. For the male spotting in the three years 1956–8 has not differed with full significance from that in the northern part of the island from 1961–68 ($\chi^2_{(2)} = 4·8$). As the spot average of the males from 1961 is still higher in the southern area, the population there is even more distinct from the situation prior to 1956.

We have here three instances, on Tean, White Island, and on Great Ganilly, in which the *M. jurtina* populations have assumed new spot-distributions. Yet they were not reduced to a few founders nor indeed were they subject to any considerable change in numbers; and this we were able to establish by direct observation. That is to say, we have complete evidence that it is unnecessary to appeal to 'intermittent drift' or to the 'founder principle' to explain the differences between the communities of this butterfly from one to another of the small islands of Scilly.

It is instructive, however, to consider what can happen when a colony of this insect is in fact reduced to a few individuals. We have not witnessed that event on the small islands themselves, where the numbers have been much too great for us to do so. However, little populations are to be found here and there round the periphery of the large islands, isolated from the main one by ecological barriers. That inhabiting the Farm Area on Tresco has already been described (pp. 20–2) in discussing numerical fluctuations. It was there pointed out that it had a female spot-frequency very distinct from that of the general population of the island. Also that after reduction to a few individuals in 1957 it returned in 1958, when the numbers rose again, to the same exceptional type of female spotting which had characterized it from 1954 (when first studied) to 1956. That is to say, selection for the conditions of the Farm Area was so powerful as to override completely any effect of 'intermittent drift' or of migration, even when the community was reduced to a few imagines only.

It has been stated (p. 64) that a single series of our own observations is sufficient to dispose of the suggestion that 'intermittent drift' or the 'founder principle' is responsible for the difference between the large and small island populations of *M. jurtina* in Scilly. Female spotting had remained stable, with approximately equal values at 0, 1, and 2 spots, on all three large islands each year in which it had been studied up to and including 1955 (see Dowdeswell, Ford, and McWhirter, 1957, for the full data). The over-all homogeneity is demonstrated by a table of general contingency with $\chi^2_{(33)} = 36 \cdot 9$, for which the expression $\sqrt{2\chi^2} - \sqrt{2n} - 1$ supplies a normal deviate. Here $\sqrt{73 \cdot 8} - \sqrt{65} = 0 \cdot 53$.

The first sign of a break-up in this stabilization could be seen on St

Martin's in 1956. It was perhaps associated with the widespread change noticed in southern English populations from 1955–7 which produced higher spotting in a number of localities (pp. 56, 81). This was probably climatic in origin. On St Martin's, however, with a spot-stabilization entirely different from that of southern England, it had an opposite effect, favouring a low rather than a high spot-average, significantly unimodal at 0.

The following year, 1957, the islands, as the rest of England, were subject to quite abnormal weather conditions. A severe drought started there in February and continued until early July. In that sandy soil it stunted the bramble and bracken and produced a poor and sparse crop of grass while the *M. jurtina* larvae were feeding. When the butterflies emerged, it was found that the St Martin's population was still more decisively unimodal at 0, while for the first time that on the Main Area of Tresco had changed, becoming low-spotted also. The vegetation had largely recovered in 1958 and the butterfly had reacquired its original 'large island' spot distribution, with approximate equality at 0, 1, and 2 spots on St Martin's. It had not done so that year on Tresco where it had swung to a different type of spotting, unimodal at 1. This is likely to represent some instability in readjusting from the new value, since its spot-distribution had become flat-topped again in 1959. The selection-pressures involved are great; indeed the selective elimination of the high-spotted females on Tresco in 1957 was 60·6 per cent, with 95 per cent fiducial limits at 80·7 and 19·3 per cent. Only upon St Mary's did the large island spot-distribution persist unaltered throughout this period (Table 3).

The uniformity of *M. jurtina* on three large islands in Scilly and its diversity from one to another of five small ones has been ascribed by Dobzhansky and Pavlovsky (1957) to the 'founder principle' and by Waddington (1957) to the very similar one of 'intermittent drift'. It seems likely that neither suggestion would have been made had it been known to these authors that large island populations, certainly involving many thousands of imagines, can in a single generation change to a different spot-distribution. Such concepts, therefore, attempt to explain what has not occurred: the permanence of the large island stabilization preserving, indeed, an original Scillonian type. Moreover, as will now be realized, the views of these authors invoke mechanisms other than those which have produced divergence in spotting. This has been witnessed both on large and on small islands and the populations concerned were not reduced to a few individuals nor were they necessarily significantly diminished at all, so that their new characteristics were the result neither of the founder effect nor of intermittent drift.

An explanation based entirely upon natural selection is indeed in conformity with all the known facts relating to the spot-distributions of *M. jurtina* in Scilly. That is to say, populations can be adapted to the special features of restricted habitats, provided these are isolated, but only to the average of the conditions found in large diversified ones. such averages will tend to be alike, while the forms adjusted to them will not provide what is needed in localities of a more distinctive kind.

TABLE 3

Spots	St Martin's				Tresco				St Mary's		
	up to 1955*	1956	1957	1958	up to 1955*	1956	1957	1958	up to 1955*	1957	1958
0	272	63	73	63	140	34	21	23	109	15	35
1	251	49	42	64	136	30	17	41	90	12	28
2	247	43	42	59	151	34	8	32	108	14	33
3	41	6	10	6	26	5	2	13	23	6	16
4	15	—	—	1	8	2	—	6	6	1	4
5	—	—	—	—	1	—	—	—	1	—	1
	826	161	167	193	462	105	48	115	337	48	117

Female spot-frequencies of *Maniola jurtina* on three large islands in Scilly.

It is especially to be noticed that even a trickle of individuals from some Main Area would hinder the accurate build-up of the gene-complex required to fit a community to the peculiarities of a small locality. In the face of such immigration, this could only be achieved by very powerful selection. Moreover, under Mendelian inheritance, a few hundred individuals reproducing bisexually in the wild possess an immense reserve of heritable variability. Consequently the *partial* isolation envisaged by Sewall Wright would be most unsuited to the type of evolution exemplified by *Maniola jurtina* in Scilly, nor does it represent the conditions which obtain from one to another of the small islands there.

The characteristic differences in spotting between the small and large island populations provide examples of the situations described under Sections 1 and 2 on pp. 44–5. They are in accord with the rapid evolutionary adjustment to changed ecology on the small islands and with the equally rapid shift to an alternative type of spot-stabilization

* These values are homogeneous for all islands over all years up to 1955. It was not possible to collect on St Mary's in 1956.

upon the large ones; also with two other facts to which attention has already been drawn. First, we have never found the large island spotting in any habitat of limited size, whether it be a small island or in the little isolated areas on the large ones which we have studied. Secondly, when during a period of exceptional drought two of the large island communities, those on Tresco and St Martin's, adopted a new type of spot-frequency, that on St Mary's was unaffected, for this has a greater area and provides much more diversified conditions than the others. Thus it is the one likely to have departed least from its normal state at such a time.

We decided to examine this hypothesis further by founding an experimental colony of *M. jurtina* in Scilly. It was difficult to find suitable islands from which the species was certainly absent. However, Great Innisvowles and Menawethan, each of them roughly circular and in the neighbourhood of 200 yards in diameter, seemed to supply most of the necessary conditions. They are very seldom visited and indeed it is difficult or impossible to land on them in certain states of wind and tide. In 1954 we made a large collection of females from the Main Area of St Martin's and liberated 120 of them on Menawethan and 117 on Great Innisvowles. These were probably all fertile, for in a state of nature it is rare to find a female butterfly which has not paired, except for very fresh individuals just beginning to fly.

The colony on Menawethan became nearly extinct two years later and, as further experience showed that it provided the less suitable habitat, we abandoned it. Great Innisvowles, on which we therefore concentrated, is divided by high rocky ground curving across the island in such a way as to leave grassy slopes on either side, some part of which is probably sheltered whatever the direction of the wind. However, very few butterflies remained in 1956, so we augmented them by a further 106 females from the same place which had provided the original founders. The colony, both before and after it was thus reconstructed, had shown a tendency to depart in the direction of higher spotting from the equality at 0, 1, and 2 spots with which it started. A control was of course provided by the characteristics of the 'parent' population on St Martin's each season. The difference was not fully significant, though it was near to becoming so, when the population died out, so that no decisive result was obtained from this artificial colony.

The study of *M. jurtina* in Scilly has illuminated various aspects of ecological genetics. Among them may especially be mentioned the differing forms of selective effect to be found in isolated populations inhabiting restricted, compared with relatively extensive, localities. It has provided observed instances of evolution in response to ecological changes of a specific kind. It has shown that extremely powerful selection-pressures are operating in nature; and it has established that while spotting is normally stabilized on the large and small islands it can, with apparently equal ease and rapidity, adjust to new types on both. Here we have proof that neither the founder principle nor random drift, whether permanent or intermittent, can be operating to produce such effects. Finally, we seem to have evidence in the Isles of Scilly that *Maniola jurtina* can adopt *alternative* stabilizations (p. 89) of a type to be discussed in the next chapter.

CHAPTER FIVE

Sympatric Evolution

It is generally held that sympatric evolution, in which distinct races arise without isolation, past or present, is hardly a possible event. It has, however, already been mentioned that the occurrence of poly-ploids and polysomics in plants, and far more rarely in animals, has always been recognized as an exception to that statement (p. 47). Moreover, in this matter we are clearly concerned neither with the formation of clines adjusted to gradations of environment nor with the existence of different polymorphisms, or polymorph-ratios within a single area of distribution, and controlled as needed by switch-mechanisms (Chapter Six). The difficulty is to envisage the subdivision of a single interbreeding and continuous population into distinct local forms. Nevertheless, this is what we have encountered in our work on the butterfly *Maniola jurtina*: true instances of sympatric evolution. These must be described here both for their own sake and because the methods used in studying them illustrate some of the techniques of ecological genetics. It is, however, unnecessary to quote our numerical findings in full since these have already been published (Creed, Dowdeswell, Ford, and McWhirter, 1959, 1962, and 1970).

It has already been explained that while the spot distribution of *Maniola jurtina* is nearly always unimodal at 2 in the male, it can take different values in the female each of which is constant in a great variety of environments. In the south of England female spotting is unimodal at o from the North Sea to west Devon; so also in such populations as have been examined from north Britain, including Caithness, Wester Ross, and the island of Raasay (Creed *et al.*, 1959 and 1962; Forman *et al.*, 1959). This arrangement is referred to for convenience as 'Southern English', from the area where it has been most fully studied. It may indeed be regarded as basic for the species. For Dowdeswell and McWhirter (1967), who have obtained extensive data on the subject throughout the greater part of the insect's range, find that females uni-modal at o characterize the population from the Pyrenees and the west coast of France at least as far as Rumania and Finland. Yet in its peri-

75

pheral regions the female spotting of *M. jurtina* breaks up into a number of distinct stabilizations. It does so in Ireland, west Devon and Cornwall, the Isles of Scilly, Iberia, the Canary Islands, north Africa, Italy, Greece, and western Asia.

The Irish race carries the southern English type to an extreme for nearly all the females are spotless, at least in the few widely scattered samples that we have been able to obtain. On the other hand, it was a surprise to find that the females are bimodal, with a greater mode at o and a lesser at 2 (Figure 6), in the far west of England. This 'East Cornish' type may spread temporarily eastwards even as far as Dorset. It was discovered in 1952 (Dowdeswell and Ford, 1953) during a survey of the Devon–Cornwall border, the results of which must briefly be described.

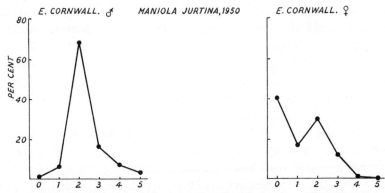

FIG. 6. *Spot-distributions of the butterfly* Maniola jurtina. *The 'East Cornish' type (to be compared with the 'Southern English', Figure 1). Males on the left, females on the right.*

In the first place, it must be mentioned that we were exceedingly unfortunate in the weather that year. In order to assess the various spot-distributions, we found it desirable to drive about the area in question, camping where we needed. Many weeks of heat and drought had brought the species out early, before we could be in the field, but these conditions changed the day we started. The insects were therefore, on the whole, few in number, worn, and difficult to obtain during the wet sunless period during which we had to work. In consequence, our samples were small and took so long to collect that our studies had to be left unfinished. Even so, we obtained the remarkable results summarized in Table 4.

Having examined samples from various places in the south of Eng-

76

land, sufficient to indicate that the general characteristics of the spotting were normal that year, we ourselves began collecting in a locality near Okehampton in mid-Devon, approximately twenty miles from the Cornish border. Thence we worked westwards in stages, on a line running roughly along the centre of these counties, our next locality being twelve miles further on upon high ground near Lydford. In both these places the spot-frequencies were typical Southern English. At that point we decided to test the situation further northwards, twenty miles away at Holsworthy, where a day's work yielded only fourteen females. As will be seen from the Table, these were, however, so suggestive of the Southern English type that we were strongly inclined to feel that they represented it.

Returning southwards we continued our western route along the original line and, crossing the river Tamar into Cornwall, obtained a sample near Lewannick, thirteen miles from Lydford. The result was surprising, for it produced our first experience of what may be called the 'East Cornish' type of spotting and we began to suspect that the stabilization which had persisted from the North Sea to west Devon had changed at last.

It became imperative therefore to determine whether or not this new bimodal spot-frequency was widespread and buffered against environmental change. We consequently moved nineteen miles westwards to Lanivet, in a different part of the county. At that place we were still roughly equidistant from the north and south coasts; but interposed between it and Lewannick is Bodmin Moor, one of the three great granite intrusions which create the wild rocky highlands of central Devon and Cornwall. They constitute formidable barriers to such an insect as *M. jurtina* which, though it follows river valleys far into them, can only maintain the continuity of its populations around but not across them, passing to the north and south between these inhospitable regions and the sea. Lanivet therefore has a different climate from Lewannick and is connected with it only by a long and indirect route habitable to the butterfly, yet the female spot-distributions in both places proved to be remarkably similar ($\chi^2_{(3)} = 0.37$, P $= 0.95$) and of the bimodal type which we now know to be characteristic of east Cornwall.

At this stage it would, in any event, have been imperative to determine whether the East Cornish spotting was a local phenomenon along the transect we had followed or one of wide occurrence. Fortunately we were this same year, 1952, supplied with collections from the coastal strip of Devon south of Dartmoor. One of these had been

D 77

made at Newton Abbot, at the head of the Teign estuary, twenty-five miles from Cornwall. The second, passing in a westerly direction, was from Noss Mayo, a peninsula at the mouth of the river Yealm, on the east side, and about eight miles from the county boundary. As will be seen from Table 6, both these were typical Southern English. So too were fifty-four females obtained the year previously (1951) in the Plymstock area, only a mile or so from the outskirts of Plymouth though still to the east. Two collections were made slightly further to the west, one just north of Plymouth and the other nearby at Roborough, about six miles from the coast. Neither was more than four miles from the county border at the river Tamar, and both were of the East Cornish type.

TABLE 4

		Spots						
		0	1	2	3	4	5	Total
E	Okehampton	36	9	5	3	—	—	53
E	Lydford	21	12	3	—	—	—	36
E	Holsworthy	10	2	2	—	—	—	14
C	Lewannick	28	17	21	7	—	—	73
C	Lanivet	30	16	18	7	1	—	72
E	Newton Abbot	47	31	10	4	1	—	93
E	Noss Mayo	15	8	1	—	—	—	24
E	Plymstock (1951)	34	14	4	2	—	—	54
C	Plymouth (north)	47	22	31	4	1	—	105
C	Roborough	21	14	18	3	—	—	56
C	Feock, Falmouth (1950)	39	24	29	7	—	—	99
C	Tavistock (1951)	17	11	13	4	1	—	46
C	„ (1952)	52	25	33	17	1	—	128

Female spot-distributions of *Maniola jurtina* in west Devon and east Cornwall (1952). The first five localities pass approximately from East to West and are midway between the two coasts; the next six, also passing from East to West, are Southern. Tavistock is half way between these two transects, approximately at the Plymouth and Roborough longitude. Two samples of the previous year and one of 1950 are also included. E = Southern English type, C = East Cornish type.

We had also been provided with a large sample from the Tavistock neighbourhood, two to three miles from the Tamar. This is seven miles north of Roborough and six miles south of our previous transect, and here the insects proved also to be decisively East Cornish. In fact we had received forty-six females from the same place the previous year (1951) and these were quite similar. We had also been given ninety-nine females caught in 1950 at Feock, near Falmouth, forty miles west of Plymouth, which were also East Cornish. Noting the bimodality of these two latter collections, we had at the time set them aside as inexplicable local variants from the normal Southern English form. They now took on a new significance, falling into line with the East Cornish spotting and suggesting its stability.

Summarizing these facts, we may say that we had in 1952 obtained samples of *Maniola jurtina* in two transects running roughly east and west, from West Devon to East Cornwall; one approximately central and the other near the south coast. The results obtained along them coincided in a remarkable manner, as did three collections made in previous years. In both regions the females in the more easterly localities were characterized by the Southern English unimodal spotting while those in the more westerly were consistently of the bimodal East Cornish form. This type was found not only close to and just into Cornwall but also much further west, beyond the granite intrusion of Bodmin Moor, at an inland locality, Lanivet, and one on the south coast at Feock, each being in a very different region from that between Launceston and Plymouth. Tavistock is mid-way between the two transects and close to the county boundary, and here also the insects were clearly East Cornish in 1952, as indeed they had been the previous year. The samples belonging to each type were respectively homogeneous and the two groups differed significantly from one another (Dowdeswell and Ford, 1953). Evidently, therefore, the two forms of female spotting are each stabilized over a wide area of very diversified country.

This matter clearly required further investigation, especially as we had encountered no populations of an intermediate kind, nor were the two types intermingled. What would happen as we passed from the Southern English to the East Cornish areas of stabilization? Considering the matter in advance, a cline appeared to be the most probable finding; we were, however, surprised that it must be as short and steep as the facts already ascertained demanded. Other possibilities of course existed, such as a confused mixture in the intervening zone. Unfortunately it was four years before we could return to this problem owing to pressure of work in the Isles of Scilly when the insect, single-brooded as it is, was on the wing.

The transition between different stabilizations of *Maniola jurtina*

In the summer of 1956 K. G. McWhirter, taking the place of W. H. Dowdeswell who could not go with us that season, and I set out to investigate the situation between the Southern English and East Cornish stabilizations; our results are shown in Table 5 and illustrated by Map 3. We decided to concentrate upon one of the two tracts already studied and we unhesitatingly chose the more northerly. For it was there that we had ourselves collected in the past so that we already knew the ecology of each habitat; while in the south it was clear that difficulties could be caused by the town of Plymouth, with its extensive suburbs, so near the line we would be following.

TABLE 5

		Spots						
		0	1	2	3	4	5	Total
E	Okehampton	45	29	20	5	—	—	99
E	Lydford	50	27	20	11	—	—	108
E	Chillaton	38	18	13	6	3	—	78
E	Tamar Valley	30	19	9	5	1	1	65
E	East Larrick	20	11	6	1	—	—	38
C	West Larrick	16	6	13	5	1	—	41
C	Lewannick, Inny Bridge	44	18	30	10	2	—	104

Female spot-distributions of *Maniola jurtina* in west Devon and east Cornwall, 1956. The localities pass from East to West. E = Southern English type, C = East Cornish type.

In the first place, we had to discover whether the conditions encountered in 1952 still persisted four years later: that is to say, we had to establish, or re-orientate, our end points before exploring the country between them. Our previous habitat for *M. jurtina* at Lewannick had been destroyed, in part by ploughing and in part by the growth of bracken. However, we discovered a place three-quarters of a mile to the east at Trekelland Bridge on the river Inny, where the species was common and could easily be collected. Here we obtained a sample of 104 females. They proved to be bimodal and completely East Cornish: the homogeneity of the 1952 and 1956 samples is measured by $\chi^2_{(3)}$ = 1·09, 0·8 > P > 0·7.

We could therefore turn to the other end of the transect and we collected again both at Okehampton and at Lydford. Here the females were still unimodal at 0. The mode was, however, somewhat less pronounced than it had been four years before, and to a significant degree. This is a 'second-order' difference and represents a shift from the Old to the New English types (p. 51) within the typical Southern English stabilization; an adjustment which, as already mentioned, had occurred widely in southern England by 1956. The females at Okehampton and Lydford were obviously homogeneous with one another ($\chi^2_{(3)} = 1\cdot14$, $0\cdot8 > P > 0\cdot7$), as they were with the nine 'New English' spot-distributions obtained east of Devon that same year.*

Clearly, therefore, it was still true that the two ends of this transect remained respectively East Cornish and Southern English, and it was with great interest that we set out to investigate the twelve miles intervening between Lydford in Devon and the Lewannick locality at Inny Bridge in Cornwall. We decided first to subdivide this region, which includes the county boundary at the river Tamar, into three blocks of approximately four miles each. Going westwards from Lydford, we came upon a suitable collecting area at approximately the right distance. This was at Chillaton, Devon, and there the population was to our surprise still normal Southern English. We therefore advanced a further four miles to the flood plain on the west side of the Tamar, just into Cornwall. Our sample there was still completely Southern English; it might have been obtained anywhere thence back to the coast of East Anglia three hundred miles away, though we were no more than four miles from the East Cornish stabilization at the Inny Bridge.

Faced with this unexpected situation we decided, as far as the terrain would allow, to halve the remaining distance. The ground is very irregular but on the whole it rises west of the Tamar, at first steeply and later gradually until, within a quarter of a mile from the Inny, it drops sharply to that river. A little more than halfway across this tract of country a road bridges the Larrick stream, and we made our way up its valley westwards from that point. We were now not quite two miles from the locality where the butterfly was known to be East Cornish. The situation therefore seemed so delicate that we decided to score our captures field by field, so far as the numbers allowed, to obtain at least something like a continuous indication of the spot-frequencies. We walked with the stream on our right hand and, crossing a lane, sat down by the hedge on

---

* Shapwick (Somerset), Middleton West (Hampshire), Worthy Down (Hampshire), Shoreham (Sussex), Canterbury (Kent), Cothill (Berkshire), Rugby (Warwickshire), Ipswich (Suffolk), and Holt (Norfolk). The female samples from these nine localities are extremely homogeneous: $\chi^2_{(16)} = 11\cdot14$.

the far side of the second field beyond it to score our captures, having by this time accumulated 82 males and 38 females. The result confirmed the estimate made as we collected, that the spotting was still completely Southern English though we were now less than a mile and a half from the Inny.

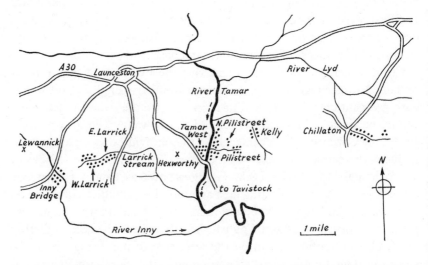

MAP 3. *The central transect across the Devon–Cornwall border where the 'East Cornish' and 'Southern English' stabilizations of the butterfly* Maniola jurtina *met in 1956. The localities where the samples were obtained are indicated by dots. (The river Tamar is here the county boundary.)* Heredity (1959), 13, 375.

We then climbed through the hedge and found the field to the west of it was a large one. As we advanced, it became clear that a change in spot-distribution had at last occurred. We continued our work in the next field and on some rough ground beyond, by which time we had obtained 59 males and 41 females. These we scored, and the result demonstrated that the female population was bimodal and completely East Cornish: a stabilization which had persisted across England, and then from western France to Rumania and Finland, had changed to another type in a few yards, at an ordinary field hedge over which the insect was flying freely (Plate 2 (2)).

We distinguished the two parts of this locality on the opposite sides of the critical hedge as East Larrick and West Larrick. The female spot-distributions in the whole series of localities from East Larrick back to Okehampton (five in all), being of the Southern English type, were extremely homogeneous: $\chi^2_{(8)} = 2 \cdot 74$, $0 \cdot 95 > P > 0 \cdot 90$. So also were those of the East Cornish form obtained at West Larrick and at Inny Bridge near Lewannick ($\chi^2_{(3)}$ = 0·49, 0·95 > P > 0·90), while the two groups were significantly different: their heterogeneity is measured by $\chi^2_{(3)}$ = 14·54, 0·01 > P > 0·001. So too are the females caught at East and West Larrick when considered alone ($\chi^2_{(2)}$ = 7·35, 0·05 > P > 0·02).

The males do not provide a definite criterion between the Southern English and East Cornish stabilizations. As expected, therefore, the series from Okehampton to East Larrick does not differ significantly in

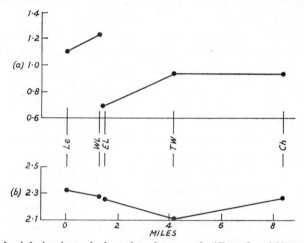

FIG. 7. Maniola jurtina: *the boundary between the 'East Cornish' and 'Southern English' stabilizations—1956. The average spot-numbers of (a) the females and (b) the males are plotted against the distance along the transect. (Le, Lewannick; WL, West Larrick; EL, East Larrick; TW, Tamar West; Ch, Chillaton).*
Heredity (*1959*), 13, 377.

that sex from the West Larrick and Inny Bridge samples combined ($\chi^2_{(3)}$ = 5·21, 0·2 > P > 0·1). When, however, the females along the whole tract (Okehampton to Inny Bridge) are compared, it is found that the difference between the Southern English and East Cornish types is greatest at the boundary between them (Figure 7). This situation, first recognized by E. R. Creed, is the reverse of a cline,* and the effect is shown most clearly by means of the average spot-numbers which, of

* A 'cline' is a continuous change in a character, from one type to another, over a considerable area. A 'reverse cline' may therefore be defined as the tendency for a character distinguishing two forms with a continuous distribution to become progressively less alike towards the geographical boundary between them.

course, are larger in the Cornish stabilization, with its second mode at 2, than in the English. No comparable effect could be detected in the males.

When at the end of the season we came to consider these results, it appeared at that stage most likely that they were due to isolation in the past. This was by no means improbable, since we had already discovered that *M. jurtina* is quite rare or absent in west Cornwall for ten miles or more east of the Marazion–Hayle isthmus where the north and south coasts are only four miles apart. This has allowed an isolated local race to evolve on the Land's End peninsula; one that is stabilized throughout that area, the spot-frequencies of the females being unimodal at 2 as in the males. Thus it is necessary only to postulate the former occurrence in east Cornwall of what can be seen today in the west in order to visualize the independent evolution of the East Cornish race to the point at which it had developed genetic isolation from the Southern English one. If the two had subsequently spread until they met, they could then remain distinct because of partial inter-sterility. This is a well-known type of phenomenon (Chapter Fifteen) which, of course, is not an example of sympatric evolution. Accordingly, we began to contemplate a combined programme of cytology and breeding work; for if it were possible to demonstrate the existence of unpaired chromosome loops in the F_1 individuals, absent from both the East Cornish and Southern English types, a past history of the kind just indicated would be established. The results of the next season, however, forced us to think upon entirely different lines.

In 1957 we found the female spot-values still completely Southern English at Chillaton and completely Cornish at the Inny Bridge, Lewannick (see Table 6 for the 1957 results of this transect). The population on the west bank of the Tamar proved, however, to have changed to the East Cornish type. Between this and the Inny Bridge lay, of course, the critical Larrick area. The butterfly was rare this year owing to continued drought up to early July, and we could obtain only four females in the parched fields of West Larrick. At East Larrick, being moister, we amassed 45 females, also East Cornish, though this site was within the Southern English region the year previously: a surprising result, consistent with the spread of the Cornish form to the Tamar two miles farther east.

Evidently we had to discover where the change from one type to the other now occurred and if it still remained a sharp one. We therefore retired eastwards and obtained a collection at Kelly, just under three miles west from Chillaton. Finding this normal Southern English we decided to advance a further mile west to Pilistreet, where the population proved to be fully Southern English also. Here we were at the top of the steep hanging woods which at this point clothe the east side of the Tamar Valley. Indeed we moved 150 yards north along the edge of them and took a further sample (North Pilistreet) which was also Southern English. Below lay the eastern flood plain, consisting of a single row of narrow fields no more than 160 yards wide; beyond these was the Tamar, about

TABLE 6

		Spots						
		0	1	2	3	4	5	Total
E	Chillaton	27	19	10	5	—	—	61
E	Kelly	50	23	15	10	2	—	100
E	Pilistreet	51	20	8	6	—	—	85
E	N. Pilistreet	26	11	8	4	1	—	50
I	Tamar East, 160 yd wide	67	36	34	5	—	—	142
C	Tamar West	84	32	41	12	—	—	169
C	East Larrick	27	3	9	5	1	—	45
C	Lewannick, Inny Bridge	80	23	43	8	1	—	155

Female spot-distributions of *Maniola jurtina* in west Devon and east Cornwall, 1957. The localities pass from East to West. E = Southern English type, I = Intermediate type, C = East Cornish type.

15 yards broad, and then the western flood plain where we had already collected this year and found the insect to be East Cornish. The spotting was homogeneous in the four localities from North Pilistreet to Chillaton ($\chi^2_{(9)} = 6\cdot24$, $0\cdot8 > P > 0\cdot7$) and in the three from the West Tamar to the Inny Bridge near Lewannick ($\chi^2_{(6)} = 7\cdot93$, $0\cdot3 > P > 0\cdot2$). The distinction between the two series, respectively Southern English and East Cornish, was heavily significant ($\chi^2_{(3)} = 18\cdot09$, $P < 0\cdot001$). Fortunately it was possible to obtain a sufficient female sample in the eastern flood plain between the Tamar and the woods, uninhabitable to *jurtina*, which cover the steep bank below Pilistreet. In this narrow belt (160 yards wide) the females were in fact intermediate (compared with the Cornish series $\chi^2_{(3)} = 8\cdot02$, $0\cdot05 > P > 0\cdot02$; and compared with the English, $\chi^2_{(3)} = 10\cdot66$, $P < 0\cdot02$); Creed *et al.*, 1959.

When the spot-numbers were studied along the whole transect, from Lewannick to Chillaton, in 1957 a 'reverse cline' was again obtained in the females as at Larrick the previous year (pp. 83–4), but its position had moved eastwards with the junction between the two races. It was, however, restricted to the English population: to the Pilistreet, but not to the North Pilistreet, sample. For there, at the beginning of the boundary, the Southern English type differed more from the East Cornish than it did farther away (Figure 8, *a*). The males, which had shown no such effect in 1956, were also involved, but differently (Figure 8, *b*); for their spot-average was higher than elsewhere at Tamar East, also at Pilistreet where the female value was lowest. Some powerful disturbance is indicated by these results: one connected with the interface between the two spotting-types, not with any particular locality (see also pp. 202–3).

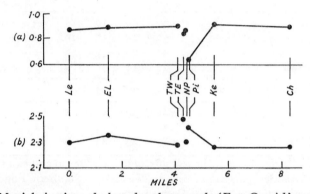

FIG. 8. Maniola jurtina: *the boundary between the 'East Cornish' and 'Southern English' stabilizations—1957. The average spot-numbers of (a) the females and (b) the males are plotted against the distance along the transect. (Le, Lewannick; EL, East Larrick; TW, Tamar West; TE, Tamar East; NP, North Pilistreet; Pi, Pilistreet; Ke, Kelly; Ch, Chillaton).* Heredity (*1959*), 13, 377.

The sharp demarcation between the East Cornish and Southern English forms persisted subsequently again at the same site as in 1957; that is to say, between the river Tamar and the eastern side of its flood-plain. The female population in that narrow belt could best be described as abnormal in 1958, with approximately equal numbers at 0 and 1 spots and substantially less at 2, while it was intermediate again in 1959.

In the latter year and in 1960, we tested once more the southern of the two transects studied in 1952; that between Dartmoor and the coast, about thirteen miles south of the line from Okehampton to Lewannick (Creed, Dowdeswell, Ford, and McWhirter, 1962). Here also we obtained evidence of a sharp switch from the Southern English to the East Cornish spotting; moreover, we found that the latter had spread two to three miles eastwards, in accord with what had happened further to the north.

In attempting to collect along two transects, respectively north and south of Dartmoor, in 1961 we found we had undertaken a greater commitment than the number of those present warranted. In the southern region the change from one type to the other occurred between Noss Mayo a promontory on the coast four miles east of Plymouth, where the population was East Cornish, and Hay Farm three and a half miles to the north, where it was Southern English. The difference, however, was not significant though it became so when three other homogeneous samples passing eastwards to Newton Abbott were included. At the latter locality the south Devon population joins the main Southern English one. Similarly in the north transect an apparently clear-cut distinction between the East Cornish and Southern English forms occurred within the tract of country from Kelly, one and a half miles from the Tamar, to Lydford seven miles farther east; but it was not significant.

Having demonstrated in previous years that the boundary phenomenon is not a local one but occurs both north and south of Dartmoor, we concentrated after 1961 upon the northern transect, while S. and E. M. Beaufoy gave us most valuable help by supplying information which demonstrated the two distinct stabilizations in the southern area. The boundary between the East Cornish and Southern English types moved one and a half miles eastwards from the Tamar in 1962, its position being identified within a few hundred yards; a further shift, but of approximately eleven miles, took place by the following year, when our results established its line within a mile and a quarter. The difficulty of finding suitable collecting grounds prevented its more accurate detection. In both these seasons the various sites within the two stabilizations were respectively homogeneous while the difference between them was significant.

The year following, 1964, when we were fortunate in having Professor Ernst Mayr with us, the line of the boundary had moved an additional eight miles eastwards and could be determined with great accuracy along the transect which we followed. We established its position within a few yards in the middle of a continuous population crossing Itton Moor along a line where no physical barrier exists. It remained at the same place in 1965.

In 1966 the East Cornish stabilization spread a further forty miles eastwards though the population did not become fully characteristic of that type except at its more westerly end. A reversal occurred the year following. Then it was for the first time that the boundary moved westwards, and for forty-four miles. We were fortunate in having Professor Th. Dobzhansky with us on that occasion and we found the two stabilizations separated only by half a mile of rough ground and agricultural land over which we could not collect specimens. The situation remained somewhat similar the following season, that of 1968, except that the position of the boundary had moved east again for a distance of between 6 and 16 miles. Within that region we could discover no concentration of the butterfly large enough for sampling; for this is the type of country over which the species ranges widely at a low density.

B. Clarke (1970b) has attempted to discount the importance of certain aspects of selection in controlling spot numbers in *Maniola jurtina*. He endeavours to attribute the boundary phenomenon to 'developmental instability' between two *races* of this butterfly. It is not clear whether he considers that these evolved in isolation and subsequently extended their ranges and met, though it seems hardly credible that he can do so in view of the similar situations in other parts of the insect's range, some being on small islands (p. 91). But instability is exactly what the boundary regions between the English and Cornish stabilizations of *M. jurtina* do not show: they would do so if mixed groups with differing spot-frequencies occurred over the distance of forty miles or so which the moving position of the interface has occupied. Normally, however, the spotting-types are highly stable in both directions up to the line of change. Clarke gives the impression that something exceptional, dependent upon the formation of local races within a band of 'developmental instability' between them occurs here. He does so by describing the occurrence of the boundary situation in *Maniola jurtina* as representing 'some of the most extraordinary phenomena in the history of population genetics'. Yet he must be well acquainted with the fact that more extreme instances of a similar kind have subsequently been detected; for example by Bradshaw and his colleagues in the grass *Agrostis tenuis* and other plant species. Thus at Drws y Coed, McNeilly (1968) shows that the change from fully copper tolerant to fully normal individuals occurs within one metre at the edge of a copper mine in spite of intense gene-flow.

When we consider the boundary phenomena of *M. jurtina*, several suggested explanations of them can be rejected. Thus it seems impossible that this barrier between the East Cornish and Southern English types represents the limit of distribution of some parasite which, like

Apanteles tetricus, destroyed the larvae destined to produce the different spotting-types to an unequal degree. The range of such a parasitic species would have to end abruptly, otherwise its effect upon *Maniola jurtina* would be gradual instead of sharply delimited, as it generally is. The distribution of such a parasite would sometimes have to enlarge or contract many miles in a season to account for the shift in spotting-types of the butterfly.

It is clear that the sharp boundary between one stabilized form and another cannot be due to a past discontinuity and subsequent extension of range, such as that discussed on p. 84. In 1956, butterflies of the East Cornish type were laying their eggs at West Larrick and those of the Southern English one were doing so at East Larrick, and on the west bank of the Tamar. Yet in both the latter localities, points on an east to west line, the spotting had become East Cornish the generation following. We cannot envisage a wave of *M. jurtina* advancing from the west and pressing back before it, along a sharp line, the insects farther to the east: a concept equally difficult whether we contemplate crawling caterpillars or imagines flying during the few weeks of their existence. On the contrary, powerful selective forces must be responsible for maintaining such forms sympatrically. Yet the facts exclude anything so simple as the idea that different genetic types are favoured in different clear-cut conditions. Rather, we are driven to suppose that this butterfly has two distinct ways of solving the problem of adjustment to certain closely similar environments.

We may compare the 1957 generation with the previous one, that of 1956, at East Larrick and the West Tamar flood-plain. This was the time when the East Cornish form replaced the Southern English in these localities, where both populations were respectively homogeneous before and after that event. The proportion of females at 0 spots was very similar in the two years (48 and 52 per cent), but in 1957 selection of 65 per cent operated against the individuals with 1 spot compared with those having 2 or more: the 95 per cent confidence limits being at 82 and 31 per cent. Yet the spotting has been maintained at approximately the same frequency in these localities subsequently, demonstrating the stability of the East Cornish type. It may be added that a further indication of the powerful selective forces acting at the boundary between the two races is provided by the 'reverse cline' effect.

This suggests, indeed, that where they meet, selection is eliminating to a disproportionate degree the less precisely adapted intermediates between them. Evidently it is powerful enough in *M. jurtina* to do so in a single generation but, if persisting long enough at the same place, it is bound to involve the two stabilizations in some degree of reproductive isolation. The parallel between this situation in butterflies on the Devon–Cornwall border and that detected where races of the snail *Partula taeniata* come into contact on Moorea, in the Society Islands (pp. 202–3), is indeed striking.

Any ecological or climatic differences along the transects must be small indeed compared with those to which the species responds not at all, judged by its spotting, further east. It is true that the seasonal change from 1956 to 1957, when the East Cornish form advanced from East Larrick to the Tamar, was a great one, but it produced no corresponding effect in the rest of southern England. Moreover, subsequent alterations in spotting have been accomplished with no great difference from one season to the next.

We are, however, familiar with other alternative adjustments in *Maniola jurtina*, though in time rather than space, exemplified by the switch from the Old to the New English stabilization in many localities in 1956 and back again in 1958. This is a 'second order' difference. Certainly we have no indication why sympatric evolution of the kind just described occurs close to the borders of Devon and Cornwall and not, so far as is known, thence to the North Sea. It should, however, be remembered that this county boundary is near the region where this species breaks up into different races: several have been found in Cornwall and many are isolated in the Isles of Scilly. The reason for this is not clear, but it occurs elsewhere at the periphery of the insects' distribution (pp. 75–6).

It is apparent that with the powerful selection-pressures now known to occur in nature, alternative adjustments of polygenes giving somewhat divergent responses may be found in one or another slightly different environment. Also that such a situation may not be achieved when the gene-complex is being built up along other lines, giving, for example, physiological plasticity.

Evidently, selection strongly eliminates intermediates between the sympatric races, the change from one to another being of a 'quantum' rather than of a continuous kind. This, as pointed out by K. G. McWhirter, is in accord with the fact that as we pass from populations with lower to those with higher female spot-averages, whether in space or time, we do not find a steady increase, with a mode first at 0, then at 1, and then at 2 spots. The passage is achieved differently from this: generally by way of bimodality at 0 and 2 spots, the smaller mode being first at 2 and then at 0. That is to say, the situation with a mode at 1 is avoided: we have detected it very seldom though we have had many opportunities of examining such transitions and have studied the spot-frequencies in a great number of localities.

We do not know how rare sympatric evolution of the kind just described may be, though in this same species we have evidence of its occurrence also in a very different environment; on the two parts of Great Ganilly, in the Isles of Scilly. These are of approximately equal areas, about 15 acres each and, as already mentioned (p. 69), they are separated by a low sandy peninsula. This is approximately 50 yards long and 90 yards wide from shore to shore. Our earliest (1951) collection here was restricted to the north-western portion of the island where the female population showed two nearly equal, and very distinct, modes at 0 and 2 spots, the latter being slightly the smaller (Figure 4). Yet our results in 1953 and 1955 gave the impression that the female-spotting was shifting progressively to a higher spot-average. This proved to be erroneous. It was due to the unexpected fact that the female spot-values differ on the southern part of the island which, seeing that *Maniola jurtina* flies freely across the central peninsula, we had in addition begun to collect.

As soon as we realized this situation, we found that Great Ganilly supports two distinct populations of this butterfly.* On the north-western part the females were bimodal at 0 and 2 spots; on the southern they were unimodal at 2 spots, with low but nearly equal values at 0 and 1. This condition had persisted from 1958, when first recognized, up to 1962. The yearly captures are still being published, but my colleagues have kindly allowed me to illustrate them by quoting those for 1962 (Table 7), which provides a characteristic picture of them.† It is true that *M. jurtina* may be somewhat commoner on the two main parts of the island than on the short peninsula between them; but this constitutes no barrier to the species, which flies freely over it. Thus we have here a further instance in which selection is sufficiently powerful to maintain two sharply distinct though continuous populations.

* A third distinct population occurs in a semi-isolated area at the south-eastern corner of the island. As this is far from the peninsula uniting the two main parts of Great Ganilly, it is not relevant to the present discussion.

† The difference between these two Great Ganilly populations only approaches significance for this one season ($\chi^2_{(3)} = 7.77$, for which P approximately $= 0.05$). Since, however, the distinction has persisted at least from 1958, it is very heavily significant over that period.

It will be seen that in 1962 female spotting in the southern area of Great Ganilly (Table 7) was equal at 0 and 1, leaving a substantial single mode at 2. These frequencies then contrasted sharply with those of the

TABLE 7

Spots	0	1	2	3	4	5	Total
North-west area	18	8	22	8	1	—	57
Southern area	7	8	27	15	1	1	59

Female spot-distributions of *Maniola jurtina* on the two main areas of Great Ganilly, Isles of Scilly, in 1962.

north-western population. That distinction has, however, become somewhat less marked in subsequent years. The female populations during the period 1961–68 are respectively homogeneous on both parts of the island so that the totals of each can be combined, with the following results:

	Female spotting 1961–8							Spot
	0	1	2	3	4	5	Total	average
North-west area	63	39	127	40	9	4	282	1·66
South area	54	44	143	62	7	2	312	1·78

It is, then, clear that both populations now tend towards bimodality, with a subsidiary mode at 0, which is more marked in the southern region, giving a significant difference between them, $\chi^2_{(4)} = 13\cdot4$.

Mention may be made of the third population at the south-eastern corner of the island, referred to in the footnote on p. 91, the total catches of which during the period 1961–8 were 32, 33, 81, 11, 7 = 164; spot average = 1·56. This is very similar to the situation in the north-western area, though separated from it by the southern one.

It seems, therefore, that the existence in this butterfly of two different but sympatric forms is not the product merely of one highly unusual set of circumstances. The fact is that the kind of information needed to demonstrate that situation in *Maniola jurtina* is at present almost lacking for other species. Perhaps such alternative adjustments under powerful selection, capable of giving a very different genetic response to slight temporal or spatial changes in the environment are, in reality, not uncommon. If so, again as suggested to me by K. G. McWhirter, they may represent concealed variation and adaptation of an important kind, capable of explaining apparently anomalous advantages and disadvantages in certain broods or strains, including, for example, farm animals and crop-plants. Such a possibility should be kept in mind, and we must hope for further studies of polygenic variability in wild populations in order to assess its significance.

Indeed Antonovics *et al.* (1967) have now detected adjustments of this general type in *Lolium perenne*. They find that natural populations of this plant respond better than cultivated varieties to high nitrogen content in the soil; a discovery which, as they point out, suggests the importance, and the possibility, of breeding directly for response to fertilizers.

In general terms, the type of evolution discussed in this chapter is strikingly in accord with the results obtained by Thoday and Boam (1959) when testing deductions reached by Mather (1955) on disruptive selection (pp. 114–15). They found that lines of *Drosophila melanogaster* selected in different directions, for larger and for smaller numbers of bristles, diverged significantly even though there was a gene-flow of 50 per cent between them.

Similarly, Jain and Bradshaw (1966) studied several species of grasses growing on mine-tips contaminated with heavy metals and on uncontaminated ground in the immediate neighbourhood (pp. 357–8). They discovered that the evolution of populations genetically tolerant and non-tolerant to those substances had taken place even against a gene-flow, as high as 50 to 60 per cent, of pollen from plants of the inappropriate type.

An especially important example of the effect of directional selection in the face of such opposing gene-flow is provided by recent work of Dobshansky and Spassky (1967). Starting with populations of *Drosophila pseudoobscura* neutral as to light and gravity, they were able to produce positive and negative lines in respect of both these characters in spite of their low heritability (8–9 per cent for phototaxis and about 3 per cent for geotaxis). Moreover, this result was achieved in contrasted lines which were exchanging 20 per cent of their members in each generation.

It will be appreciated that work on *Maniola jurtina* has proved important in the experimental study of evolution. K. G. McWhirter, in a lecture delivered to the British Genetical Society in November 1966, has further extended its application in terms of 'quantum genetics' (McWhirter, 1967). This takes into account several facts which became evident during our studies on this butterfly.

93

Heritability of spotting, estimated from mid-parent to progeny, is extremely different in the males and females, yet evidence has accumulated to show that the spot-averages of the two sexes are closely correlated in the wild, though they generally fail to be so when the frequencies are responding to changes in the environment. Also at such times the spotting does not always take up its new value immediately. It may in the process of doing so correspond for a season or two with recognized quantum steps (modal at 0 or at 2, bimodal at 0 and 2, and others) already well known in different populations, or even show interquantal patterns (McWhirter, 1957).

The spot distributions of *M. jurtina* exist in *quantized* groups stable within geographical areas (pp. 51-2, 75-6). In western Asia, this condition passes over into a related species *M. telmessia* and, in general, it does so though in a less variable form in the whole family Satyridae; much as Dobzhansky has found for the chromosome inversions of *Drosophila* (Chapter 11).

McWhirter pointed out that ancient gene-systems adjust by taking discontinuous frequencies and strongly resist minor changes in the environment: this because they will have evolved gene-complexes appropriate to each stage that they may adopt. On the other hand, new genic systems, not working in such accurately built-in genetic arrangements are not so committed to quantum steps, and prove sensitive to small environmental and genetic changes. McWhirter suggests for these two situations the terms *palaeogenes* and *neogenes* respectively. He defines the former as genes recognizable by allelic differences of relative antiquity which are often carried over intact during the evolution of new species; in relation to the distinction between genes and supergenes (pp. 110-16), palaeogenes would be expected often to exhibit the qualities of the super-gene, while the neogene would tend to be simple.

As Mourant (1954) has indicated, most of the blood group polymorphisms of Man are ancient and stable, even when the populations carrying them are dispersed (as in the gypsies, p. 123). As he points out, this is in striking contrast with the variations in kind and frequency of the human anti-malarial polymorphisms when directed against what seems to be the relatively recent spread of *Plasmodium falciparum* (Darlington, 1964, p. 263). We see the one situation as palaeogenic and the other as neogenic.

It appears likely that the selective values of neogenes may interact with and be governed by the palaeogenic systems in which they operate.

It may well be that these latter can be shifted to new quanta only by strong and sustained selective changes.

McWhirter has shown that *M. jurtina* larvae collected both before and after the phase of slow growth in winter were consistent, in different localities and years, in producing butterflies with spot-values remarkably different from the flying insects in each population; nearly always in the direction of being higher spotted. Two possible sources of error must be eliminated before that fact gains a meaning. First, the 'sweeping' of larvae is now known to produce a random sample of the population; for by removing all the grass from quadrats, the whole of the larvae present in them can be obtained for comparison. Secondly, some colonies show a shift in males only from higher to lower spotting when reared in the laboratory compared with wild insects, others in females only and some in both. That either sex may produce homogeneous samples in the laboratory and in the field seems to limit environmental effects on spotting due to the transfer of larvae to artificial conditions.

It follows that a large proportion of prospective higher spotted specimens must be selectively eliminated in late larval or pupal life; in some places in either males or females, in others in both sexes. The degree of selection may amount to 80 per cent, yet the populations are stable in spotting. Thus a corresponding amount of counter-elimination must take place in the earlier stages of the life-cycle. We have here an alternating sequence in which the total selective process is, as McWhirter says, *endocyclic*. Such a situation, compared with the alternative possibility of super-gene formation (pp. 110–16), could be an efficient one for certain types of evolutionary adjustment. For it allows a situation to evolve in which selection can operate to produce variation demanding different quantum levels in a palaeogenic system. It also enables a species to control its numbers efficiently. The importance of built-in palaeogenic systems, shifting by quantum stages, combined with the more labile neogenic ones, is likely to have a widespread impact on the organism, and signs of it are to be detected in the most diverse groups.

CHAPTER SIX

The Theory of Genetic Polymorphism

The work on *Maniola jurtina* described in the last two chapters aimed at analysing the evolution of polygenic characters in wild populations. It is necessary also to consider how those controlled by major genes can be used for similar evolutionary studies. This can in fact be done in all instances of genetic polymorphism. A strict definition of that situation had proved a fundamental necessity. It was provided in the following terms: Genetic polymorphism is the occurrence together in the same locality of two or more discontinuous forms of a species in such proportions that the rarest of them cannot be maintained merely by recurrent mutation (Ford, 1940*a*).

The implications of this definition must be assessed. Evidently it excludes geographical races, as well as continuous variation controlled by polygenes and falling within a curve of normal distribution, as with human height. It excludes also the segregation of rare recessives, or heterozygous conditions, eliminated by selection and maintained only by mutation-pressure. Thus the occurrence of Huntington's Chorea does not constitute a polymorphism in Man.

Discontinuous variation, which is nearly always genetic (Ford, 1965, pp. 11–12), must be maintained by some form of switch-mechanism, to which certain general conclusions are applicable whatever the nature of the controlling unit may be: whether a major gene, a super-gene or a chromosome reconstruction such as an inversion. Indeed the distinction between these is somewhat arbitrary and usually not ascertained.

It has already been pointed out that the major genes, as indeed the other units of genetic switch-control, seem always to have multiple effects influencing the body in a variety of ways (pp. 16–17). Yet the view is often taken that many of them are of negligible importance for survival. That is not a correct conclusion when they control a polymorphism. Fisher (1930*b*) has shown that the balance of advantage and disadvantage involved must be remarkably exact if a gene is to be of neutral survival value compared with its allele. He demonstrated, more-over, that it can then spread only at an extremely slow rate: such indeed that if derived from a single mutant the number of individuals which possess it cannot greatly exceed the number of generations since it arose (Fisher, 1930a). Moreover mutation takes place so seldom that its recurrence cannot hasten the process very materially. That is to say, not only will such 'neutral' genes be very rare but, long before they have advanced in frequency to any appreciable extent, the delicate equipoise required for their neutrality will have been upset by changes in the environment and in the genetic outfit of the organism.

When, however, genes are actually disadvantageous, they tend to be eliminated and their progress through the populations checked at an early stage. Consequently each one of them will be very uncommon, being maintained only at a level determined by their respective muta-tion-rates on the one hand and the counter-selection to which they are exposed on the other.

In view of these considerations it is clear that if any unifactorial character is at all widespread it must be of some value. Indeed it is probably true to say that even if it occurs at as low a frequency as 1 per cent, it must have been favoured by selection.

There is a possible source of error which could perhaps affect our judgement of what appear to be polymorphic phases when these are decidedly uncommon. That is to say, distinct genes sometimes have apparently identical effects (as with several of those responsible for *retinitis pigmentosa* in Man). In such circumstances, a rare condition may give the impression of occurring slightly more often than it would if the different major genes which can give rise to it were phenotypically distinguishable. However, when such 'mimic' genes are eliminated by selection and maintained merely by mutation it would require an unreasonably large number of them, compared with any such situation ever detected, even to raise the character for which they are responsible to the 1 per cent level.

Yet genes which gain and preserve an over-all advantage will spread until they displace their 'normal' alleles, reducing them to the status of rare mutants. While that process is taking place, though not of course before or afterwards, the population manifests a 'transient poly-morphism'.

This, however, must be an uncommon situation, for advantageous genes will usually have been already incorporated into the genetic constitution of the organism. Only when one of them has recently become an asset, owing to a change in the environment or the gene-complex of the population, will it be possible to study its spread through the community.

Most genetic polymorphisms, then, are of a different type: that is to say, they are 'balanced', being maintained by contending advantages and disadvantages at a level determined by the relative strength of the opposing selective forces to which they are subject. Thus they ensure permanent diversity, and of this the existence of the two sexes provides the most universal example.

The distinction between transient and balanced polymorphism was first made by myself (Ford, 1940a). It will be appropriate to deal with the latter type first, since this is much the more frequently met with. Its evolutionary aspects have been reviewed in detail by Huxley (1955) in an article of outstanding importance. Transient polymorphism will be reserved for discussion in Chapter Fourteen.

Balanced polymorphism

There are a certain number of situations which automatically promote discontinuous variation, and polymorphism may be maintained by any system that does so. The proportion of the sexes, based upon a funda-mental arrangement which generally gives an approximate 1:1 ratio, has been adjusted to the needs of each species. It is obvious that any tendency for the males to increase at the expense of the females, or the reverse, would be opposed by selection. Thus the sexual situation in itself constitutes a polymorphism. This is maintained by a 'built-in' genetic switch-mechanism which, of course, is generally chromosomal, though it may occasionally depend upon a single major gene as Winge (1932) has shown in the fish *Lebistes reticulatus*, and it can even be environ-mental (in *Bonellia*).

The same general concept applies to the heterostyly of plants (Chapter Ten) worked out most fully in the Primulaceae and in the tri-stylic *Lythrum salicaria*. Here the morphology of the flowers favours diversity through out-crossing, while this is further ensured on physiological grounds by the 'illegitimacy' mechanism. The whole double-acting system is controlled by a genetic switch taking the form of a super-gene (pp. 110–16).

Another instance of the kind is Batesian mimicry (Chapters Twelve and Thirteen), in the sense that the ecology of this situation is itself, and for obvious reasons, sufficient to maintain distinct and contrasted forms at high frequencies in the same population. Yet this is reinforced by the reactions of the genetic mechanism, sometimes a major gene, sometimes a super-gene, controlling it. A number of examples could also be cited of conditions which promote diversity, though not for evident ecological reasons. Thus Sheppard (1952a) has shown that females of the three genotypes, each phenotypically distinct, constituting a polymorphism in the moth *Panaxia dominula*, all prefer to mate with males of another rather than their own form. This obviously opposes uniformity and we may take two further instances which do so; one which favours dissimilarity in Man. The incompatible children of individuals incompatibly mated on the O,A,B blood-group system are better protected than those of compatibles against haemolytic disease of the newborn when due to the Rhesus and the Kell groups, and probably to other groups also.

Indeed, B. Clarke and Kirby (1966) point out that transplant antigens appear to be polymorphic. Apart from matings within a closely bred line, the mammalian placenta must be judged a homograft because it differs genetically from its maternal 'host'. It seems that the agency which prevents the actual rejection, through the relevant foetal antigens reaching the mother unchecked, is at least in part anatomical: at any rate in mice, it has been identified as the fibrinoid material in the placenta (Kirby *et al.*, 1964). Billington (1966) finds, none the less, that in this respect there is a sufficient antigenic reaction between mother and foetus to produce some effect, its extent being directly proportional to the immunological dissimilarity involved. The greater it may be, the larger the placental and foetal size at the end of prenatal life, and this exceeds that of pure strains in the F_1 hybrids between them. Moreover, by means of transplanting eggs from one pure mouse strain to another, Billington (*l.c.*) has shown that this difference is not due to heterozygous advantage (see the next Section). Clarke and Kirby (*l.c.*) point out that increased size in foetus and placenta probably carries with it an increased chance of survival and, it may be added, this is particularly likely to be true in the mouse in which there is powerful intra-uterine competition between embryos (Ford, 1950, p. 164); that situation which seems thus to favour diversity, will tend to promote balanced polymorphism.

A further important matter must be raised here. Kojima and Tobari (1969) produce evidence for density-dependent selection against the commoner forms of an enzyme difference, so giving rise to polymorphism: a situation which may well be fairly widespread.

Apart from the types of situation just mentioned, the majority of polymorphisms advertise no reasons for their existence. We find them controlled in stable equilibrium by super-genes (or other switch-mechanisms) at all frequencies up to the maximum, when the two alleles of a diploid are in equality. Genes cannot have reached such values unless they confer some benefit; they cannot maintain such values were that benefit to persist unabated as their frequencies increase.

Attention needs to be drawn to a fundamental principle underlying the study of polymorphism and evolution. That is to say, the subject should be analysed by the techniques of ecological genetics, not merely by a mathematical approach to population genetics or by laboratory or ecological work alone. The parameters employed in the construction of theoretical analyses of population-structure and evolution must be derived from studies of species in their natural habitats including, of course, the selective forces operating upon them at different stages of their life history and in different environments. Too often simplifying assumptions have had to be made in order to construct mathematical models: assumptions which falsify the situation as, for example, in those of Haldane discussed on pp. 140, 323–4.

Moreover, wide generalizations on selection and evolution have often been drawn from a study of *Drosophila* populations maintained in culture bottles or population cages. These provide an environment which bears no relation to the natural one and the selective forces operating on them are of an entirely abnormal kind. As one aspect of that situation, the stocks do not undergo the long- and short-term numerical fluctuations arising from the succession of the seasons and of favourable and unfavourable conditions of various kinds, which they must endure in the wild. A revealing comment upon that type of experimental genetics is provided by the fact that we are still in almost complete ignorance regarding the ecology of *Drosophila melanogaster*, though it has been the species used more extensively than any other in genetic work. To this, the brilliant studies of Dobzhansky himself on other Drosophilidae, especially *D. pseudoobscura*, provide a contrast striking indeed.

The evolution of heterozygous advantage

It may be said at the outset that the most general basis of genetic polymorphism is a balance of opposed advantage and disadvantage such that the heterozygote is favoured compared with either homozygote, as originally suggested by Fisher (1927). This situation, which has been regarded as particularly baffling and mysterious (e.g. Medawar, 1960, p. 52), tends to arise automatically in polymorphism and in either of two ways. (1) It may result from dominance modification. When a gene previously rare becomes an asset owing to changed conditions, we can be sure that the advantage will be an average one rather than absolute. That is to say, if one of its effects chances to fit in with the complex adjustments of the body so as to promote harmonious working, it is very unlikely that all the others will do so too. For mutations take place at random relative to the needs of the organism, though they will not be random changes from other points of view, for instance the chemical. Selection will tend to make the favourable effects of the gene dominant and the unfavourable ones recessive. Accordingly, the heterozygote will have nothing but advantages, and be superior to the homozygotes which will have both advantages and disadvantages (Sheppard, 1953). A good instance demonstrating this tendency is provided by the work of Caspari (1950) on the moth *Ephestia kühniella*. A pair of alleles responsible for either red or brown testis colour control a number of other characters in addition. Some of them are advantageous and prove to be dominant or nearly dominant, while others are disadvantageous and these are recessives. Similarly, in human sickling erythrocytes (pp. 120–1), the dangerous anaemia is recessive while the formation of the abnormal haemoglobin, which is at an advantage in certain areas, is not. Such conditions ensure that a balanced polymorphism will arise, the frequencies of the forms depending on the relative strength of the selection and counter-selection involved.

Gustafsson (1946) and Gustafsson *et al.* (1950), working primarily upon chlorophyll deficiencies in barley, gave instances in both plant and animal material in which mutants known to have arisen in experimental stocks produce heterozygotes superior in viability to the wild type though their homozygotes are markedly weak or lethal. This situation must represent the recurrent mutation of genes which are in fact polymorphic. Doubtless their physiological advantages have been enhanced by selection while their morphological effects, being too unfavourable to improve, have become recessive; a process which should be easy, compared with those having a heavy over-all disadvantage, owing to the relatively high percentage of loci occupied by such genes. Indeed their frequency could be very considerable without being detected in wild populations, unless a special search were made for them, considering that their homozygotes, which alone can be visibly distinguished, are eliminated. Indeed Gustafsson recalls that Masing was able to demonstrate recessive lethals in 22 per cent of second chromosomes in natural populations of *Drosophila melanogaster* in a botanic garden in the Crimea.

As Sheppard (1953) points out, heterozygous advantage due to dominance-modification can also result from an inversion. This will generally carry many genes and (even if long enough for counter-turned loops to form, allowing a chiasma to arise and produce a dicentric and acentric chromatid) it will behave as a single genetic unit. Some of the included genes will be dominant in effect and others recessive, respectively favoured and opposed by selection. Thus the inversion may at once produce stable polymorphism owing to its genetic content; alternatively, if not of a type to do so at first, polymorphism can evolve owing to dominance-modification along the lines already indicated. The great advantage of the inversion is that it can hold together co-adapted groups of genes in a considerable segment of a chromosome (see Chapter Eleven).

(2) The occurrence of recessive lethals and semi-lethals can also contribute to the unequal viability of the three genotypes, favouring the heterozygote. Such recessives are a common type of mutant, though individually rare because eliminated by selection. However, heterozygotes for the unit controlling a polymorphism must be relatively frequent, and in them disadvantageous recessives can accumulate. They do so when situated so close to the locus of the switch-gene concerned that they are seldom separated from it by crossing-over or, more effectively still, when they occur within the length of an inversion forming a super-gene. They are then sheltered from counter-selection because held in single dose, in which they cannot exercise their deleterious effects.

The gene or super-gene responsible for a polymorphism spreads initially because it has some advantage; as already indicated, it would not progress beyond the status of a rare mutant otherwise. But it may accumulate recessive lethals, as just described, and these are harmful only to the corresponding homozygote. For an additional reason, therefore, polymorphic heterozygotes tend to become the most favoured genotype.

Williamson (1960) states the fact, which is clear enough, that when a polymorphism is maintained by the advantage of the heterozygote this genotype will generally exceed its expected frequency in the population, on the assumption of equal viability with the two homozygotes (though it will not necessarily do so, pp. 145–6). He adds 'but the converse is not true'. This is a little obscure. The word 'converse' suggests a situation in which the frequency of the heterozygotes falls below the Hardy-Weinberg expectation. This, presumably, is not the meaning that Williamson intends to convey, but rather that a significant excess of heterozygotes cannot be taken as a proof of their relative advantage. If so, that view is not valid: with random mating, an excess of heterozygotes does indicate their selective advantage if the gene-frequencies are not substantially changing.

Thus the establishment of heterozygous advantage constitutes the normal course of events to be expected in polymorphism and, although relatively seldom looked for so far, the examples to be described in the succeeding chapters will indicate that it is generally realized. Indeed it is an impressive fact, extremely relevant to the present point of view, that it can be detected in instances in which the ecological situation provides all that is needed to maintain diversity (pp. 188–9, 251, 287). It is, however, an important matter to decide how widespread such heterozygous advantage may be.

On the one hand, the genetic structure of organisms with their great number of rare mutants, whose alleles have the status of 'normal' major genes, demonstrates that many of these must have passed through a phase of transient polymorphism which has not become a balanced one. On the other, there are a few balanced polymorphisms in which no evidence of heterozygous advantage has been obtained even though it has been sought. An instance of this is provided by the moth *Panaxia dominula*. This is a colony-forming species widespread in southern England and western Europe (p. 129). Yet it is known to be polymorphic for the *medionigra* gene in a single locality only: one of 15 acres, isolated by agricultural land and woods, at Cothill in Berkshire (Chapter Seven). It looks very much, therefore, as if this polymorphic situation is quite recent: a view supported by the fact that the heterozygote is still visibly distinct and becoming more dominant, p. 144. Moreover, the rarer of the two genes maintains a relatively low frequency; it has not exceeded 4·6 per cent of available loci in the last twenty-four generations, of which there is one a year, and has only once reached 11·1 per cent since it was first assessed in 1939. There appear, in addition, to be ecological grounds for the diversity (pp. 139–40). In these circumstances, it seems probable that the heterozygous advantage has not yet had sufficient opportunity to become established.

We may contrast this situation with another of those to be discussed in Chapter Nine in which the polymorphism is universal in a widespread species, and in its near relations, and in which we possess direct evidence of its existence since the Pleistocene period (Diver, 1929); that is to say, the colour and banding of the snail *Cepaea nemoralis*. Here the physiological advantage of the heterozygote is incontestable (p. 188).

That condition is in fact likely to arise whenever selection begins to favour a major gene provided sufficient time has elapsed, which it may not have done if the spread is rapid. Moreover, a very considerable period may be required if one allele (or unit of the switch-mechanism, in general) be much less common than the other.

It is evident therefore that many of the major genes which acquire an over-all advantage must become polymorphic instead of being incorporated into the normal gene-complex. It is still true that genetic polymorphism is often regarded as a somewhat exceptional phenomenon, in spite of the indication to the contrary given by sex itself. This is quite incorrect, as a study of variation in many groups of organisms tends to show. Even so, the condition is frequently masked by the fact that the physiological effects of the switch-mechanism concerned may not be associated with any visible qualities. Such cryptic diversity must, in the vast majority of instances, pass unrecognized, but see below.

For example, the hydrogen-cyanide polymorphism of *Lotus corniculatus* (pp. 166–9), the widespread homomorphic 'illegitimacy'-mechanism (p. 218), the ability to taste phenyl-thio-urea in Man, together with the blood-groups and the human secretor factor, and the beta-globulin and haptoglobulin polymorphisms already found in many mammals, are all quite without visible indications of their existence. Moreover, the chromosome polymorphisms of *Drosophila* are associated with no gross morphological characters. It is inconceivable that these and other instances like them could have been detected unless situations of this kind were very common indeed.

In recent years methods have been developed for estimating the relative frequency of heterozygosity and of polymorphism in higher organisms. These are provided by the technique of electrophoresis (pp. 28–9, 173–7) in which, if indeed applicable, all three genotypes of a diploid can usually be identified. Lewontin and Hubby (1966) studied eight gene loci controlling different enzymes, and ten for larval proteins, in wild populations of *Drosophila pseudoobscura* from five localities in the western U.S.A. Seven of the eighteen (39 per cent) were polymorphic in more than one population. Furthermore, from the frequencies of homozygotes and heterozygotes at different loci, including those showing no polymorphism, the percentage of genes heterozygous in an average individual could be assessed. It amounted to 11·5 per cent. Both these calculated values are obvious underestimates, for a proportion only of variant proteins are discriminated by electrophoresis. What that proportion may be is quite uncertain. Lewontin and Hubby themselves suggest 50 per cent, but this is little more than a guess.

Before considering the implications of these high frequencies, two general problems obviously call for solution. In the first place, do the genes that have been chosen for study in this way represent a fair sample of those in *Drosophila pseudoobscura*? There seems no reason to think the contrary, for particular care was taken to avoid selecting exceptionally variable loci. Secondly, is the situation in that species fairly typical of higher organisms in general? This we simply do not know, but there do seem to be grounds for believing that it may be so. Johnson, Kanapi *et al.* (1966), have obtained corresponding results when examining *Drosophila ananassae* by similar methods. They found 40 to 50 per cent of the loci controlling five enzymes and one other protein to be polymorphic in light strains of the fly in three islands, and for dark strains in two islands, in the Samoa Group.

Results quite remarkably consistent with these have been obtained in the Mammalia. Selander *et al.* (1969) examined electrophoretic variation in thirty-six proteins determined by forty-one genetic loci in two subspecies of the House Mouse, *Mus musculus musculus* and *M. m. domesticus*, though unfortunately the samples were small. Sixteen of them (44 per cent) proved to be polymorphic at one or more localities in Jutland, Denmark, as did seventeen (41 per cent) of the controlling loci. The individual mice could, on the average, be assessed as heterozygous for 8·5 per cent of their alleles; certainly indeed an underestimate. Moreover, Harris (1966) applied the electrophoresis technique to Man. He investigated ten loci affecting enzymes in blood serum, without bias in favour of variable or invariable ones, and found three of them to be polymorphic. In discussing seven other human polymorphic enzymes, Harris points out that their gene frequencies may vary quite widely from one population to another: as examples of this he cites red-cell acid phosphatase and placental alkaline phosphatase. He reasonably inferred that selection is presumably responsible for such effects. In general terms, we may say that when a similar situation obtains in higher Insects (*Drosophila*) and in Mammals as dissimilar as Mice (*Mus musculus*) and Man, it is presumably one of wide occurrence. Yet the variation here envisaged may have to be reduced if the results obtained by Gillespie and Kojima (1968) in *Drosophila ananassae* prove to be of general application. In that species, they find much less polymorphic variability in enzymes involved in energy-production (the commoner allele having a frequency of not less than 0·92) than in non-specific enzymes.

Taking everything into consideration, it certainly appears that a considerable proportion of genetic loci are polymorphic in animals, and probably in plants. Such diversity could be maintained either by random genetic drift, by mutation, or by balanced selective forces usually operating to produce heterosis. The first of these (drift) cannot be sustained owing to the arguments in Chapter 3, and to those of Fisher on the rarity of alleles whose members are of neutral survival value (see pp. 96–7,). The second is excluded by the consideration that if the diversity of polymorphism were frequently supplied by mutation, the genes would be too impermanent to provide the necessary stability of the heritable material (p. 361).

King and Jukes (1969), as well as Kimura (1968), have argued that evolution in DNA and proteins is largely due to selectively neutral mutations and random drift. Their view depends upon the supposition that some base pair changes in DNA do not result in the substitution of a different amino-acid into the encoded protein. That is to say, such mutations are 'synonymous', because they do not alter the functions of the genes. But if they do not alter the functions of the genes, they do not bring about evolution.

Apart from the latter fundamental consideration, B. Clarke (1970a) demonstrates the fallacy of the view of King and Jukes. Thus they had tabulated the distribution-numbers of amino-acid changes in variants of globulins, cytochrome-c and the variable (S-) regions of immuno-globulins. When plotting the number of such changes per stite against the number of sites having the specific number of changes, they claim that the results are distributed in a Poisson series. As Clarke points out, their grounds for that conclusion are statistically unsound. Before the figures will fit, they have to exclude an arbitrary number of sites from the zero group: but such manipulation of data would allow them to adjust their results to almost any distribution. Clarke further shows that the Poisson series obtained by King and Jukes for haemoglobin is made up of several distinct components which invalidates the conclusion to be drawn from it: while even if the figures did fit the Poisson series, this does not provide a sound basis for neutrality.

King and Jukes have to support their contention by some curious arguments. For example Primates and Guinea Pigs cannot convert

2-keto-L-gulonolactone to ascorbic acid and therefore develop scurvy in the absence of Vitamin-C. These authors hold that this inability is of selective neutrality; an improbable and quite unsupported supposition.

Clarke (*l.c.*) is able to produce four reasons why synonymous mutations may not be selectively neutral, while Richmond (1970) gives further grounds based on chemical considerations for doubting such neutrality. He also cites other lines of evidence directly opposed to it. Thus two alleles which determine enzymes in *Drosophila melanogaster* reach an equilibrium frequency at which they are selectively neutral; yet on departing from that equilibrium value, they produce differences in viability (Kojima and Yarborough, 1967). Moreover, the genes controlling esterases examined by Burns and Johnson (1967), Koehn (1969), and others in a variety of organisms (*Drosophila*, Lepidoptera and Fish) show clines in frequency, indicating selection.

Beardmore (1970) has demonstrated that in general there is a positive correlation between ecological diversity and genetic variability. He has also shown that the genes controlling the Esterase-6 polymorphism of *Drosophila melanogaster* must be of powerful selective importance since changes in temperature and population density strongly affect their frequencies. As he remarks, if this system be typical of protein polymorphism in general, 'it seems unlikely that "effective neutrality of genotypes" is a meaningful notion'.

In addition, B. Clarke (1970a) rightly says that even if the majority of amino-acid substitutions that have occurred in evolution were to affect the action of the genes but in a manner selectively neutral, we should expect to find, but do not, such neutrality in protein polymorphisms today. Its presence or absence can be determined accurately by observation both of natural and artificial populations. As Clarke stresses, that evidence, which excludes it, is not considered by King and Jukes.

Yet it seems likely that a small change in an amino-acid at the end of a polypeptide chain may sometimes be insignificant in its effects and so produce a mutation approximately neutral in value. There would not then be the pressure to reduce such occurrences to the normal low mutation-rate, so that they may possibly be taking place with considerable frequency. So long as they remain 'neutral', their evolutionary consequences are potential only: that of providing genetic diversity which, in changing circumstances, would become of significance for the organism if selection were to begin to operate upon it. Such alleles would then qualify to fall within the class of ordinary mutants. Whether or not their effects on viability are detectable to us, whenever we find an excess of heterozygotes above expectation we can be sure we are dealing with genes upon which selection is operating *through their effects upon the individual*: whenever we find genes exposed to selection, we can be sure also that there is pressure to minimize their mutation-rates.

One of the great needs in genetics at the present time is research upon the evolution of mutation-frequency. This is likely to be beset with difficulties since so many mutants will antedate the speciation of the forms that we can investigate. Thus a failure in the balance of the gene-complex sufficient to disclose an increase in mutation-rate would probably require hybridization too wide for fertility. A search for such effects is never the less very desirable, perhaps in material so far unexplored in genetic research. One thinks, for example, of the extreme yet successful out-crosses possible in the Echinodermata. For in that phylum genuine hybrid offspring (excluding facultative parthenogenesis, that is to say) have survived at least through early development even in crosses involving distinct Sub-Classes (Morgan, 1927, p. 635).

The work of Lewontin and Hubby (1966) demonstrating the high frequency of polymorphism in *Drosophila pseudoobscura* is brilliant and of great importance. In their interpretation of it, however, they were led to exclude heterozygous advantage as a mechanism for producing and maintaining such diversity. They did so owing to the intrusion into their argument of two errors, well exposed by Milkman (1967). These are the concepts, first, that the genes controlling distinct polymorphisms act independently; and, secondly, that the unit of selection is the gene, whereas it is the individual. As Milkman remarks, 'artificial selection has shown us nothing if not the cumulative effects of genes at many loci'; one may add, their frequent interaction also produces characters that may be qualitatively distinct.

The reason why Lewontin and Hubby (*l.c.*) maintained that allelic variation at many loci cannot be due to selection operating by way of heterozygous advantage is a point that should receive brief attention. To remove 10 per cent of homozygotes at one locus at each generation reduces the relative fitness of the population by 5 per cent; assuming equality in the action of the two alleles and an initial heterozygote frequency of 0·5. That is to say, such fitness declines to 0·95 among the progeny of the initial generation. When we consider 2,000 loci, Lewontin and Hubby state that it would become $(0·95)^{2,000} = 10^{-46}$, which is an entirely impossible situation: see also Hubby and Lewontin (1968).

Milkman (*l.c.*), however, points out that this conclusion is wrong, since it incorporates the error of regarding the loci as acting independently, consequently with fitness combined as a product, though it is the individual not the locus upon which selection acts. He calculates the relation between selection-pressure on loci and on individuals and uses a model in which the population is divided into classes on the basis of the number of alleles that are heterozygous. This compares with the situation in which they have a frequency of 0·5, and no assumptions are made except that the effects of many genes are cumulative. The size of each class is given by the binomial distribution. Milkman considers what proportion of a population must be selected (selection-pressure) for heterozygotes at a number of loci, acting cumulatively, to exceed by a certain percentage (selection-differential) the number of heterozygotes in an unselected population. The selection-differential proves to be equivalent to the average selection-pressure at individual loci, and the result demonstrates that heterosis is acceptable as a major cause of heterozygosity in nature.

A similar difficulty to that envisaged by Lewontin and Hubby will be found in a much quoted paper by Haldane (1957) who points out that the substitution of one gene or inversion by another results in a number of genetic deaths, and that the cost of these occurring at many independently acting loci may be too great for the organism to bear; that is to say, its fitness or its rate of evolution may be seriously reduced. It might thus become impossible to maintain many polymorphisms in an organism by means of heterozygous advantage. Consider the genotypes A_1A_1, A_1A_2, A_2A_2 having fitnesses of 0·9:1:0·9. Assuming equality of the alleles, their mean fitness in the population will be 0·95. Here, the organism sacrifices 5 per cent of its zygotes to obtain complete heterozygous advantage. This would be quite reasonable for a single locus though not for many, if their effects are multiplied as a product as assumed by Haldane. But that is precisely what is not allowable, as shown by Milkman (*l.c.*), and Haldane's dilemma is an unreal one.

Moreover, as P. M. Sheppard has pointed out to me, Haldane ignores density-dependent factors. Where these are acting, an increase in deaths from other causes will not much affect the total number of progeny produced. Only when density-dependent factors are not operating, or the species in question have very low reproductive rates, will genetic deaths be of serious consequence.

This matter has been further discussed in a valuable article by Turner and Williamson (1968). They stress that most species produce far more offspring than are needed to maintain a constant density. Many individuals die before maturity of starvation, infection, predation, genetic diseases and other causes. Thus a population will be able to tolerate what seems a heavy genetic load without being in danger of extinction. Such genetic load is indeed largely an expression of the fact that not all genotypes are equally viable; a consideration of special importance in crowded conditions.

Others (e.g. Sved, 1968), in addition to those mentioned, have indicated the misconceptions inherent both in the views of Lewontin and Hubby (1966) and in Haldane's (1957) article. Further examples in which Haldane's theoretical models fail because unrelated to the relevant facts of ecology and, in one instance at least, owing to mathematical imperfections, are discussed on pages 140, 323–4.

The formation of super-genes

A further important aspect of balanced polymorphism remains to be noticed. That is to say, the switch-mechanisms controlling it frequently take the form of super-genes. Since genes interact or reinforce each other to produce their effects, there must be many occasions in which the simultaneous presence of particular alleles at two, or more, loci would be encouraged by selection; while the complex polymorphic adaptations of organisms, seen for instance in mimicry (Chapters 12 and 13) must often require the co-operation of several pairs of alleles. Evidently something is required to hold these together, and that end is attained, here as elsewhere, by close linkage.

If two major genes co-operate in an advantageous way, selection will favour rare structural interchanges, as well as translocations, bringing them on to the same chromosome and then the means of checking crossing-over between them. That may be done by reducing chiasma-formation in the region concerned or by moving the two loci nearer together as a result of small chromosome reconstructions. This will continue until the two genes so seldom break apart that they act effectively as a single switch-mechanism; that is to say, until they have become a super-gene. In addition, this can be achieved at a single step by the inclusion of several major genes within an inversion which, therefore, itself constitutes a super-gene also.

The occurrence of rare interchanges is of course well known but, as already pointed out (p. 10), very little work has been undertaken on the adjustment of chiasma-frequencies between genes on the same chromo-some, whether or not they have always been linked. However, sufficient information is available to demonstrate that super-genes can arise in the manner indicated here. Thus Parsons (1958) selected for a greater amount of recombination between black (b) and purple (pr) in *D. melanogaster*, and succeeded in raising it from 5 to 8 per cent. More-over, Cain, King, and Sheppard (1960) have demonstrated that the cross-over value between the genes for shell-colour and presence or absence of banding is subject to significant differences in the snails *Cepaea nemoralis* and *Arianta arbustorum* (pp. 179, 199); this is of special relevance since both the characters concerned are polymorphic. In general terms also, the evolution of crossing-over can be deduced from the fact that the distribution and frequency of chiasmata vary within and between species, also that they are genetically controlled and so must be selective and adaptive.

R. A. Fisher (1930a, pp. 102 *et seq.*) had envisaged such selective adjustment of linkage-values more than thirty years ago, while at about the same time Haldane (1930) pointed out the high frequency with which polymorphism is associated with close linkage. To mention a few instances only: apart from long inversions, we find this in the control of heterostyly in the Primulaceae, and doubtless in that of other plants, in Batesian mimicry, repeatedly in the human blood-groups, in the snails *Cepaea nemoralis* and *hortensis*, in the fish *Lebistes* and others; even though genetic analysis of the kind necessary to reveal it has seldom been carried out.

The formation of super-genes is well illustrated by the work of Nabours (1929), and of Nabours *et al.* (1933), on polymorphism for colour and pattern in Grouse Locusts (Tettigidae). Darlington and Mather (1949, pp. 335–6) make an important comparison between the linkage-values of the species he studied. In *Acridium arenosum* the colour-patterns are controlled by thirteen genes which, though carried in the same chromosome, reassort easily. They are recognizable also in *Apotettix eurycephalus* in which, however, they form two tightly linked groups, between which there is 7 per cent of crossing-over. A further step has been taken in *Paratettix texanus*. Here there seems to be com-plete suppression of crossing-over among twenty-four out of twenty-five of the colour-pattern genes, which nevertheless remain distinguish-able since their effects correspond with those found in the other species. As Darlington and Mather point out, there can be little doubt that the various genes responsible for these phases in *P. texanus* have been gradually aggregated into a group which can act as a single switch-mechanism.

Fisher (1939), moreover, demonstrated heterozygous advantage in the polymorphism of that insect. He showed that the minimum selection in favour of homozygotes for the single dominants controlling it is, in three large samples, 6·6, 10·4, and 14·2 per cent per generation. He found also that individuals in which two of the dominants are combined are at a disadvantage of 40 per cent or more in the six wild populations from the U.S.A. which could be tested: a fact which indicates the importance of bringing these genes together into one unit, the alternative super-alleles of which can be selected to meet the requirements of each locality.

As crossing-over decreases between two linked loci controlling a polymorphism, a smaller selection-pressure suffices to maintain the combination of alleles most advantageous to each phase, or does so more effectively. When the loci are so close together that a cross-over between them is extremely uncommon, so that they form a super-gene, the effect will simulate that of multiple allelism: unless, of course, a recognizable inversion be involved, when the difference may be detected cytologically. Otherwise it is only when mutation, or a rare chiasma in the right place, occurs that the distinction may become evident.

No doubt the formation of a super-gene may on occasion have to wait for appropriate mutations at linked loci, rather than unlinked ones as discussed on p. 110. Whichever alternative is used, presumably sometimes one and sometimes the other, it favours few and large chromosomes (Ford, 1965).

Turner (1967b) produces a theorem for the evolution of super-genes and (1967a) considers the question why the chromosome sets of organisms do not pass to one extreme or the other: to a single large unit or else fragment to small particles of DNA. His theorem shows that in a stable environment a two locus polymorphism is, under any mating system, subject to selection for the formation of a super-gene. This tendency augments with an increasing number of loci. Thus if continued unchecked, the whole genotype would tend to condense to a single large chromosome-pair. This of course does not prevent crossing-over, though additional steps could then be taken to do so. As Turner points out, such a situation is opposed by the side effects of structural changes in the chromosomes and by complex genetic interactions in a fluctuating environment.

We may consider the point a little further. Without any crossing-over, each advantageous gene is threatened by the disadvantageous mutants linked with it. That difficulty is resolved not only by the existence of multiple chromosomes but by chiasma formation; indeed the limit (rarely, and but little, exceeded) of 50 per cent of crossing-over is genetically equivalent to independent assortment. But this, and the approach towards it, is effective only for genes far apart on a long chromosome. It is probable that great meiotic difficulties are introduced by considerable chromosome lengths, so that the situation with numerous chromosomes is favoured. It may be added that, as Darlington (1956a) shows, it is much easier for organisms to move towards higher than towards lower polyploid values. Also, although chromosome reduction has often taken place in animals, in *Drosophila* for instance, it is yet true, as Darlington further remarks, that it is much less dangerous for an organism to gain genes than to lose them, or merely to subdivide the chromosomes carrying them.

On the other hand, an upward trend in chromosome numbers will be opposed by the stress, powerful and universal as it must be, to build co-adapted genes into super-genes. For the difficulty of doing so must be much reduced when the units comprising them are linked initially: a situation which will be promoted if the chromosomes are large in size and few in number. On the whole, the balance between the opposed tendencies thus involved will be a complex one, varying with the genetic adjustment to the environment and therefore with the species.

In addition, there can be little doubt that super-genes have sometimes evolved through duplication when, and only when, their members are on the whole similar in action. Thus, for example, one cannot imagine duplication to have been responsible for the combination that determines the colour-pattern of the wings, the presence or absence of mimetic tails, and the body-colour in *Papilio memnon*, p. 284. On the other hand, it seems likely that the genes controlling the β and δ chains of human haemoglobin arose by duplication. For they are congruous in their action, differing only by ten residues, and are closely linked. Though less nearly related, those for the α and β chains are very similar in function. Thus they too probably originated as duplications from the same locus. In that event, they seem to have done so in the remote past, for they are no longer linked. That is to say, if this view be correct, duplications can be separated. We may argue that those still held closely together, as are the genes for the α and δ chains (Boyer *et al.*, 1963), must be co-adapted or else have duplicated recently. We think here of the fact that cross-over values can increase, both spontaneously and as a result of selection experiments (pp. 111, 179), also of much information on chromosome reconstructions. I am greatly indebted to Dr J. D. Weatherall for information on the human haemoglobins; his survey on the subject, with special reference to the thalassaemias (1969), has also proved most helpful.

In this connection, it is relevant to find that a form of thalassaemia exists in which the synthesis of both the β and δ chains is suppressed in the homozygotes. This fact may well mean that these are to an important extent determined by the same mechanism and therefore there is pressure to preserve as a super-gene the units involved in their control.

Mather (1955) has made a penetrating analysis of disruptive selection. He points out that this process operates when several optima are favoured in a population. That situation must, as he indicates, lead either to genetic isolation, which a free flow of genes no longer takes place between the forms, or to polymorphism if they continue to share a common gene-pool.

Of these two alternatives, isolation must result when three conditions are all fulfilled. (1) The optimal phenotypes must be independent, so that none constitutes a necessary part of the environment for the others. (2) They must be sufficiently distinct for different selective forces to operate on them. (3) The environmental conditions to which the various optima are adjusted must be persistent. Alternatively, if all these requirements are not met, disruptive selection tends to produce polymorphism, as must occur when effective viability depends upon co-operation between the different phases: for instance, in sex, or in Batesian mimicry in which the success of each mimetic phase depends upon the frequencies of the others.

Darlington (1958) has shown that once a super-gene has been established, it will tend to grow by accumulating additional units, owing to the need for increasing co-adaptation as the alternate forms of a complex system evolve. Pursuing this line of thought, Mather points out that the longer polymorphism survives the more likely are its component features to be interdependent. Any breakdown in the genetic system will then affect the whole complex of phenotypes, not a few of them only.

Thoday and Boam (1959) have tested Mather's conclusions experimentally. They maintained stocks of *Drosophila melanogaster* represented in each generation by four single-pair cultures, two selected for high and two for low sterno-pleural chaeta-number.

Each male selected for a high value was mated to a female selected also for high chaeta-number, but taken from a low-value male-line. Similarly each male selected for low chaeta-number was mated to a female chosen for low chaeta-number also, but from a culture of a high-value male-line. Therefore cross-mating ensured that the stocks for high and low chaeta-number all shared in a common gene-pool.

Though there was a 50 per cent gene-flow between the lines selected in different directions, these diverged significantly. This was due to the evolution of appropriate super-genes. One of these, in chromosome II, favoured low chaeta-number and one, in chromosome III, favoured high chaeta-number. Both had intermediate heterozygotes. The population therefore became polymorphic even though there was bimodality between the phases rather than discontinuity; which indeed could evidently arise later by a continuation of the same process. Thus the results fully justify Mather's contention that disruptive selection can, in the right circumstances, give rise to polymorphism. His concept is indeed in full agreement with that developed here, though differing somewhat in point of view.

Mather (1955) also considers the possibility of treating as a polymorphism the occurrence of distinct seasonal phases, such as the very dissimilar spring and summer types of the butterfly *Araschnia levana*. He does so owing to 'the dependence of each on the other for continued propagation under conditions which would be adverse to itself'. I would not myself favour stretching the polymorphism concept in this way; for it is one which essentially involves a balance of contending advantages and disadvantages between the forms, absent from seasonal differences. Further, as Mather himself rightly remarks, the latter form of variation must be achieved environmentally and cannot depend upon simple genetic segregation, as with true polymorphism. Moreover, the fact that the frequency of polymorphic phases may change from one time of year to another (pp. 221, 226) cannot be regarded as an intermediate stage between the occurrence of seasonal forms and definitive polymorphism. For, in the latter condition, the proportions are determined by opposed selective forces which are differently balanced in different environments; as with the seasonal fluctuations of the ST and CH inversions in *Drosophila pseudoobscura*, the relative advantages of which change with temperature (p. 221).

It will be realized from what has already been said that any polymorphism, be it transient or balanced, must be built up and maintained by selective forces. These, as previously mentioned, turn out to be far more powerful in a state of nature than was supposed twenty years ago, while the existence of such genetically controlled diversity must itself be a sensitive indicator of changes in the contending advantages and disadvantages involved, or in the spread of a gene. The existence of polymorphism therefore always advertises a situation of importance, and is especially well suited to detecting evolution in progress (Ford, 1953a, 1961; Huxley, 1956). It is for this reason that it was originally picked upon as one of the conditions which could be used effectively for that purpose (p. 8).

Human polymorphism

It seemed appropriate to apply this generalization to instances in which it had been ignored. Among these the human blood-groups were outstanding. They were first treated as examples of balanced polymorphism by Ford in 1942a. Though that step received emphatic approval from Sir Ronald Fisher, it was some time before it attracted general attention; Dobzhansky regarded the blood-group genes as of neutral survival value up to 1951 (his p. 156), though he was among the first to see the importance of altering that view. Yet theoretical considerations clearly indicate that they must be maintained in the population by a balance of advantages and disadvantages. Indeed it had been suggested by Ford in March 1940a that they must have unknown effects for which they are selected, though this was before Levene and Katzin had demonstrated Rhesus incompatibility between mother and foetus, which they did in the autumn of that year. It thus became essential to analyse the blood-groups from this aspect.

The genes and super-genes controlling them in Man have no effect upon bodily structure nor upon the choice of husband and wife. Their selective importance therefore means that they must influence either fertility, general viability, or susceptibility to disease. Considerations along these lines led to the prediction that the blood-groups, the presence or absence of the A,B,O antigens in the saliva, and the ability or inability to taste weak solutions of phenyl-thio-urea are associated with liability to develop specific diseases (Ford, 1945). The first instance of the kind was reported after six years, when Struthers (1951) showed that a larger proportion of Group A than of Group O babies died of bronchopneumonia during the first two years of life. This result has not been confirmed in later work, a discrepancy possibly due to the much more extensive recent use of antibiotics. At present, therefore, the correlation reported by Struthers must be regarded as suggestive only. Since that date, however, various other such associations have been detected and fully established and these, increasing in numbers, are opening up a new branch of medicine. Its importance for diagnosis, and conceivably for treatment, is yet to be explored (Clarke, 1961). As these discoveries are dependent upon the polymorphism concept, one so largely analysed by techniques of ecological genetics, it is appropriate to mention some of them here since they have implications for experimental work in the field which will be mentioned in subsequent chapters. It is also possible to indicate a few further lines of research to which they may lead.

One comment must be inserted at this point in parenthesis, as it were. The notation generally used for the human blood-groups is in a chaotic state. It is not only completely out of accord with that otherwise universally employed in genetics but it is self-contradictory and inconsistent from one group to another. The confusion it introduces has prevented many geneticists from interesting themselves in serology; I suspect also it has prevented many serologists from understanding genetics and the genetic unification of their work. Accordingly, I have devised a uniform notation for the blood-groups (Ford, 1955a) which is in accord with normal genetic practice and in which the symbols for genes, antigens, and antibodies are related in a simple and consistent way. This being a first attempt is likely to be improved upon but it is the only one of its kind yet available and, though it has not so far come into general use, it is employed in this account in the few instances in which any blood-grouping notation has to be introduced.

Aird, Bentall, and Fraser Roberts (1953) have shown that a significantly high proportion of those who develop cancer of the stomach belong to blood group A. Moreover, Clarke et al. (1955) find no excess above expectation of A Secretors, compared with A non-secretors, among them. Thus the carcinogenic stimulus is likely to be some other effect of the determining gene (G^A) rather than irritation of the stomach-wall by large quantities of the G(A) antigen secreted by the salivary glands and swallowed. It should be noticed that at the present time the frequency of cancer of the stomach could not be altered by selection since the disease occurs almost entirely after the period of child-bearing. There are, however, reasons for thinking that in past ages most cancers occurred earlier in life than they do today (Ford, 1949), while the gene in question (G^A) must have other effects in addition to those influencing malignancy and controlling blood-group A. One of these is already known: it increases the liability to pernicious anaemia.

Clarke et al. (1959) have proved that group O individuals are about 35 per cent more likely to develop duodenal ulcers than are others, and that the tendency to do so is still greater, by about 40 per cent, in those who fail to secrete the G(A) and G(B) antigens into their saliva. This is itself a polymorphism, with the non-secreting type recessive and affecting approximately 22 per cent of Europeans.

Another polymorphism involves the ability to taste phenyl-thio-urea. This is a dominant trait found in 70 to 80 per cent of the population in Europe and the Middle East, and it affects also the pathology of the thyroid: toxic diffuse goitre (Graves' Disease) being much commoner in the 'tasters' who, however, are significantly less subject than 'non-tasters' to the multiple adenomatous form. The latter condition is rarer, but more closely associated with the taste polymorphism, in men than in women.

What may prove to be a highly important human polymorphism is that controlling rapid or slow inactivation of isoniazid, which is one of the drugs now used in the treatment of tuberculosis. Those reacting to it in these two distinct ways are in approximate equality and this was true for the two Caucasian and the two Negro populations which comprised the 484 people studied by Evans *et al.* (1960). He and his colleagues find that the control is autosomal and the slow inactivators are recessives. We do not yet know whether tuberculosis patients belong disproportionately often to one or the other phase; nor do we know whether the two types respond differently to isoniazid treatment, as that substance is not generally used alone. Clarke (1962) makes the very reasonable suggestion that this particular 'concealed' polymorphism may exist in the human species because some substance which is normally ingested may be metabolized in a similar way to isoniazid.

Infections were included in the association between the human polymorphisms and disease when this was originally predicted, a feature which has also been confirmed. Thus those who respond to type-A haemolytic streptococcal infection by developing rheumatic fever are significantly deficient in blood group O and in Secretors (Clarke *et al.*, 1960).

Of far greater importance in human affairs is the fact that individuals belonging to blood groups A and AB are much more liable to develop smallpox, and to do so in a more severe form, than those of the O and B types. When first reported, this was denied on experimental grounds (Harris, Harrison and Rondle, 1963 and others). However, the connection between that disease and the blood groups has been fully established by Vogel and Chakravartti (1966) working in Western

Bengal and Bihar, and confirmed by the studies of Bernhard (1966) both in India and Pakistan. This correlation which seems to be limited to severe epidemics in unvaccinated populations, must have been an important factor in determining the OAB frequencies of the human races.

When the connection between the blood groups and disease was originally established after its prediction, it chanced that the pathologies involved were associated with the alimentary canal and, accordingly, it was suggested that they only were inter-related with the blood groups. No-one can maintain this now, nor indeed was any such restriction probable on theoretical grounds.

Some confusion has arisen in regard to the prediction that the human blood groups must be associated with specific diseases. It has repeatedly been pointed out that the effect involved is too small to be an important agent in controlling the frequencies of these serological types; while it certainly could not do so if operating after the age of child-bearing, as with cancer of the stomach. One may of course remark that there is no limit below which a selection pressure is ineffective. On the other hand, the association between the blood groups and diseases is one predictable effect of the polymorphism involved, one indeed of special medical interest, but others much more powerful may exist. Thus it is now known that differential elimination of incompatible types of foetus takes place: that involving O mother and A foetus amounts to about 10 per cent (Chung and Morton, 1961). We have at present no information on differential survival at the other end of the age scale (from 90 years upwards, for example), partly owing to the technical difficulty of providing appropriate controls. The powerful effect of Rhesus is well-known, so also is its interaction with the A,B,O, groups to produce haemolytic diseases of the new-born, as already mentioned (p. 99). We can be sure that the human polymorphisms have evolved heterozygous advantages, but in many of them that situation is obscured owing to dominance: for instance we cannot at present discriminate between the homozygotes and heterozygotes of groups A and B of the O,A,B, series, but we can be confident of finding an excess of the latter when that distinction can be made. However, all three genotypes are separable among children of AB × AB matings and an excess of the AB group among them is fully established (Chung and Morton, 1961). A similar distinction is possible in another series: for when we consider the parents both of whom are of the MN type, they too produce a significant excess of the MN group (Chung et al., 1961).

The human polymorphisms are, like any others, constantly associated with super-genes and with multiple alleles either apparent or real. Thus the MN,L, Hunter, Henshaw system and the Rhesus factors certainly provide instances of super-genes. It will probably be found that the genes of the O,A,B series (the principal members of which are G^{A1}, G^{A2}, G^B, G) do so too. A genetic study of these conditions from this point of view may make it possible to obtain further evidence on the evolution of linkage. An additional instance of the type to be examined is provided by the Lutheran and Secretor genes. These, both affecting the blood-groups, are fairly close to one another on the same chromosome: Race and Sanger (1958, Addenda) give 9 per cent as the best estimate for the recombination-fraction in Western Europeans. This association may be due purely to chance. It is possible, however, that there is some advantage in keeping these genes together at least in certain populations. Further studies on their linkage should be carried out in very different races, Hindus, Chinese, and American Indians for example, so as to provide a considerable contrast with the information now available. It is conceivable that significant differences in crossing-over between these two loci would then be detected.

The last time that I recommended this type of investigation (Ford, 1957), it may well have appeared less hopeful than it does today. For work on the snail *Cepaea nemoralis* has demonstrated that significant differences can occur in the linkage-values between two major genes each controlling polymorphic characters (p. 179).

Another form of heterozygous advantage is provided by a human polymorphism associated with the blood but not, so far as is known, interacting with the blood-groups or secretor to influence human disease, though itself of importance in that connection. In Man the foetal haemoglobin is normally replaced by the ordinary juvenile and adult type 'A' within a year from birth. In certain parts of the world another haemoglobin, 'S', partly or even wholly replaces A. These two adult forms are controlled by a pair of alleles Hb^A and Hb^S. In the homozygous condition, $Hb^S Hb^S$, haemoglobin A is absent but, in addition to S, there is a small amount of the foetal type normally lacking in adults. Such individuals are at so great a disadvantage that four out of five die before the age of reproduction. They do so because in the absence of the normal haemoglobin the erythrocytes become curved and distorted (the 'sickle' shape). These tend to block the capillaries and are then phagocytozed, causing severe anaemia and other symptoms, generally fatal. In the heterozygotes $Hb^A Hb^S$ about half of the haemoglobin is of form A and the erythrocytes are normal in the blood-stream. They only sickle if the oxygen-tension of a drop of blood is reduced, by ascorbic acid or other agencies, by which means the heterozygotes can be distinguished from the ordinary homozygotes $Hb^A Hb^A$. These heterozygotes are not only healthy but the proportion of haemoglobin S which they carry renders them relatively resistant to subtertian malaria due to *Plasmodium falciparum* (Allison, 1954, 1956). Moreover, even when infection occurs, the course of the disease is then short and mild. Thus in malarial regions, but not elsewhere, the heterozygotes are at an advantage over both homozygotes. They occupy 17 per cent of the population in some parts of Greece and even up to 40 per cent of some East African tribes, in spite of the almost lethal character of the $Hb^S Hb^S$ condition; and here, where malaria is hyperendemic, Allison (1954) finds that the fitness of the heterozygotes is 1·26 times that of the normal homozygotes.

This neogenic situation is more complex in West Africa owing to the presence there of yet another haemoglobin-type, 'C', controlled by a gene Hb^C. That this behaves as a multiple allele of Hb^A and Hb^S is suggestive. It is possible that a super-gene has been evolved, for there is interaction here. Those with the genotypes $Hb^C Hb^C$ and $Hb^S Hb^C$ suffer from anaemia since they have no haemoglobin A, but it seems that $Hb^A Hb^C$ heterozygotes are at an advantage over normal people, probably because they are partially resistant to a blood parasite, though not necessarily malaria. In such a situation as this, where all three adult haemoglobins occur, one would expect the Hb^C gene to be rare where Hb^S is common and the reverse, and there is some evidence of this. It is noteworthy that in the United States, where malaria is now uncommon, the frequency of Hb^S has been much reduced among the Negro population. It is clear, then, that these haemoglobin polymorphisms are maintained by heterozygous advantage.

Furthermore, the two thalassaemias can conveniently be classified as alpha and beta, since these conditions consist in a failure in the production of the α or of the $\beta N + \delta$ globin chain. In the heterozygotes of either, the interference with haemoglobin synthesis is not serious while there is considerable protection against malaria (thalassaemia minor). On the other hand, the homozygous state (thalassaemia major) is usually fatal.

Summarizing, we may say that the abnormal haemoglobins are due to a change in the amino-acid sequence while the thalassaemias are due to a defect in the production of normal haemoglobin. Yet both, in different ways, protect against the same parasite.

In addition, a deficiency of the enzyme-system glucose-6-phosphate-dehydrogenase, which leads to haemolytic anaemia, provides partial immunity to *Plasmodium falciparum* (Allison and Clyde, 1961). The abnormal enzyme is controlled as a sex-linked dominant, in which the heterozygote may be intermediate.

It is very desirable to detect as many as possible of the concealed poly-morphisms which must exist in Man. For this there are two main reasons. The first is that just given, the fact that they are influenced by powerful selection-pressures. The second is that they could act as markers of the human chromosomes. A good linkage map of Man would be a great benefit in certain types of medical work; making it possible, for example, to detect carriers of hereditary diseases, or those who will or will not develop them later in life if they are members of a family in which such conditions occur; as with Huntington's chorea.

The blood-groups and other human polymorphisms supply the physical anthropologist with decisive evidence. This has been recognized for forty years, but for the wrong reasons. Commenting upon this matter some time ago (Ford, 1957) I remarked:

> By a curious inversion of logical thought, it was held that their occurrence in distinct and characteristic proportions in the different races of mankind was especially important because the variation involved was selectively neutral. Precisely the contrary is true. The fact that the genes concerned are balanced by selection at optimum frequencies, which differ from race to race, is the one which gives them significance as a criterion of relationship. It does so because in these circumstances their proportions are influenced by the average genotype of the population in which they occur.

The matter is very relevant to the study of ecological genetics, for in no other organism have so many polymorphisms been analysed as in Man. Consequently their adjustments are a guide to what may be expected in lower forms. It is so far only in human polymorphisms, and in certain butterflies (Chapters 12 and 13), that we are able to examine the frequencies of many different phases in many different races. These indicate that genes stabilized by selection to maintain certain values in one community may be maintained with equal exactitude at quite different proportions in another: a situation determined partly by genetic adjustments to the balanced gene-complex of each population and partly by the need to produce an optimum reaction with the external environment.

To take a relevant instance, Group B of the O,A,B blood-group series becomes commoner as we pass south-eastwards from north-west Europe to India, where it reaches its highest value. We have good reason for thinking that this is an adjustment to certain aspects of the environment since the exceptionally high values of Group B in southern Asia are attained by people of both Caucasoid and Mongoloid stocks. On the other hand, we know that the different frequencies are partly determined by the genetic structure of the various populations concerned, for a Hindu race, the Gypsies, has preserved its North-Indian blood-group characteristics unimpaired after living for hundreds of years in north-western Europe (Table 8). Thus a palaeogenic situation has been built up in South Asiatic peoples which interacts advantageously with a high frequency of the gene G^B, responsible for producing the B-antigen and the other characteristics associated with it.

The human blood-groups thus supply valuable information to the student of ecological genetics. In certain respects they illustrate more fully than lower organisms have so far done both the stability and the varying adjustment of genetic polymorphisms.

TABLE 8

	O	A	B	AB	Total sample
France, Bouches-du-Rhône	41·5	50·4	6·0	2·1	752
Germany, Dresden	38·1	41·6	14·1	6·2	6,341
Hungary, Debrecen	31·1	38·0	18·7	12·2	1,500
Persia	37·9	33·3	22·2	6·6	1,000
United Provinces, Hindus	30·2	24·5	37·2	8·1	2,357
Gypsies: French Riviera	22·1	25·7	38·1	14·1	113
Gypsies: Hungary (Debrecen)	28·5	26·6	35·3	10·0	975

The percentage-frequencies of O,A,B blood-group phenotypes, with special reference to group B, passing approximately from west to east. French and Hungarian Gypsies are added to demonstrate their North-Indian characteristics (extracted from Mourant, et al., 1958).

Selection for sensory appreciation and physiology, presumably heterozygous advantage, seems to be combined in human red–green colour-blindness. It has been said that this provides a simple example of a recessive controlled by total sex-linkage in the X-chromosome. Actually a super-gene of two units seems to be concerned: one determining red-perception and the other green. The disability is maintained at a remarkably high frequency in Europeans. It affects about one man in eight, and is therefore a polymorphism. Consequently if completely recessive and if the three genotypes were of equal survival value, one woman in sixty-four would be red–green colour-blind.

Post (1962) provides evidence that the condition is slightly less frequent among primitive peoples who live or have recently lived by hunting and food-gathering in whom, apparently, it is a greater handicap. Presumably therefore the super-gene responsible for it has increased in civilized populations until the selective balance between the two sets of alleles has been restored. Post himself attributes that increase entirely to mutation. This is an error for indeed mutations could have no such effect (p. 361 and see Ford, 1965) on a polymorphism. The gene must even in primitive peoples have been associated with advantages as well as disadvantages; the balance between which has merely shifted owing to an alteration in the conditions of society.

I have in the past pointed out the remarkable effect that a similar abnormality in perception might have if occurring in any species which appreciates colour and hunts by sight (Ford, 1955c, pp. 95–6). Owing to the fact that colour-blind individuals of this type can discriminate shades which appear alike to normal people, a reaction used in the Ishihara tests, a single mutation occurring in a predator could annul the advantage of a protective colour-pattern which had been evolved in its prey. It is indeed an instructive experience to accompany a trained but colour-blind naturalist in the field and to notice the preternatural powers of detecting cryptic forms which he will occasionally display.

Criticisms of the polymorphism concept

There has been considerable opposition by H. J. Muller and some others in the U.S.A. to the view that polymorphism is a widespread and important phenomenon and to the concept that it is generally associated with and maintained by heterozygous advantage. The points raised by Muller in his 1950 and 1958 articles are characteristic of these objections and it will be useful briefly to comment upon them.

Muller speaks of 'the recent claims, harking back to pre-Mendelian views, that heterozygosity in itself tends to be beneficial'. But, as explained in this book, students of the subject claim nothing of the sort. They find that heterozygous advantage tends to be *evolved* in polymorphism (pp. 228, 302), which is an entirely different contention from the one which Muller is criticizing. He continues by saying that

the evidence from inversions in *Drosophila* is quite beside the point since they carry with them intact multigenic complexes established by selection. A situation comparable to this cannot exist in most organisms since crossing-over in the male (unlike that in the female), when it occurs within heterozygous inversions, results in meiotic bridges that lower the reproductive potential and thus eliminate the inversions before they can accumulate complexes that possess a net advantage when heterozygous.

Here Muller was misled by applying to other animals his great knowledge of *Drosophila melanogaster*, and in consequence he was mistaken. Thus such inversion polymorphisms have been subjected to critical study in *Chironomus* species, in which they are common and widespread in the presence of male crossing-over (Acton, 1957) (see also p. 232). Moreover, Acton (1956) has in fact demonstrated the formation of meiotic bridges within the heterozygous inversions of *C. dorsalis*. The truth is that the reproductive potential of most organisms is so great that they can well support some reduction in fertility without prejudice to their evolution: a fact pointed out by Darlington (1956*a*) when discussing the frequent successful establishment of polyploids in nature in face of their necessarily reduced fertility.

Muller also states that

even if heterozygosity did confer a *per se* advantage [which, as already mentioned, modern students of polymorphism do not claim] it would be expected that each species having large effective population numbers would in consequence of natural selection have tended to increase its degree of heterozygosity nearly to optimum level. Thus, any further increase in heterozygosity, such as would be brought about by the application of mutagens, would not be to the species' advantage.

In the first place, when a major gene spreads in nature and becomes polymorphic, as it will often do, heterozygosity is in fact increased up to the optimum level: that resulting from the contending advantages and disadvantages involved. In these circumstances, the application of mutagens is indeed most unlikely to be to the advantage of the individuals since, as Muller himself points out, random changes relative to the needs of the organism can rarely be beneficial. One may also remark that Muller is here involved in the well-known confusion between the idealized concept of advantage to the *species* and the working of selection, which operates upon *individuals*.

It is generally and reasonably held that the possession of many heterozygous mutant genes by a diploid population gives it a more useful store of variability in case of need than would be available in a haploid one. This Muller denies, remarking a little analysis shows, however, that the diploid accumulates mutant genes only to the point where it has the same amount of phenotypic variance per genome as it would if it were haploid, and that, if conditions changed, no more material would be available for selection in the diploid than in the haploid.

There is here a confusion between phenotypic and genotypic variation. It is approximately true that the diploid only possesses the same amount of phenotypic variation as would the haploid. It does, on the other hand, possess tens of thousands of times more genotypic variation. Here, contrary to what Muller says, is its immense advantage, for it consequently possesses vast opportunities for trying out genes in new genetic and environmental situations in some of which they may react advantageously, a resource denied to the haploid.

There are many instances in which diversity itself is an advantage, as with the wider range of conditions tolerated by the phases of some *Colias* species (pp. 163–4). Apart from these, the maximum adaptive values are reached in balanced polymorphism when two alleles, supergenes, or inversions, $A1A2$ are in stable equilibrium. Yet in these circumstances the less well adapted homozygotes continue to be produced. That is to say, the population may not be as well adjusted as it would were it monomorphic for another allele $A3$ equal or superior in fitness to the heterozygote $A1A2$.

It has therefore been suggested by Muller that natural selection must lead to the latter state. Thus he remarks that balanced polymorphisms 'are to be regarded as temporary makeshifts that arise in the stress of comparatively rapid evolutionary flux and that they are due to be rectified ultimately, where a long-term natural selection repairs its short-term imperfections and miscarriages'. This fallacy has been exposed by Dobzhansky and Pavlovsky (1960) who point out that the flaw in it lies in the assumption that whatever level of adaptation may be reached by the heterozygote it can always be equalled, if not exceeded, by another allele arising by mutation. They give an excellent analysis which shows that balanced polymorphisms are not merely 'temporary makeshifts' but that they may be very stable. Their evidence is derived both from long-continued laboratory tests, in which the heterozygous advantage is maintained, and from situations found in a state of nature. Thus as they say,

in *Drosophila willistoni*, which is a species not closely associated with Man, some inversion heterozygotes occur over vast territories, including much of the inter-tropical zone of the Western Hemisphere. A genetic variant must have existed assuredly for a long time to have spread to, and become incorporated in, the gene pools of populations so numerous and so remote.

One might well add here some of the human blood-groups and the taste-test distinction, for which the Great Apes are also polymorphic.

It is especially to be noticed that genes equal or superior in their adaptive value to the heterozygotes of a balanced polymorphism are very unlikely to arise by mutation. This is due to the fact that the switch-control of a balanced polymorphism is so often a super-gene, which is itself the product of evolution by selection (as of course is the gene-complex in which it acts).

Speaking of the benefits derived from heterozygosity at individual loci (contrasted with inversion heterozygotes, that is to say) Muller says, 'the critical cases in point are so rare and special (although they tend to make themselves conspicuous) as almost to prove that they are contrary to the rule'. In the first place, it is difficult to draw a distinction between point-mutations, super-genes, and inversions, all of which are frequently used as switch-mechanisms in polymorphism. Cytological evidence indicates that small inversions are common in the chromosomes of organisms in general. Indeed a comparison of the detailed morphology of human chromosomes and those of the Great Apes proves that small as well as large chromosome rearrangements have been frequent in the ancestry of Man.

Muller is misled in suggesting that polymorphism is a rare phenomenon: an erroneous conclusion to which the student of *Drosophila melanogaster* is almost necessarily subject. Here we have a species in which disadvantageous mutant genes have proved an invaluable tool; the one by which the physical basis of heredity was first revealed. Yet its genetics in a state of nature, its ecology, and its adaptations are virtually unknown. The laboratory mutants, by which the triumphs of Drosophilosophy have been achieved, leave the experimenter who works in that field ignorant of the fate of advantageous genes in wild populations, of the polymorphisms so often associated with them, and of the frequency with which heterozygous advantage is generated. Those subjects have, however, been studied with outstanding success in other members of the Genus, especially *Drosophila pseudoobscura* and *D. persimilis*, to which the methods of ecological genetics have been applied. The results of that work, due so largely to Dobzhansky, will be discussed in Chapter Eleven.

CHAPTER SEVEN

Balanced Polymorphism in
Panaxia Dominula

Until 1964 the genetics of balanced polymorphism had been studied in
relatively few instances, although more often in the Lepidoptera than
in any other group. Yet by now the available information upon this
subject has become too extensive to summarize effectively in the space
available in this book, supposing indeed it were desirable to do so. However, this is a work designed to develop principles, with examples
sufficient to illustrate them but no more. A proportion only of the
known genetic polymorphisms will therefore be discussed here. These
will be described in some of the succeeding chapters. In the present
one a single example will be considered more fully to indicate some of
the methods used in analysing polymorphic forms.

No natural population of animals in the world has been so fully
quantified as that of the Scarlet Tiger moth, *Panaxia dominula*, at
Cothill in Berkshire; a species which seems to belong to the family
Arctiidae, though it has sometimes been placed in the related Hypsidae.
The locality is an unusual one; a marsh of about 15 acres (Map 4; Plate
3(2)), consisting of highly calcareous peat which attains a maximum
depth of fourteen feet, filling what were once several small converging
valleys. It is drained by a little stream which flows into the river Ock and
thence into the Thames. The whole area is completely isolated from other
swamps by woods, a village, and encircling agricultural land. Its flora
and fauna are exceptional. The plants include Butterwort, *Pinguicula
vulgaris*, and Grass of Parnassus, *Parnassia palustris*, while the predominant species comprise the Great Reed, *Phragmites communis*,
Comfrey, *Symphytum officinale*, and Hemp Agrimony, *Eupatorium
cannabinum*. Among animals, the Grasshopper Warbler, *Locustella
naevia*, is abundant and there is a striking population of Fenland moths,
some among them being unexpected, such as the Scarce Burnished Brass,
Plusia chryson. The glow-worm, *Lampyris noctiluca*, is particularly
common here and so are several of the rarer Coleoptera. The charac-

teristics of this locality are indeed more like those associated with the 'Broads' of East Anglia; while the alkaline peat which, surprisingly enough, is of the earliest Atlantic Age almost to the surface, gives it a quality of its own.

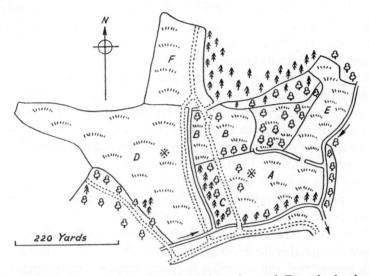

MAP 4. *The marsh at Cothill, Berkshire, where the moth* Panaxia dominula *is polymorphic for the* medionigra *gene. The colony formerly occupied areas A to E. It has now spread into F but no longer extends to E, C, or the eastern two-thirds of A.* * = *releasing sites; formerly in A, now in D. Except where other indications are given, the locality is bounded by agricultural land.* Heredity (1947), 1, 146.

This remarkable environment is doubtless responsible for the fact, previously mentioned, that Cothill is the only place where *Panaxia dominula* is known to be polymorphic for the *medionigra* gene, although the species is widespread but local in southern England and in Europe south of the Baltic. It seems likely indeed that opportunities for the same situation have occurred elsewhere but that conditions have not favoured it, for single specimens of what appear to be *medionigra* have been found occasionally as rare aberrations in other places.

The Scarlet Tiger is a day-flying moth (Plate 5(1), Figures 1 and 2), always colonial in habit. Its colours are brilliant. The fore-wings are blackish with green iridescence, their spots are white except for the two basal ones which are yellowish. The hind-wings are scarlet with black spot and markings. Thus the insect looks almost tropical on the wing. The males are active and difficult to catch in the sunshine, the females are more sluggish and less readily disturbed. Both sexes can be captured more easily in dull weather and in the rain, when they sit on the vegetation.

The colour-pattern of this insect provides it with three main means of defence. At rest, when only the fore-wings are seen, it is cryptic, for it may well be mistaken for a dark patch of leaves with light shining through them, while on a flower-head the white spots resemble buds (Ford, 1967, Plate 15). On the wing it exhibits flash-coloration, for the undulating flight, with flickering scarlet which vanishes suddenly when the moth alights, makes it a difficult species to follow. Moreover, it is also warningly coloured. If a resting specimen be disturbed in dull weather it thrusts it fore-wings forward, exposing the brilliant hind pair in a characteristic attitude with the body curved downwards, while two large yellowish drops of fluid well up from the cervical glands on the prothorax just behind the head. It has been shown that this is extremely repellent to a number of animals (bats and other mammals, various birds, reptiles and amphibia), which reject the insect vigorously as soon as they begin to eat it (Fraser and Rothschild, 1960a and b). Indeed, starlings refuse meal-worms, to which they are greatly addicted, when moistened with this liquid, wiping their beaks in violent disgust. It is worth noting that the secretion is nearly tasteless, and certainly not repellent, to human beings, though it produces a disagreeable local irritation of the throat if the moth be chewed (Lane, 1957). This indicates how incorrect our own judgement in such matters may be and how unreliable experiments can be in which a group of people assess the flavour of animals or their eggs in order to detect whether they are protected by nauseous qualities. Furthermore, Lane and Rothschild (1959) have shown that some species of moths are so poisonous that it may be decidedly dangerous to taste them.

In Britain *Panaxia dominula* is found in two very different types of habitat: (i) marshes and marginal agricultural land, damp and usually copsed; (ii) hedgerows. Here and in France the favourite food is Comfrey, *Symphytum*, and for this the larvae show a preference even if they have come from a locality where it is absent (Cook, 1961). They feed also upon *Filipendula ulmaria*, *Urtica dioica*, *Rubus salix*, and many other plants; indeed in most European countries they seem to be polyphagous. The female lays on the average about 200 eggs. Since these are without any adhesive substance and are scattered broadcast among the herbage, the preference for *Symphytum*, already mentioned, is a larval one not due to selection by the imago.

Panaxia dominula has only one generation in the year in England. The imagines first appear in early July and are to be found for about three weeks, with stragglers well into August. The larvae hibernate among litter when young and resume feeding in the spring. They are then quite conspicuous: black and somewhat hairy with yellow marks along the back and sides. Their length when fully grown is about 3·5 to 4 cm. (Ford, 1955*c*, Plate XI). They are, indeed, often present at Cothill in immense numbers and although they are eaten to some extent by cuckoos, and possibly by pheasants and thrushes, they seem almost free from attacks by other predators. Very curiously, they are principally eliminated when nearly fully fed and during the pupal period. Their death seems due to some unknown factor such as bacterial or virus disease, not to insect parasites, from which in Britain at any rate the larvae are almost free. The pupae, which lie among vegetation on the surface of the ground, are occasionally parasitized by a Calcid, *Pteromalus puparum*, but not to a serious extent. The Hon. Miriam Rothschild tells me that captive mice and voles will not eat them. It is, I believe, not known if shrews constitute serious enemies of *P. dominula* while in the pupal stage.

It is worth noting that elimination may take a very different pattern in other Lepidoptera. It has been most fully studied, by Dempster (1967), in the Small White, *Pieris rapae*. In that butterfly he showed that there is a mortality of 45 per cent or more during the first two larval instars, due to predacious Arthropods (the bettle *Harpalus rufipes* and the harvestman *Phalangium opilio*). Birds take about 20 per cent of the larvae but, effectively, they are a danger only to the larger specimens, in the last two larval instars. Virus disease destroys no more than 1 to 4 per cent of this species.

Estimates of larval populations

Our estimates of larval and pupal mortality in wild insects are based entirely upon conjecture and, save in a single species, those important facts have never been assessed with an approach even to their correct order of magnitude. That sole exception is provided by *Panaxia dominula*, in which these values have indeed been calculated in two colonies. This has not been done at Cothill but at the neighbouring, though fully isolated, marsh of Sheepstead Hurst, a mile away (pp. 7–8, 143) and formerly in a garden at the University Museum, Oxford, where a small population of the species, originally introduced there, had long maintained itself. In both these places the nearly full-grown larvae were labelled with a radio-active tracer (sulphur-35) which could still be recognized, by means of a Geiger counter, in the imagines to which they give rise; see pp. 7–8 (Kettlewell, 1952; Cook and Kettlewell, 1960), and these latter were themselves marked in the ordinary way with quick-drying cellulose paint (pp. 6–7, 134–7).

The results confirmed the high late-larval and pupal mortality predicted from observations in the field and evident in breeding experiments. In May 1953 Kettlewell returned to the Sheepstead Hurst colony 1,227 radio-active larvae of *P. dominula* which had been removed from it and fed on deadnettle, *Lamium*, grown in a water culture, containing sulphur-35. In the following July Sheppard (1956), using the ordinary marking, release, and recapture technique, estimated the total population of imagines as 20,000 to 26,000 and, in all, 18 radio-active moths were found among samples which amounted to 3,418 specimens. The larval population can be calculated as: $1,227 \times 3,418/18$ (see p. 8) giving 232,994 as a mean figure. The moths resulting from radio-active larvae, however, hatched slightly earlier than usual because they had been kept at a somewhat higher temperature during the period of treatment. The scatter of the results indicated that about five or six might have been present in the first samples and these had not been scored with a Geiger counter. An adjustment for this produced a mean larval population of 198,185 to 225,210 which, with a total imaginal population of approximately 23,000, showed an intervening mortality of 88 to 90 per cent.

In 1959 Cook worked upon the small colony near the Museum at Oxford. His control experiments gave no sign of any differential mortality between the normal and the radio-active larvae, nor of any difference in their time of emergence, for those feeding on the radio-active *Lamium* had been kept at the ordinary temperature. The data he obtained indicated a population of 1,210 nearly full-grown larvae, and a mean mortality of 94 per cent between that stage and the imaginal one: a figure which agrees well with Kettlewell's previous estimate at Sheepstead Hurst.

There is obviously a great need for work of this kind, which opens up the possibility of calculating the mortality of holometabolous insects in natural conditions and of assessing the relative viability of different polymorphic forms during development. The error involved in that comparison would be large. Against this, however, must be set the powerful selection-pressures operating to maintain the distinct phases, and these must generally affect the early stages.

Estimates and analysis of imaginal populations

The polymorphism of *P. dominula* at Cothill is controlled by a gene whose effects can be detected in all three genotypes; consequently its frequency is susceptible of direct calculation. The rare homozygotes, known as *bimacula* (Plate 5(1), Figure 7) are very distinct from the normal form already described. All the pale spots on their fore-wings are replaced by black except for the two basal ones. The small white dots near the outer margin that are to be seen in some specimens are, however, under separate polygenic control and are not necessarily affected. The black spots and markings on the hind-wings are much enlarged and confluent, largely obscuring the scarlet ground-colour. *Bimacula* is more lethargic than ordinary *dominula* and, owing to the absence of the white spots on the fore-wings, its cryptic coloration fails and it is decidedly conspicuous at rest.

The heterozygotes, known as *medionigra*, though visibly separable from both homozygotes, are much more like the wild type, *dominula*. From this they can be distinguished by a reduction in the size (Plate 5(1), Figures 3 and 4) or by the absence (Figures 5 and 6) of the central white spot in the fore-wings, and by the presence of an additional black one in the middle of the hind pair. This may be quite noticeable (Figures 4 and 6) or reduced to a few black scales (Figure 3). These, however, are ringed by a narrow circle of yellow, which seems to persist even when all trace of the black has disappeared. Doubtless it occasionally happens that the heterozygote is indistinguishable from typical *dominula*, but this must be an exceptional event since the three genotypes approach so closely to their expected proportions (p. 139).

I noticed a number of *medionigra* in the *P. dominula* population at Cothill in 1936 and 1938, suggesting that the form was more frequent than it had been in 1921 when I had visited the site previously; for at that time no examples of *medionigra* were seen, though the moth was common. This impression later received some support from a study of old collections. A considerable search in museums and elsewhere made it possible to examine 168 specimens caught at Cothill in various years prior to 1929 and only 4 of them (2·4 per cent) were *medionigra*. Since no homozygous mutants (*bimacula*) were included, this represents a gene-frequency of 1·2 per cent. Now variety-hunting has been one of the chief interests of those who collect Lepidoptera. It is usual therefore for them to catch a short series only of the typical insects in each locality but to secure as many varieties as possible. Thus ordinary collections greatly exaggerate the variability of the species. There is certainly no tendency to reject abnormal specimens: a statement which anyone acquainted with butterfly- and moth-collecting in Britain would emphatically endorse. Consequently the sample of 168 *P. dominula* obtained before 1929 will tend to over-estimate, certainly not to under-estimate, the frequency of *medionigra*.

Bearing these facts in mind, and having regard to the potential advantages of the situation for evolutionary study, it seemed to me worth while to take a random sample of *P. dominula* at Cothill on the next possible occasion: the summer of 1939. This amounted to 223 specimens, which included 37 *medionigra* (16·6 per cent) and 2 *bimacula*; thus the frequency of the *medionigra* gene had now risen to 9·2 per cent. In a similar random collection the year following this had become 11·1 per cent. Owing to pressure of other work in war-time conditions, it was only possible to obtain a total of 117 insects on that occasion, but the relative abundance of the mutant compared with the earlier records, apparent in 1939, was corroborated.

It was therefore decided to obtain information in future on the annual frequency of the *medionigra* gene together with the number of insects flying in the Cothill colony. The totals of the three forms, which have been recorded yearly since 1939, are given in Table 9, in which the corresponding gene-frequencies and population-sizes are also shown. The latter calculations are based upon the method of marking, release, and recapture (pp. 6–7), the results being set out in a triangular table showing the number of captures and releases per day, together with recaptures related to dates on which they were originally caught and subsequently recovered. One of these is illustrated here (Figure 9). The details of this technique can be obtained from Fisher and Ford, 1947, and Ford, 1953*a*, in which both the necessary field-work and the required calculations are described. Thus a few points only need be mentioned now.

Special precautions are taken to ensure that the marked specimens rejoin the flying population and that any which fail to do so on account of injury sustained in the marking process are correctly accounted for. The insects are liberated at a more or less central point and, in *P. dominula* and the other Lepidoptera so far used, they scatter well though relatively few of them reach the more distant parts of the colony. This is compensated for by collecting to something like an equal extent in the whole area. Provided that these precautions are taken, approximately random sampling can be achieved and it is found that marking the wings

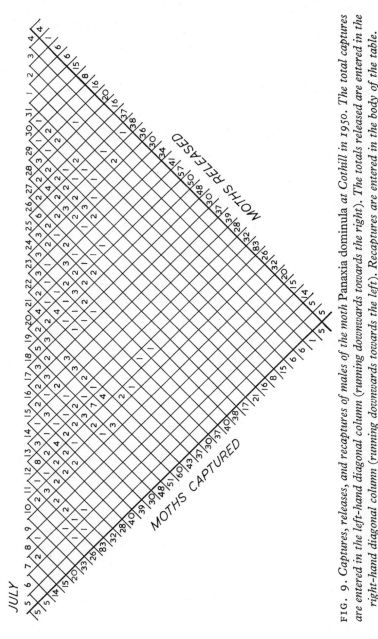

FIG. 9. *Captures, releases, and recaptures of males of the moth* Panaxia dominula *at Cothill in 1950. The total captures are entered in the left-hand diagonal column (running downwards towards the right). The totals released are entered in the right-hand diagonal column (running downwards towards the left). Recaptures are entered in the body of the table.*

with dots of cellulose paint does not affect ease of capture or expectation of life (Sheppard, 1951a). As is almost universal in the Lepidoptera, the males are on the wing earlier in the season than the females, but the three polymorphic forms (the normal one, *medionigra*, and *bimacula*) do not differ in their dates of emergence.

TABLE 9

Year	Captures dominula	medionigra	bimacula	Total	Population Size of total colony	Gene-frequency per cent of medionigra
1939	184	37	2	223	?	9·2
1940	92	24	1	117	?	11·1
1941	400	59	2	461	2,000–2,500	6·8
1942	138	22	0	205	1,200–2,000	5·4
1943	239	30	0	269	1,000	5·6
1944	452	43	1	496	5,000–6,000	4·5
1945	326	44	2	372	4,000	6·5
1946	905	78	3	986	6,000–8,000	4·3
1947	1,244	94	3	1,341	5,000–7,000	3·7
1948	898	67	1	966	2,600–3,800	3·6
1949	479	29	0	508	1,400–2,000	2·9
1950	1,106	88	0	1,194	3,500–4,700	3·7
1951	552	29	0	581	1,500–3,000	2·5
1952	1,414	106	1	1,521	5,000–7,000	3·6
1953	1,034	54	1	1,089	5,000–11,000	2·6
1954	1,097	67	0	1,164	10,000–12,000	2·9
1955	308	7	0	315	1,500–2,500	1·1
1956	1,231	76	1	1,308	7,000–15,000	3·0
1957	1,469	138	5	1,612	14,000–18,000	4·6
1958	1,285	94	4	1,383	12,000–18,000	3·7
1959	460	19	1	480	5,500–8,500	2·2
1960	182	7	0	189	1,000–4,000	1·9
1961	165	7	0	172	1,200–1,600	2·0
1962	22	1	0	23	216	(2·2)
1963	58	1	0	59	470	(0·8)
1964	31	0	0	31	272	—
1965	79	2	0	81	625	1·2
1966	37	0	0	37	315	—
1967	50	0	0	50	406	—
1968	128	3	0	131	978	1·1
1969	508	38	0	546	5,712	3·5
1970	444	31	0	475	4,493	3·4
	17,062	1,295	28	18,385		

Captures of three forms of *Panaxia dominula* at Cothill, Berkshire; population-size of the colony, and frequency in it of the *medionigra* gene.

An essential step in assessing population-size consists in determining the death-rate of the imagines, and this in itself provides the student of ecological genetics with information of great value (pp. 229–31). In the *P. dominula* colony at Cothill, the average daily mortality up to 1946 proved to be 16 per cent, giving an expectation of life of 6·25 days. It was consistent with a constant death-rate which, indeed, has been assumed in the calculations. Yet it seemed likely that this may not be strictly correct and that the value may increase towards the end of the season: a conclusion so far demonstrated for certain only among females of the Sheepstead Hurst colony in 1953. However, it was found that the error introduced by assuming a constant in place of a changing mortality was slight and merely decreased the total female population by 129 in 10,000. Lees (1970), who obtained the 1969 results, showed that the survival-rate at Sheepstead Hurst (0·79) was higher that year than at Cothill (0·71).

It is possible to calculate the extent to which the frequencies of a pair of alleles can alter from generation to generation by chance, provided their proportions and the size of the population in which they occur are known (Fisher, 1930a). Such information had not previously been available for any wild species, but the Cothill colony of *P. dominula* has now provided it. Thus it has become feasible to determine whether the changes in gene-frequency observed there were of a magnitude consistent with random survival. They have proved much too large to be so. Even by the end of the 1946 season, when only eight independent observations of frequency were available, their heterogeneity could be measured by $\chi^2_{(7)} = 20\cdot81$ for which $P < 0\cdot01$, assuming a total of only 1,000 imagines in each generation, while P would be still less were the population greater. Actually, it was only once as small as a thousand in the six years for which estimates were then available. Thus the observed changes in gene-frequency were not of a random nature and must have been due to selection (Fisher and Ford, 1947).

This was the first occasion on which a comparison of this kind was possible in natural conditions. Wright (1948), who has consistently stressed the importance of random drift, challenged the accuracy of these results, which showed that process to be relatively ineffective. He did so on three issues, one of which was clearly inadmissible even at the time. That is to say, he questioned the deductions because no estimates of population-size had been made in 1939 or 1940. Yet to invalidate them the total for 1940 would have had to be no larger than the number of moths actually caught. In that year, only a small sample could be taken and at a period when the insect was flying in considerable numbers, while subsequent data have demonstrated that at no time has more than a quarter of the population been captured in a season.

Wright also pointed out that the selective value of the *medionigra* gene may fluctuate from year to year and that no attempt had been made to estimate such changes (in fact because the observations had not then continued long enough to do so). Indeed he calculated from the records that in half a century the heterozygotes might vary from a semi-lethal condition to one in which they possess a 50 per cent advantage over *dominula*. In reaching this conclusion he had to assume that their selective value 'has no trend but varies according to non-secular fluctuations in conditions from year to year'. Time has proved him wrong. We now know that the *medionigra* gene was consistently subject to a 10 per cent disadvantage compared with *dominula* from 1941 to 1955; its frequency increased significantly, however, in the following two years but has since undergone further trends (Table 9).

Thirdly, Wright suggested that the real and apparent populations of *dominula* at Cothill might be very different. He pointed out that whole broods might be subject to similar environments and be killed off or survive if the females tend to lay the majority of their eggs in one place. But this is precisely what they do not do, as Sheppard (1951a) has been able to demonstrate; for each female scatters her eggs widely over the habitat as she flies. We have here an example of a well-known type of error: that into which it is easy to fall if theoretical deductions on evolution are unsupported by a study, or any knowledge of, the ecology of the situation concerned. Wright indeed seems to have made no attempt to justify his objections though it is more than twenty years since their erroneous nature was demonstrated (by Sheppard, *l.c.*).

It is clear that although the 10 per cent disadvantage of the *medionigra* gene had been maintained as a long-term trend, its selective value must previously have been reversed; for it must have possessed a powerful advantage, which can hardly have been less than 20 per cent, between 1929 and 1939. An alternative possibility would be the reduction of the Cothill population to an extremely small size on several occasions during this period, so allowing the *medionigra* gene to increase by random drift. Cothill is a well-known habitat of great interest only five miles from Oxford and attracts entomologists from the University. It was possible therefore to get into touch with a number of people, at least one of whom had collected there each year from 1929 to 1935 and in 1937 (the species was already known to be common in 1936, also in 1938 when the initial observations leading to this work were made, see p. 133). None of them had difficulty in obtaining a series of specimens for their collections in any of these years, and they would certainly have done so if the population had been reduced even to two or three hundred individuals. That a sharp reduction in population-size would have been noticed has now been proved during the period 1962–8.

The principal purpose of studying the *P. dominula* population at Cothill, to determine whether or not its polymorphism is controlled selectively, has already been achieved. There are, however, important reasons for continuing the work. One especially must be noticed. It is to be hoped that the situation which occurred for twelve generations or less up to and including 1940 will be repeated: namely, that in which the effect of selection upon the *medionigra* gene is reversed so that it gains an advantage once more and begins to spread. If it were to do so as rapidly as before, it might be possible to detect and examine the causes of that occurrence, which must certainly have been a remarkable one. The fact that *medionigra* was actually favoured in 1956 and 1957 after being at a disadvantage for fifteen years suggests that at any time a return of the former situation may take place (see also 1969). It is indeed by watching for such opportunities as this that some of the problems of ecological genetics are to be solved.

The total catches of *medionigra* amount so far to 1,295 specimens; assuming equal viability of all three genotypes 1,404·4 were to be expected when 17,062 *dominula* and 28 *bimacula* (the rare homozygote) have been obtained. The latter number is so small that a large error is involved in this estimate, but *medionigra* is certainly not in excess. Thus in this instance there is no indication of heterozygous advantage (and see Williamson, 1960). This might of course be detected if the frequencies could be studied over a period when the *medionigra* gene is heavily favoured. It seems more likely, however, that even then such an effect would not be apparent since it has probably not had time to evolve at Cothill, the only colony in which the polymorphism concerned is known to occur.

It is, nevertheless, established that the *medionigra* gene has other effects in addition to colour-pattern upon which selection must operate, favourably or unfavourably according to circumstances. These influence mating-habits, fertility, and survival between egg and full-grown larva.

Differences in mating-behaviour between geographical races have often been recorded (e.g. pp. 345–6) but they are also known to be produced by a single gene. Thus Rendel (1951) showed that in *Droso-*

phila melanogaster ebony and vestigial males both tend to pair preferentially with females of the opposite type; and a similar situation has been discovered in the bird *Zonotrichia albicollis* (see p. 231). A comparable condition occurs in *Panaxia dominula*. Sheppard (1952a) has shown that the females of *dominula*, *medionigra*, and *bimacula* all prefer to mate with males belonging to a different genotype from their own. The males on the other hand show no such preferences. In this species the sexes are in approximate equality and the males can pair many times but the females once only, except when the previous mating has proved infertile or the store of sperm has been used up. Numbers of males are attracted to the scent of a virgin female so that there is competition between them, giving an advantage to the one which stimulates her most rapidly: a situation which, in view of the mating-behaviour just mentioned, must favour whichever gene is the rarer, in this case *medionigra*, so tending to establish and maintain polymorphism. It should be mentioned that a calculation by Haldane (1954) gave no indication of assortative mating within the wild population of *P. dominula* at Cothill; but his test assumed that no other selective forces were operating. As we have clear evidence that this assumption is invalid as one would anticipate, his conclusion was not justified and has subsequently proved incorrect.

Sheppard (1953) has also demonstrated that though there is no detectable difference in the number of eggs laid by *dominula* and *medionigra* females, the heterozygous (*medionigra*) males are slightly the less fertile. The available numbers of *bimacula* were too small to extend the comparison to that form (the rare homozygote). Moreover, results obtained in his artificial colonies (pp. 141–6) prove that in certain circumstances the early stages of *medionigra* can be heavily eliminated compared with those of normal *dominula*.

It has been suggested that there is a tendency for the gene-frequency of *medionigra* to change inversely with population-size. It is, however, clear that no such simple relationship as this connects these two variables, for an inspection of Table 9 will indicate that their fluctuations are sometimes in phase and sometimes out of phase with one another; a point made by Sheppard (1951a). Williamson (1960) is in agreement with this conclusion. As he further points out, there seems some indication that these two functions (gene-frequency and population-size) are each subject to two-year cycles independently of one another, but of this there is at present no clear proof. In addition to the effects of the *medionigra* gene which we already know, there must be others which contribute to maintaining its polymorphism, and we are not yet fully informed of the selective forces which act upon them. A detailed analysis of the conditions which favour, and which check, the spread of *medionigra* at Cothill should involve a study of the causes of mortality in that colony and of the micro-climate of the habitat. Research into these subjects would be particularly rewarding in an instance such as this in which the importance of selection has already been fully established.

Experimental populations

No evidence of differential survival in the early stages had been detected in the extensive experiments in which *medionigra* had been bred in the laboratory. However, Sheppard (1951*a*) analysed this matter further by exposing segregants to the impact of selection in natural conditions. In 1951 he scattered 4,000 eggs from fertile back-cross matings between *dominula* and *medionigra* on Comfrey (*Symphytum*) in a suitable locality near Hinskey, Oxford, from which the species was known to be absent. The following April, 35 larvae were found on the same plants while none was seen in other parts of the area. They gave rise to 30 imagines of which only 9 were *medionigra*, a deficiency which constitutes a significant departure from the expected 1:1 ratio (P < 0·05). It seems therefore that the two forms were subject to differential elimination during the early stages.

Larvae were found in this artificial colony in small but gradually increasing numbers each spring up to 1956, but no adults were seen. The area was not visited in 1957 or 1958, but by 1959 larvae were common and a sample of them was taken and reared, followed by a much larger one the next year when they had actually become abundant (Sheppard, 1961*a*).

TABLE 10

	dominula	medionigra	bimacula	per cent of medionigra gene
July 1951	—	—	—	25·0*
April 1952	21	9	—	15·0
April 1959	20	3	—	6·5
April 1960	269	32	3	6·2
April 1961	217	35	1	7·3

The frequency of *medionigra* in P. M. Sheppard's artificial colony of *Panaxia dominula* at Hinksey near Oxford.

* The population in July 1951 consisted of the 4,000 eggs from the back-cross *medionigra* × *dominula* which founded the colony.

The results are shown in Table 10. Evidently the chief alteration in the frequency of the *medionigra* gene occurred in the first generation, although its reduction from 1952 to 1959 was significant ($P < 0.01$). The departure from the original value of 25 per cent need not be calculated after 1959 for by that time its significance had obviously become very great.

These facts may reasonably be explained in one of three ways. (1) The selective value of the genotypes had altered greatly between 1951 and 1959, owing to environmental changes. For, as Sheppard points out, the ecology may have been considerably modified by the larvae eating out the young shoots of the Comfrey in the spring. (2) *Medionigra* was initially disadvantageous here, as it is almost everywhere (p. 140), but selection acting upon the gene-complex adjusted its effects so as to fit them to the conditions of the locality, a possibility which will be considered further on p. 145. It certainly looks as if, after a period of adjustment, the gene-frequency has stabilized in the new population. (3) The selection against the heterozygotes during the early stages is later balanced by the mating-behaviour, which favours *medionigra* when that form is rare.

This experiment demonstrates one aspect of an important technique in ecological genetics: that of founding artificial colonies. It not only allows us to assess the effects of the normal environment upon the forms which we study but makes it possible to detect differences which can only be recognized under such stringent elimination as occurs in the wild. For it must be remembered that even in comparison with our more unsuccessful attempts at rearing animals or growing plants in the laboratory, the arts of Man prove immensely superior to the works of nature. We recall that the average output of *P. dominula* is the equivalent of two breeding individuals from 200 eggs: almost all of our own results, other than total failure, are better than this. Consequently a differential which escapes attention in the laboratory may be very evident in the field. Thus when Gordon (1935) liberated in a locality in South Devon 36,000 *Drosophila melanogaster* carrying in equality the gene for recessive ebony body-colour and its normal allele, the relative elimination of the mutant proved to be exceedingly severe; for it had fallen from 50 per cent to 11 per cent in 120 days, representing five to six generations. This was equivalent indeed not only to the destruction of all homozygous ebony flies before they could breed but involved a bias against the heterozygotes as well. Yet ebony is one of the few *Drosophila* mutants whose heterozygous viability in experimental cultures equals or even surpasses the wild type, for it generally establishes a stable polymorphism in the stock bottles. A somewhat similar technique consists in subjecting segregating families, compared with controls, to severe conditions in the laboratory: one which successfully demonstrated differences in viability indetectable in stocks maintained to secure maximum survival (p. 296).

Another way in which an experimental colony can be founded is to use an already existing one and add new forms to it. This technique has advantages. We know that the locality must be suitable, while a sufficient breeding-community is ensured from the start. Indeed if the introduced mutant or polygenic character fails to maintain itself in such circumstances we may presume that it has different requirements from the type already established, provided the experiment has been conducted on a relatively large scale. This method was employed in studying the Sheepstead Hurst colony of *Panaxia dominula*.

That locality has already been mentioned in other connections (pp. 18–19 and 131–2). It is a marsh rather similar in size to Cothill, approximately 20 acres, but with a very different ecology, being much damper and more wooded. However, the food-plant of *P. dominula* is Comfrey here also, and *Phragmites communis* and *Eupatorium cannabinum* again provide shelter and flowers of a kind favoured by the imagines. The whole area is completely surrounded by agricultural land and a neighbouring aerodrome and the insect, though sometimes present in great numbers, seems to be closely restricted to this habitat and completely isolated from any of its other colonies, the nearest of which is that at Cothill. This is rather over a mile away and, in contrast with the conditions there, the Sheepstead Hurst community is not polymorphic for *medionigra*: not one had been found among the 11,102 specimens caught from 1949 to 1954. Moreover, half the releases were given a distinctive mark and none of them has ever been seen at Cothill (Sheppard, 1956).

In view of these facts, Sheppard (*l.c.*) decided to introduce the *medionigra* gene at Sheepstead Hurst, where it was clearly absent except, presumably, as a rare mutant. In 1954, when the population-size was approximately 11,100, he therefore scattered the eggs from 50 heterozygotes back-crossed to normals in suitable places, and the following year 2 of this form were caught among 875 insects. Subsequently the introduced *medionigra* gene has been maintained in this colony at a low frequency, up to about 2·3 per cent of available loci. Moreover, it survived, without considerable alteration in proportion, the great reduction in numbers which this population suffered in 1959 (p. 136).

Thus *medionigra* individuals amounted to 9 out of 407 in 1961, and 1 out of 233 in 1962.

It might have been supposed that this form would gradually be eliminated at Sheepstead Hurst where it has never established itself by normal means. It seems, however, that starting from a considerable number of individuals instead of from a single mutant, there must have been opportunities for the effects of the *medionigra* gene to be adjusted to the requirements of the habitat. There is, moreover, evidence for this, for contrary to the situation at Cothill, its appearance has become less extreme and more like the *dominula* form which it now closely approaches (Ford and Sheppard, 1969). The heterozygotes may thus be able to exploit the advantage which they gain, while rare, from the differential mating-preferences of this species.

In general terms, a heterozygote always produces the most variable phenotype if distinct at all, while that to which the common homozygote gives rise is the least so. For the heterozygous condition is not buffered against changes by a safety-value, as in complete dominance or recessiveness: one sufficient to produce the correct effect throughout the range of environment to which the organism is ordinarily exposed. That is to say, intermediate heterozygotes are particularly likely to be labile and easily influenced by selection acting upon the gene-complex. This might then simply modify their visible features, but much more likely it would increase or diminish their expression; that is, to change our terminology, it would make them more dominant or more recessive. Consequently it has proved a relatively simple matter to alter the colour-pattern of *medionigra* by selecting respectively the more and the less extreme specimens for four generations, until the condition closely approached dominance in one line and recessiveness in the other (Ford, 1960*b*). A similar process has evidently been taking place in the wild population at Cothill (the reverse of that at Sheepstead Hurst) for, though very variable in expression, the heterozygotes (*medionigra*) have on the whole become more extreme there (that is to say, more easily distinguishable from homozygous *dominula*) than they were during the early part of our study there and prior to 1929 (Ford and Sheppard, 1969; Lees, 1970). This seems to be the first occasion on which it has been possible to forestall in the laboratory subsequent evolutionary changes taking place in nature.

In a book of this kind, which assumes a knowledge of genetics, there is no need to give an account of the evolution of dominance, by means of selection operating on the gene-complex, as illustrated indeed by the facts just mentioned. It is necessary only to recall that this concept is due to the genius of Sir Ronald Fisher, who suggested it in 1928 and developed it from that date onwards (Fisher, 1928*a* and *b*, 1929). He himself provided the first experimental proof of it by crossing the Jungle Fowl with breeds of domestic poultry (Fisher, 1935), and its operation was demonstrated for the first time in wild material in the moth *Abraxas grossulariata* (Ford, 1940*b*).

It will be realized that dominance is not a property of genes but of individual genetic characters. For each major gene has multiple effects some of which can be dominant and some recessive, while some may have an intermediate degree of expression in single dose (p. 101).

It is important to notice that the production of dominance and recessiveness is only one aspect of a wider principle: the selective evolution of the effects of individual genes, by which the characters of a given mutant can be gradually adjusted to the needs of the organism. Fisher, of course, fully appreciated this. Indeed he supported his argument with reference to the change which takes place in the expression of the gene for 'eyeless' in *Drosophila melanogaster* owing to selection operating in the stock bottles (Morgan, 1929); also by citing the numerous complex adaptations of butterfly mimicry which are controlled by single genes (Chapters Twelve and Thirteen). Instances of a similar type will repeatedly be encountered in the succeeding chapters of this book.

Sheppard and Cook (1962) give a valuable discussion of the *medionigra* polymorphism in *Panaxia dominula*. Three artificial colonies have now been established in which the gene for this form started at a high frequency, and in all of them it was rapidly reduced to a much lower value. In that at Hinksey (pp. 141–2), the proportion fell from 25 to 7·3 per cent between 1951 and 1961; while it declined from 50 to 43 per cent in a single generation (1960 to 1961) in a population started at Ness, Cheshire. In the Genetics Garden at Oxford (p. 132), where the species was accidentlly introduced in 1958, the *medionigra* gene initially occupied about 25 per cent of available loci, but only 14 per cent by 1961. The viability of *medionigra* seems, indeed, to have been about 50 per cent that of the normal form at all three places. We now have information on two components of this differential. For it is clear that, compared with *dominula*, *medionigra* has about a 75 per cent survival-rate from egg to imago, while its low male fertility has become heavily significant, with $\chi^2_{(1)} = 12·90$, for which $P < 0·001$ (similar information on *bimacula* is still lacking owing to its rarity).

When, on the other hand, the *medionigra* gene was introduced at a very low frequency into the natural population at Sheepstead Hurst (pp. 143–4), it actually increased there (from about 0·2 per cent of available loci in 1954 to about 2·3 per cent in 1969). We thus have a rather clear indication that *medionigra* maintains a stable polymorphism because it is disadvantageous when common but advantageous when rare. That is to say, when at a small proportion its relatively poor viability is balanced by the value of its disassortative mating (p. 140).

As Sheppard and Cook (*l.c.*) point out, these facts are in accord with most of the events that have taken place in the Cothill population. The logarithmic regression of the *medionigra* gene-frequency on year did not show a significant departure from a straight line until 1952. It seems clear, therefore, that the frequency of the heterozygotes starting, strangely enough, at a high level in 1939–40, was for twelve years or so subject to a reduction attributable to their low viability. When these became uncommon, the advantage gained from mating-preferences checked the decline and established a stable polymorphism. Since the differential elimination of *medionigra*, so clear in the wild, is not apparent in laboratory stocks, the fluctuations in gene-frequency actually observed at Cothill could be due to changes in the selective values of the phases consequent upon the varying conditions of a natural habitat.

These considerations throw no light on the conditions existing between 1929 and 1939 when *medionigra* greatly increased in frequency. That form must have had an advantage of 20 per cent or more during this period. Evidently entirely different conditions were operating then from any seen subsequently. We can hardly expect to elucidate their nature until they are repeated. It is necessary therefore to continue the yearly study of the Cothill colony in the hope that the extraordinary changes which it must have undergone during the 1930's are re-enacted: an event which might well throw light upon certain aspects of micro-evolution in general; those in which an uncommon phase appears suddenly to have gained an outstanding advantage.

CHAPTER EIGHT

Polymorphism and the Effects of the Switch-Gene

Selection for the effects of the switch-gene

Considering the mechanisms by which the effects of the genes can evolve we may suspect that, even where the same genetically controlled character is concerned, dominance may sometimes be attained by distinct adjustments in different isolated populations. That situation must often arise in polymorphism and the Lesser Yellow-underwing moth, *Triphaena comes*, seemed to provide an opportunity for examining it. This species, which belongs to the family Agrotidae, is widespread in Britain. It has a single generation in the year and feeds upon a variety of low-growing plants. It is monomorphic* throughout England and a considerable part of Scotland: the fore-wings being ochreous-brown and the hind pair light yellow with a black border. But in Central and north-west Scotland and in most of the Scottish islands it is dimorphic owing to the presence there, in addition, of a dark phase *curtisii*, which is nearly dominant. The heterozygotes have fore-wings of a chocolate shade while the yellow of the hind-wings is, in addition, clouded with black. This form is very variable, but is always clearly distinguishable from the normal, which is relatively constant in its coloration. Homozygous *curtisii* are on the average considerably darker than the heterozygotes, the fore-wings being blackish to coal-black while the yellow on the hind-wings is largely, and may be almost entirely, obscured by black scales. However, the darkest heterozygotes and the palest homozygotes overlap, so that though we can be sure that the two *curtisii* extremes belong to different genotypes, the intermediate range cannot be classified genetically by inspection. This difficulty is enhanced by the fact that the correlation between the amounts of black pigment on the fore- and hind-wings is far from complete. There is, moreover, a separate form, under distinct genetic control, with dark red fore-wings with which *curtisii* must not be confused.

* See p. 148 for an exception.

147

The frequency of *curtisii* evidently differs greatly from one part to another of its distribution. It may in some districts amount to 50 per cent of the population or even more, but the normal *comes* (*cc*) never becomes merely a rare variety. The forms from the island of Barra in the Outer Hebrides and from Orkney are indistinguishable. The two localities are one hundred miles apart, though the Scottish mainland on which *curtisii* also occurs lies between them. However, Barra is separated by fifteen miles of sea from Skye (which is effectively a peninsula); while between Orkney and the north coast of Scotland runs the Pentland Firth, six miles wide at the narrowest point, in which is the small wind-swept island of Stroma.

Evidently the populations of *T. comes* on Barra and in Orkney are very isolated from one another, for the species is not a regular migrant. Thus specimens of *curtisii* do not scatter from north and central Scotland, where they are common, to be caught in the Lowlands or northern England. Moreover, a quite exceptional and perhaps now extinct population of *T. comes* on the sandhills of the Lancashire coast was the only one in which blackish specimens resembling *curtisii* were known to occur in England. However, the genetic identity of this local melanic with *curtisii* has not been proved.

It was possible to secure specimens of *T. comes* from Barra and Orkney in order to compare their genetics. In each of them the dominance of *curtisii* was nearly but not quite complete, while the normal *comes*, save on rare occasions, was fully recessive, as already described. Indeed in the segregating material reared within the Barra and Orkney stocks only 6 specimens out of 749 were unclassifiable intermediates between *curtisii* and *comes*, of which the totals were respectively 511 and 232.

Having thus established the recessiveness of the normal form in both strains, crossings were made between them so as to allow the *curtisii* gene to segregate in an F_2 ratio in a gene-complex of mixed origin. Accordingly Barra and Orkney heterozygotes were interbred. Three families were successfully reared from them (several others were failures), producing totals of 74 (43 males, 31 females), 23 (9 males, 14 females), and 63 (38 males, 25 females), the parent of Orkney ancestry being the male in the first two of these and the female in the third. All three produced results of a similar type. Segregation in a ratio of $1CC:2Cc:1cc$ was of course taking place, but it no longer led to a clear discontinuity between *comes* and *curtisii* as it had done previously. There was indeed an unclassifiable range from the palest to the darkest specimens: that is to say, the dominance of *curtisii* had broken down, while further crossings demonstrated that the same gene was responsible for that form in the two islands (Ford, 1955b).

In this instance, therefore, we know what meaning to attach to the statement that 'the same' polymorphic phases of a species occur in two isolated localities. In these places, identical switch-genes control similar effects, but that similarity has been reached by distinct means, for the sets of 'modifiers' which adjust the dominance of *curtisii* have been built up differently: an example which illustrates rather clearly the way in which polymorphic forms can evolve.

In *Triphaena comes* it was possible to break down the dominance which had developed under natural conditions. H. B. Williams (1946–7) working on slightly different lines was able to extend the action of a gene in the moth *Angeronia prunaria* much beyond the normal limits of its expression. This is a polymorphic species belonging to the family Selidosemidae. It is confined to woods and is rather local but with a wide range in the Palaearctic region, extending to southern England and Ireland. The larva feeds upon *Prunus, Ligustrum, Crataegus,* and some other plants. In the common form, *prunaria*, the wings are uniformly orange in the male and yellow in the female, marked with minute greyish speckles. A second phase, *corylaria*, seems to be found wherever the species occurs but is much the rarer, comprising, perhaps, 5 to 10 per cent of the population. The wings are chocolate-brown crossed by a band, rather variable in width, of the normal male or female colouring (Plate 7, Figures 1 to 4). It is a simple dominant except that the homozygotes are generally distinguished by lacking the greyish speckles on the pale areas, though a few of these may sometimes be present, especially on the nervures, so that homozygous and heterozygous *corylaria* are occasionally difficult to recognize (Doncaster and Raynor, 1906). By means of his own selective breeding and that of H. D. Smart, from whom he received some of his stocks, Williams obtained broods in which the band was reduced to a spot in the centre of the fore-wings and others in which it was greatly extended (Plate 7, Figures 5 to 8). We have here an illustration of the way in which selection operating on genetic variability can alter the effect of a gene so as to make it dominant or recessive or adjust it as may be needed in different types of environment.

Heterozygous advantage and multiple-phase polymorphism

There appears to be no evidence that the heterozygotes of *corylaria* are at a physiological advantage compared with the two homozygotes, though it is probable that they are so. There does not seem any clear indication of this in bred families, but it has already been pointed out (p. 142) that, owing to the great reduction in selection to which they are exposed, laboratory stocks are but a poor guide in this matter unless the departure from expectation is a striking one. Nor have we, to my knowledge, the necessary information on the genotype frequencies in wild material; and these comments apply also to *Triphaena comes*.

However, in another Selidosemid moth, *Cleora repandata*, these deficiencies are to some extent overcome. This is a woodland species which normally has a cryptic colour-pattern which makes it extremely inconspicuous when resting upon the bark of trees. It is about one and three-quarter inches across the wings, brownish grey and mottled. In Britain it is monomorphic in the east (with the exception of an industrial melanic, p. 296) but dimorphic in the south-west of England and the north-west of Scotland, owing to the presence in addition of the black-banded form *conversaria* which has also been found both in the east and west of Ireland. This is an incomplete dominant: the ground colour (that excluding the band) being brownish grey in the heterozygotes but white in the homozygotes (Williams, 1950). It would be a matter of much interest to discover what requirements of *conversaria* are satisfied by the ecology or climates of the regions to which it is restricted. An initial approach to that problem should be attempted by studying the predation to which it is exposed there and elsewhere; also by rearing segregating broods in diverse environmental conditions combined, in some instances, with powerful selection through starvation or other means. An investigation of the species should also be made in those parts of its range where it is dimorphic and monomorphic respectively, to determine whether or not it tends to feed on the same plants there or has other distinguishing features.

Most of the families of *C. repandata* bred by Williams were of the back-cross type and the few F_2 broods were small. None showed irregularities except one F_2 family consisting of 19 imagines: 8 *repandata* and 11 heterozygotes only. This looks like a back-cross. If indeed it represents an F_2 generation, and of this Williams seems to be confident, it strongly suggests that the viability of homozygous *conversaria* may sometimes be reduced. Moreover, Cockayne reports a large count of wild specimens in Ross (north-west Scotland) consisting of 480 *repandata*, 141 heterozygotes, and 5 homozygous *conversaria*: a result in which the heterozygotes much exceed expectation on the assumption that all three genotypes survive equally well ($P < 0.001$), though the error in such a calculation is great when one homozygous class includes only 5 specimens. The moths were obtained during the day. Had they been found by searching the tree-trunks, on which they normally sit, the result would be subject to strong sampling errors since it is almost certain that the three genotypes would not be equally distinguishable to the human eye. However, those with experience of this species in natural conditions know that it is extremely restless and easily disturbed. Most of Cockayne's specimens would therefore have been caught flying. Even so, Kettlewell has shown that melanic specimens of this moth are much less conspicuous on the wing than are those of the relatively pale normal form (p. 317). Heterozygous *conversaria* is on the whole lighter than *C. repandata repandata*, and homozygous *conversaria* is the palest of the three forms. Therefore, if the melanic situation be a guide, any sampling error introduced by catching this species as it flies would operate against the typical specimens and consequently tend to diminish, rather than spuriously inflate, the recorded departure from expectation on the assumption of equal viability.

If indeed the observed excess of heterozygotes be a real one, as it seems to be, there are three possible explanations for it. (1) It could result from non-random mating of which, however, there is no indication. (2) It might be due to the fact that the rare homozygote is at a severe disadvantage. The date of Cockayne's sample is in doubt but it was taken before (and probably not long before) 1940. If the frequencies could be ascertained again and it were found that they had not changed appreciably (indeed a small amount of information tends to support the view that they are in fact stable) this interpretation could reasonably be excluded. (3) The observed departure of the three classes from expectation could be due to heterozygous advantage and, in the circumstances, this is the most probable interpretation.

The unequal viability of the genotypes in polymorphism has sometimes been detected solely by means of a disadvantage of the rare homozygote. An example is provided by the Grass-Finch, *Poephila gouldiae* (Ploceidae), which is distributed over the tropical and subtropical parts of Australia (Southern, 1945). Two common forms of this bird are known; in one the facial mask is black, like the rest of the head, but in the other red. The latter condition is a sex-linked dominant (the female, of course, being heterogametic). Both sexes are affected, the red colouring is, however, brighter and less variable in the cock birds. In these, as expected, it is about twice as common as in the hens, while its total frequency in all populations is about 20 per cent. This value does not vary throughout the whole distribution of the species. Thus the diversity does not relate to environment and is presumably maintained by heterozygous advantage. Indeed, it seems that the homozygous cocks are pathological; they suffer from twisting movements of the head and inability to judge distance.

It may not be easy to detect the superiority of the heterozygote in natural conditions, owing partly to the large samples that are needed when the frequency of this genotype is rather low, and partly to the difficulties of obtaining random samples. However, such 'single-gene heterosis' may be demonstrated by means that are indirect though conclusive, as in *Cepaea nemoralis* (p. 188), or by breeding results in the laboratory. It will be useful to introduce two instances of the latter type at this stage, though many examples establishing the widespread occurrence of heterozygous advantage will be discussed in subsequent chapters. Thus Lerner (1954) and Muller (1958) are misled in regarding this as an uncommon phenomenon, quite apart from the fact that its existence is to be deduced from the general theory of polymorphism, as explained in Chapter Six.

Drosophila polymorpha, a Brazilian species, is polymorphic in nature, occurring in three phases in which the abdomen is dark, light, or intermediate in colour. Da Cunha (1949) showed that they are controlled by a single pair of alleles without dominance (*EE* dark and *ee* light), the intermediates being the heterozygote (*Ee*).* The forms are not distinguishable at hatching and the abdomen does not darken completely for a week, though the light and dark types can usually be separated after twenty-four hours. The experimental flies were, in fact, classified on the sixth day after emergence or later.

Da Cunha made a study of this phenomenon using a combined approach of genetic analysis in the laboratory and of sampling the phenotype frequencies in wild populations. He found that the F_2 ratios contained in his bred material differed very significantly from the expected 1:2:1 segregation ($\chi^2_{(2)} = 134.25$) due to an excess of heterozygotes; for these survived better than the homozygotes in the environment provided by his cultures. His normal technique involved some overcrowding and, if the result were really due to differential mortality between egg and imago, a closer approach to expectation should be obtained when the larvae are reared in improved conditions in which

* Since there is no dominance in the characters studied, E and e (indicating dominance and recessiveness) is not, in fact, the correct notation, which should be E_D and E_L.

there is little or no competition. That conclusion was amply justified. In these circumstances, the excess of heterozygotes was almost, though not quite, obliterated and was indeed no longer significant $(\chi^2_{(2)} = 3\cdot73)$. If the viability of the heterozygotes be taken as unity, then that of the three genotypes under normal and under optimum laboratory conditions may be expressed as follows:

	EE	Ee	ee
Normal conditions	0·85	1·00	0·70
Optimum conditions	0·98	1·00	0·95

It will be noticed that while the heterozygotes are the most viable, the pale form is the least so; also that the survival of the latter is disproportionately poor in the less favourable conditions, as already discussed in general terms on p. 142. In addition, da Cunha obtained evidence of another widely applicable proposition previously mentioned (p. 144): that the heterozygotes are the most variable of the three classes.

Two other important pieces of laboratory work upon *Drosophila polymorpha* must be mentioned. In the first place, unlike the situation encountered in the *medionigra* polymorphism of *Panaxia dominula*, da Cunha was able to show that mating is at random relative to the genotype of the individual. Secondly, he devised an apparatus for maintaining experimental populations for a number of generations; the well-known population cages developed by Dobzhansky being, for technical reasons, unsuited to this species. In some lines he started with equality of the two alleles (50 per cent *E*) and in others with 20 per cent of *E*. In both of them the frequency of *E* rose to about 64 per cent at which value it stabilized, so giving rise to a balanced polymorphism in the stocks.

Da Cunha made large counts in seven localities in southern Brazil. The species proved to be polymorphic at all of them, with the pale form constantly the rarest phase: the gene-frequency of *e* apparently ranged from 18 to 33 per cent, but it differed significantly from place to place and from one season to another. In view of his experimental results, the distribution of the three forms in nature appeared most surprising, for the heterozygotes always fell below, and generally significantly below, expectation.

This apparent contradiction seems to have been resolved by Heed and Blake (1963). They have identified a third allele at the locus in question. This produces a dominant pale abdomen-colour which is homozygous in the northern part of the species' range (Trinidad, the Sierra Nevada mountains of Colombia, and elsewhere). If this allele exists also in small frequencies in southern Brazil, as it probably does, the phases would in fact be in equilibrium there: indeed its occurrence at 1 to 7 per cent of available loci would be sufficient to achieve that result. As already indicated, da Cunha established the advantages of the heterozygotes in his laboratory cultures and this he confirmed by several lines of work. Thus such 'heterosis', though complicated by the segregation of three genes at the controlling locus, doubtless balances the polymorphism in nature as it has been shown to do experimentally.

The second example of heterozygous advantage which it is useful to quote at this stage is provided by the work of Battaglia (1958) on the marine Copepod *Tisbe reticulata*. This species has colonized the brackish waters of the lagoon of Venice, where it is found in four polymorphic phases differing in intensity of pigment. They are controlled by three alleles, V^V, V^M, and v. The first two, giving rise to the forms *violacea* and *maculata* respectively, are both fully dominant over the last which, when homozygous, produces *trifasciata*. Individuals of the constitution $V^V V^M$ are, however, phenotypically distinct, showing the characteristics both of *violacea* and *maculata*. At least in the laboratory, they survive better than the corresponding homozygotes, to an extent which becomes accentuated in crowded conditions but is always heavily significant. A corresponding situation seems to apply to the $V^V v$ and $V^M v$ heterozygotes, but the point has not been fully established. It will be noticed once more that the advantage of the favoured genotype becomes greater as the environment deteriorates, a fact which demonstrates its selective importance.

Both here and in *Drosophila polymorpha* we have a polymorphism associated with a series of alleles (see p. 112), and it is probable that these in fact represent the units of a super-gene: as in examples to be discussed subsequently (pp. 169–73). Of this, however, there is no decisive evidence in *Tisbe reticulata* and the co-existence of multiple phases which it displays can be examined in more detail in material of a different kind, the insect *Zygaena ephialtes*.

This is a day-flying moth, the forms of which, and their distributions, have been analysed in much detail by Bovey (1941). The species is protected by poisonous qualities like other Zygaenidae (Lane and Rothschild, 1959). It is found throughout Europe, including southern Russia but excluding the Scandinavian countries, Britain, and apparently the Iberian peninsula. Its pattern is of two distinct types, giving a very different appearance:

1. *Ephialtes*, which is blackish with five or six spots on the fore-wings and one or two on the hind. Red or yellow pigment is restricted to a band on the abdomen and to the anterior of the two basal spots on the fore-wings, the other spots being white.

2. *Peucedani*. In this phase the red or yellow colour is extended to all the fore-wing spots. It occupies the basal area of the hind-wings also on which the spots, which stand out clearly in the dark form, can still be detected but only by transmitted light. There is in addition the red (or yellow) abdominal band.

Peucedani is dominant to *ephialtes*, except that the heterozygotes are decidedly variable. Also red coloration is fully dominant to yellow. These two pairs of alleles are said to assort independently and this is probably correct, though it must be admitted that the data on di-hybrid crosses are not sufficiently extensive to exclude loose linkage.

The fore-wing spots are placed in pairs, the members of which are anterior and posterior. The proximal and median ones seem almost invariable in size but the more posterior of the distal pair may be reduced or absent. The variation of this sixth spot is under polygenic control. On the other hand, the presence of one or two spots on the hind-wings is unifactorial, the two-spotted condition being a simple dominant to that with one.

Bovey (1941) published a back-cross family for the genes controlling hind-wing spotting and the *peucedani-ephialtes* pattern. The total was only 53, consisting of 7 twin-spotted *peucedani*, 17 single-spotted *peucedani*, 18 twin-spotted *ephialtes*, 11 single-spotted *ephialtes*. The departure from expectation approaches significance ($\chi^2_{(3)} = 6.93$; $0.10 > P > 0.05$) and is in a direction which could be the result of linkage. Of this, however, there is no sign in a di-hybrid segregation with a total of 94 (47, 24, 15, 8: scored in the previous order).

There are regions in which either of the main forms occurs alone and others in which they constitute a polymorphism. For instance in central and northern France, northern Germany and north-west Switzerland, only *peucedani* is found; and in south-west France, peninsular Italy, and Cyprus, only *ephialtes*. However, in central and southern Germany the population is mixed in varying proportions, as it is in southern Switzerland where *ephialtes* in fact predominates, and in many other places. So too, there are races composed entirely of insects with red pigment, as in Cyprus and south-western France, and others entirely of those with yellow as in peninsular Italy. Yet there are extensive areas where the red and yellow phases fly together: south-eastern Germany and the south-eastern Balkans for instance.

Evidently, therefore, either pattern or colour can be at a decisive advantage or disadvantage, or the alternative forms may be balanced by opposed selective forces. In addition, these pairs of alleles clearly demonstrate factor interaction, for yellow colouring is extremely disadvantageous when combined with the *peucedani* pattern, so that no populations of the kind have been found. The form is known only in places where yellow (or red and yellow) *ephialtes* and red *peucedani* constitute mixed communities and the yellow *peucedani* have never established themselves outside the area where they are produced by recombination and as constantly eliminated. Bovey (1941) describes three (homogeneous) di-hybrid families, amounting to 82 in all, in which yellow *peucedani* is unduly rare (he obtained 5 when 15·5 were expected). He crossed specimens from the same F_1 brood with yellow *ephialtes* to give 2 back-cross families, together producing a total of 149 in which yellow *peucedani* actually exceeded expectation to a slight extent (39 individuals). Evidently there may be considerable variation in the extent to which the yellow *peucedani* are eliminated in laboratory conditions.

A similar type of interaction occurs between the genes controlling spot-number on the hind-wings and the *peucedani-ephialtes* pattern. Though two-spotted *peucedani* can be produced by the appropriate crosses, they are unknown in nature for that form is invariably single-spotted while *ephialtes* may belong to either type. However, Bovey remarks that two-spotted *ephialtes* are geographically rather isolated from populations containing *peucedani*, though they do occur together in Bohemia and Moravia. Two-spotted *peucedani* fall a little below expectation in the results quoted on p. 155 which are relevant to this matter. Yet, as already remarked, this deficiency might be occasioned by linkage in the back-cross broods though there seems no evidence of this in those of the dihybrid type.

Bovey had done extensive and valuable work on the genetics of this insect and on its geographical distribution. Yet, as pointed out in the first edition of this book, much more ecological information was needed to throw light upon the conditions which favour one rather than another of its forms. This has now been provided by the researches of Bullini *et al.* (1969). They have shown by means of feeding experiments with captive birds that *Zygaena ephialtes* is a highly distasteful insect; also that its two morphs, so distinct in appearance, gain additional protection by falling respectively into separate associations of Müllerian mimics, the other members of which are also poisonous and warningly coloured. The more southerly form, *Z.e. ephialtes*, is extremely similar in appearance to three species of the genus *Amata*, Syntomidae: *A. phegea*, *A. marjana* and *A. ragazzii*. Farther north, where these models become rare or absent, protection is correspondingly enhanced by the very different *peucedani* colour-pattern, which accords perfectly with that of the ordinary Zygaenidae, comprising an assemblage of species extremely difficult to distinguish from one another (see Lane and Rothschild, 1959, for the dangerous qualities of one of them, *Z. trifolii*).

Although red pigment can be replaced by yellow in all species of *Zygaena*, it never occurs, save as a rare variety, in the insects typical of that genus. On the other hand, yellow coloration enhances the resemblance of the *ephialtes* form to *Amata*. We thus have an explanation for the absence of yellow *peucedani* save in special circumstances, to which attention has been drawn.

It looks as if the Syntomids provide the preferred model for *Z. ephialtes*, since even where the *ephialtes* form occupies 100 per cent of the population, as in peninsular Italy, normal Zyganidae are common and they are always highly protected. Only where the *Amata* species are relatively rare and then absent does *peucedani* appear, becoming first polymorphic and finally monomorphic. There must indeed have been strong selection to produce the *ephialtes* colour-pattern differing, as it does, profoundly from that of related Zygaenids, which *peucedani* so exactly resembles. It is to be noticed, in addition, that in the region of polymorphism the latter form may gain some advantage from heterosis.

The life-cycle of *Z. ephialtes* generally occupies two years though some individuals complete it in one. Lane and Rothschild (1961), who encountered a similar type of variation in the related *Z. lonicerae*, point out that this may be important in enabling a colony to survive a season in which the numbers are greatly reduced.

In the majority of instances it is not clear to what type of ecological situation a polymorphism is adjusted. In some it is evidently related to certain specific conditions, and in others again we know *how* it is adapted to them. Of these latter, mimicry and industrial melanism, to be discussed in subsequent chapters, provide noteworthy examples.

The second of the above groups is that in which different phases are favoured in certain types of environment or habitat: a situation which was indeed suggested by the work of da Cunha on *Drosophila poly-*

morpha. It is illustrated in the next two examples and in a number of those that follow.

Ricinus communis, the Castor Bean, is widespread in Peru, where it is cultivated and occurs also as an introduced weed. It is subject to a dimorphism in the presence or absence of a waxy bloom on the stems. This character is controlled by a pair of alleles. The phase with the bloom, which is dominant, increases with altitude because it is the more viable in conditions of considerable sunlight and absence of fog; its frequency, in fact, changes from 0·15 per cent of the population at sea level to 30 per cent at 2,000 feet and 100 per cent at 7,700 feet. The disadvantage of the waxy, compared with the waxless form, consists in its reduced power of fruiting in a sunless and foggy climate (Harland, 1947).

Sex-controlled polymorphism

The moth *Parasemia plantaginis* (family Arctiidae) has cream-coloured fore-wings with a pattern of black blotches and longitudinal stripes upon them. The hind-wings are deep yellow, sometimes orange or red in the females, and they also are marked with a varying amount of black. The males fly rapidly in the sunshine but the females become active only at dusk. The larvae feed upon *Viola*, *Plantago*, *Senecio vulgaris*, and other low-growing plants, and the imagines appear in May and June, with a partial second brood in August.

Parasemia plantaginis is widely distributed throughout Europe, including the British Isles, and northern Asia, and it inhabits moors, hillsides, and open woods. There is an alternative phase, *hospita*, completely restricted to the males, in which the normal cream and yellow colouring is replaced by white. It occurs only on mountains. Even there it is generally much the rarer, though it constitutes a polymorphism, occupying 5 to 10 per cent of the male population in the Lake District of north-west England: it is, however, relatively commoner at high altitudes in some parts of Continental Europe. *Hospita* is a sex-controlled dominant (Suomalainen, 1938) so we have here a form adapted to certain genetic and ecological situations: those provided both by the male sex and an alpine or semi-alpine environment. In these conditions the heterozygotes are doubtless favoured, though not sufficiently for the distinction to have become apparent in the laboratory. Yet it must be admitted that the breeding so far carried out has not been designed in a way to detect differential mortality. If, as is probable, homozygous *hospita* are normally of poor viability, this seems to have been overcome in at least one area, by selection acting upon the gene-complex in the way indicated on p. 121. For Schulte (1952) reports that all the males are of the *hospita* form in a locality near Abisko in Swedish Lapland. However, unequal survival of the genotypes has been established in some other similar instances in which the genetic work has been conducted on a more extensive scale. Two examples of this will suffice.

The palaearctic butterfly, *Argynnis paphia* (Nymphalidae) is normally golden-brown, rather duller in the female, with black pencillings, while the underside of the hind-wings is green washed with metallic silver. The males are always monomorphic but the females may be dimorphic, owing to the·existence of the form *valezina* in which the brownish colouring is replaced by blackish-green. This is a sex-controlled dominant, the expression of the gene being limited to the female (pp. 165–6, 239); that is to say, the condition is the reverse of that found in the *hospita* phase of *Parasemia plantaginis*.

The genetics of *A. paphia* are, nevertheless, subject to complications which have not yet been resolved. Goldschmidt and Fischer (1922) bred the species on a large scale and obtained most curious results. Their back-crosses from genetically normal males (*vv*) × heterozygous *valezina* females (*Vv*) produced, in all, 759 males, 357 *paphia* and 353 *valezina* females. These were the progeny of eleven families showing no heterogeneity, and it will be noticed that the expected equality of the two female forms is almost perfectly realized. The reverse cross, heterozygous males (*Vv*), which are identical in appearance with the *vv* genotype in that sex, × normal females (*vv*) gave 189 males, 104 *paphia* and 68 *valezina* females. They were the results of three large and homogeneous broods, and of two others contributing only 7 (2:5) and 5 (3:2) of the females. The deficiency of the *valezina* females is here significant ($\chi^2_{(1)} = 7.51$ with P < 0.01). Goldschmidt and Fischer also raised thirteen broods from double heterozygotes. These show no heterogeneity and the combined totals gave 526 males, 143 *paphia* and 367 *valezina* females. The departure of this ratio (2.57:1) from expectation approaches but does not reach formal significance ($\chi^2_{(1)} = 2.51$, and P just exceeds 0.10). Among the large numbers bred from other types of pairings, males homozygous for the *valezina* gene (*VV*) were repeatedly obtained and shown to be indistinguishable from the normal *vv* type. One family from the cross *VV* male × *Vv* female was also raised. It produced 19 males, 0 *paphia* and 21 *valezina* females. In the broods in which the *valezina* females are relatively rare, the difference in sex-ratio is not significant ($\chi^2_{(1)} = 1.64$, P = 0.2). Thus it seems probable that those effects of the *valezina* gene which produce a disturbance in segregation are expressed in the male as well as in the female.

Fischer (1929–30) reports the existence of a very few male *valezina*. These may be the result of a rare genetic condition which allows the blackish-green colouring to appear in the sex which usually suppresses it, in much the same way that the effect of a normally complete recessive can occasionally be detected in a heterozygote. Alternatively, the male *valezina* may not be sexually normal; they might in fact be females largely converted to the male sex, like some of the well-known intersexes studied by Goldschmidt in *Lymantria dispar*. There is, however, a further and more probable interpretation of them. That is to say, they may genetically be entirely distinct from *valezina*, being controlled by a different gene from that form. This would be a parallel situation to the rare white males reported in certain *Colias* species which are autosomal recessives without limitation to sex, in contrast with the ordinary dominant pale phase restricted to the females (p. 162).

The unequal results among female progeny from the two types of backcross obtained by Goldschmidt and Fischer can be due to either of two causes. (1) Sperms carrying V may be at a relative disadvantage compared with those carrying v, or at a relative disadvantage only when the two are in competition: that is to say, they might, on the one hand, survive less well, or on the other, they might swim the more slowly or penetrate the egg less effectively. (2) Sperm carrying V, or indeed the early stages of *valezina* development, might be at a disadvantage in normal but not in *valezina* cytoplasm. We are, of course, acquainted with maternal effects of this kind in which the genes of the parent modify the egg cytoplasm so as to influence segregation in the next generation (pp. 207–8).

If, in fact, the deficiency of *valezina* females in the F_2 broods is a real one, as seems likely, the first but not the second of the above alternatives could account for this also. On the other hand, the F_2 situation might well be a distinct phenomenon in which heterozygous females are at an advantage over homozygous *valezina*.

The female dimorphism of *Argynnis paphia* is restricted to certain areas. In parts of Hampshire and Somerset *valezina* constitutes perhaps 5 to 15 per cent of the females but it is merely a rare variety (below 1 per cent) elsewhere in the British Isles, where the range of *paphia* includes the southern half of England, Wales, and Ireland. A similar situation is encountered in Continental Europe, the females being dimorphic in some districts but monomorphic in others, such as parts of northern and of south-western Germany respectively. However, it is said that *valezina* becomes much commoner in the south-eastern part of the Palaearctic region. Indeed it has been asserted, but on insufficient evidence, that all the females are of that form in southern China.

This is essentially a woodland butterfly but, owing to its tendency to wander and its powerful flight, it may be found ranging along sheltered lanes and hedgerows. It is generally agreed among entomologists that the habits of the two female forms are distinguishable. The normal one is to be found particularly in sunny clearings and pathways while *valezina* is more restricted to the shade of overhanging trees. This has on the whole been my own experience, and it is probable that we have here a genetically controlled behaviour-difference. Unfortunately, however, no ecological work capable of establishing this firmly has yet been under-taken.

Differences in habit certainly open up the possibility of maintaining forms as a balanced polymorphism. Moreover, Magnus (1958) has discovered a situation which opposes the spread of *valezina*. He has demonstrated experimentally that the optical stimuli which must be provided by a female in order to be recognized by a distant male consist in the normal brown colouring of this butterfly when kept in rapid motion by the fluttering of the wings. He showed in addition that the greenish shade of *valezina* is wholly ineffective for this purpose, so that such females must be at a reproductive handicap. He points out, how-ever, that in favoured localities this species is very common and that in such circumstances *valezina* females are likely to meet some males ready for copulation when they are feeding on flowers. These females might then be recognized on account of some sexual scent which they emit. At any rate, when caught in the wild it is generally found that they have in fact been fertilized, though they may well have had to await copu-lation longer than those of the normal *paphia* form.

It is evident that the disadvantage of the *valezina* colouring in court-ship must be counterbalanced in some way in those regions where that phase is polymorphic. No doubt this is due in part to a physiological or behavioural superiority, ensuring that it survives better than the normal one in certain environments.

A very similar situation, combining unisexual polymorphism with unequal viability of the genotypes and a genetically determined difference in habit, occurs widely in the butterfly genus *Colias*. It is the subject of a valuable survey by Remington (1954). Many of these species are lemon-yellow or greenish in the male and more or less cream-coloured in the female. On the other hand, large numbers have a monomorphic orange or bright yellow male but two forms of the female, the majority being of the male coloration while a minority is very pale yellow to white.

This dimorphism differs greatly in its proportions among the *Colias* butterflies in which it occurs. The light-coloured specimens usually amount to between 5 and 15 per cent of the females in the Palaearctic *Colias croceus*, though they may sometimes be considerably commoner (perhaps up to 30 per cent) early in the emergence (p. 164). In the North American *Colias eurytheme* they are relatively more numerous, 15 to 50 per cent, and Hovanitz (1944*b*) has even recorded instances in which they are slightly the commoner form. There are, however, other species in which less than 5 per cent of the females are of this type.

As far as is known, the pale phase is always a sex-controlled dominant and seems to be due to an autosomal gene ineffective in the male even when homozygous (Hovanitz, 1944*a*) (see pp. 165–6).

The light-coloured females are of a primrose shade in some species, white in others; alternatively, they may range between these extremes, as in *Colias croceus*. Such variation within the ambit of the major gene is at least partly genetic, for the pale offspring of some females may include a significantly higher proportion of the whiter specimens than those of others even when reared in similar conditions. That is to say, these minor differences are due to segregation within the gene-complex. Doubtless, therefore, their frequency and degree could be altered by selective breeding from the yellowest and the whitest forms respectively. Yet this would be a much slower process than, for example, was the corresponding one affecting the expression of the *corylaria* gene in *Angeronia prunaria* since in *Colias* it could be practised on one parent only.

A very few white or light-coloured males have been recorded in some *Colias* species (e.g. *C. myrmidone*, *C. eurytheme*, *C. philodice*, *C. croceus*). They prove to be entirely distinct from the normal white females, being due to an autosomal recessive which is nearly lethal (Remington, 1954). This seems to affect both sexes and influences a number of other characters in addition to wing coloration. The legs, antennae, and body are white, the two parts of the proboscis fail to unite near the tip, the wings do not expand properly, and the affected individuals appear to be sterile. They have blue-green eyes and blood, and bluish larvae and pupae (see also p. 2). Thus they constitute a rare abnormality, not a polymorphism.

Hovanitz (1948) has made important discoveries by studying the ecology of the two female colour-forms of several of the *Colias* species, in particular *C. eurytheme*, for he finds that the gene controlling this dimorphism affects also the habits of the imagines. If specimens are collected throughout the day and the times of capture be recorded, the frequency of the pale form proves to be highest in the early morning. It subsequently declines but may rise again towards evening. The activity of both the female colour-phases in fact increases towards noon but the whitish females are less affected by this tendency than the orange (or deep yellow) ones, so altering the proportions in which they are caught.

As Hovanitz points out, the most likely physical features to determine these diurnal changes are solar radiation, temperature, and humidity. However, in the San Joaquin Valley, California, where he principally worked, and indeed in many other localities where he studied these phenomena, relative humidity was so near to being inversely proportional to temperature that it was not possible to discriminate between the two components. Neither the curve for solar radiation nor for temperature alone fits that for total female activity (both colour phases taken together), but a combination of these two environmental effects does so. That is to say, total female activity declines as the conditions pass towards low temperature and low solar radiation or towards high temperature and high solar radiation: it increases with a transition in the direction of low temperature and high solar radiation or the reverse. It seems reasonable to conclude, therefore, that since a combination of these climatic factors influences female activity as a whole, they are important in controlling the differential behaviour of the two phases. The Hon. Miriam Rothschild, however, points out that this situation must be complicated by the way in which the quality of solar radiation varies with altitude.

A strong indication that a similar situation exists in *Colias croceus* was obtained in England in 1947 (Dowdeswell and Ford, 1948). This was the only occasion in recent years in which the species has been sufficiently common in that country for such a comparison to be attempted; but even so, the numbers were not quite large enough to show that an observed increase in the pale phase at the beginning and end of the day was in fact significant. It is likely, however, that a real difference in habit is indicated here for it is in the same direction as that occurring in the American *C. eurytheme*.

In view of these findings, it seems that the behaviour of the white females lengthens the period, but reduces the degree, of the insects' activity while that of the yellow ones concentrates it towards the middle of the day. The optimum balance between these trends will depend upon the geographical and ecological conditions of each locality and may be important in adjusting the various species to different parts of their ranges. Thus Hovanitz (1944b and c) finds that white females are relatively commoner in northern than southern populations and in those at high compared with low altitudes, as he demonstrated in the equatorial Andes (1945). Another feature controlled by the gene for white coloration is likely also to be important in this connection. That is to say, the pale females on the average emerge in advance of the orange (or yellow) within each brood. Evidently they have a more rapid development and this may well be important in cooler localities.

The female dimorphism of *Colias* must be maintained by a selective balance the nature of which has, however, aroused considerable discussion. Gerould (1923) at first suggested that the homozygous pale form *AA** is lethal. However, he later obtained undoubted specimens of that genotype. In the light of this and of further information from bred families he substituted the theory that *A* is linked with a recessive lethal from which it is sometimes separated by crossing-over. It will be realized from the discussion in Chapter Six that this is in itself a reasonable hypothesis, but it does not seem applicable to the *Colias* polymorphism. On the contrary, Remington (1954) provides evidence to show that each of the three genotypes can be relatively more, or less, viable than the others in certain genetic or environmental conditions, especially temperature. Lorkovic and Herman (1961), who have also studied this subject by means of extensive breeding work on *Colias croceus*, confirmed Remington's suggestion and could find no evidence of a recessive lethal linked with the '*alba*' locus. However, as Remington (*l.c.*, p. 141) himself points out, the maintenance of such a polymorphism as that of the *Colias* species seems essentially to require heterozygous advantage, the frequency of the *alba* gene being controlled in each population by the varying relative advantage of the genotypes as just indicated. As already stressed (p. 142), the elimination taking place in nature, reducing the average output of one pair of butterflies to two only, is very much stricter than that studied in the laboratory. Thus the greater superiority of the heterozygote might well be detected by breeding the species in unfavourable conditions, though it is difficult so to adjust such treatment that the experimental broods are not completely destroyed.

* *Alba* in *Colias philodice*.

It has generally been assumed that the female polymorphism of butterflies is determined by ordinary sex-controlled inheritance in which the effects of the switch-gene are suppressed in the environment provided by one of the two sexes: in this instance by that of the male. However, Stehr (1959) has interpreted this situation differently. He finds that the haemolymph is always yellow in the males and green in the females of the moth *Choristoneura pinus*, also that the related species *C. fumiferana* is dimorphic for these two characters but that their frequencies are often different in the two sexes. As a result of extensive genetic work, he concluded that when the polymorphism of the Lepidoptera is restricted to the female it is controlled by the interaction of genes at two loci, one autosomal and the other sex-linked: a suggestion previously made by Cockayne (1932), though Stehr does not refer to this. Hybrids between the two *Choristoneura* species show that the haemolymph dimorphism is controlled by the same two loci in both, the alleles having different potencies in *C. pinus* from those in *C. fumiferana*. Thus he says,

> Where the determination of a character is divided between an autosomal protagonistic, and an X-chromosomal antagonistic gene, the conditions for a sexual dimorphism are given on the basis of the X-chromosomal/autosomal potency balance, in perfect parallel to the determination of sex as such. However, if there exist, as in *C. fumiferana*, alternative alleles of different relative potencies on one or the other of these loci, many forms of sex-controlled polymorphism and even some that do not seem to be sex-controlled, may be created. Such polymorphism thus becomes a phenomenon akin to intersexuality.

However, as Sheppard (1961a) points out, polymorphism differs from intersexuality in one important respect: that it involves clear-cut segregating classes. This must on Stehr's hypothesis mean that the gene-complex has been so adjusted by selection as to produce an all-or-nothing effect, as with sex itself.

There seems no doubt that Stehr's view of two interacting pairs of genes at autosomal and sex-linked loci respectively provides a correct interpretation of the situation in *Choristoneura*. However, he extends this concept to the control of unisexual polymorphism in general. There are several reasons for rejecting that deduction.

1. It would be almost impossible to obtain a series of potencies to account for the control of the female colour-pattern by a large number of alleles as in some mimetic butterflies (more than twelve in *Papilio dardanus*, see Chapter Thirteen). This could be done only by a

complicated system especially arranged to sustain this particular hypothesis.

2. If Stehr were right, the control of such polymorphism on an auto-somal and sex-linked basis should be about equally common, whereas the latter system is very rare indeed.

3. Sex-controlled polymorphism is far more frequent in butterflies than in moths.

4. Sex-controlled polymorphism is almost invariably confined to the female in butterflies but not in moths, in which it may be confined to the males.

The two latter points indeed require a selective interpretation based upon the ecology of butterflies (p. 239). Thus, though Stehr's hypothesis certainly seems applicable to some instances, including *Choristoneura*, it cannot be extended to unisexual polymorphism in general, for which the ordinary control by a pair of switch-genes selected to operate in one sex only provides a valid explanation, as for the accessory sexual characters.

Polymorphism in hydrogen cyanide production

It will have been noticed that the work of Hovanitz on female dimorphism in *Colias* has taken into account the multiple effects of a single gene, or conceivably of a super-gene. He has thus been able to relate a genetic polymorphism to the ecological requirements of the species in a way which helps us to understand its adaptations to the conditions in which it flies.

Such multiple effects are well known to control not only variations in colour, pattern, or structure, but at the same time differences in physiology and behaviour. This is a less easy aspect of their function to study but it is surely the more important for the organism. Doubtless, it often happens that these rather elusive qualities are subject to polymorphism alone without involving others that are more obvious and easily recognizable, as in protein polymorphism (pp. 28–9, 104–10). Various other instances of this kind have come to light and the occurrence of substances producing hydrogen cyanide in plants provides an example of them. I am indebted to Mr D. A. Jones for information on this subject, which he has himself investigated.

Cyanogenic* glucosides occur very widely. They have been identified in at least fifty orders of flowering plants, in six ferns, and in several

* Often, and very confusingly, miscalled 'cyanogenetic'.

Basidiomycetes. The genus *Prunus* (Rosaceae) contains numerous species which possess them, for instance the almond *P. amygdalus*. In some forms (e.g. *Phaseolus lunatus*) they have disappeared under cultivation. In others they are present only in the seedlings, becoming weaker and vanishing as maturity is reached, as in *Sorghum vulgare* in which their concentration may be affected also by the environment, for it seems to be increased by drought.

Several species are dimorphic for the presence or absence of the glucoside, and of these *Lotus corniculatus* (4n = 24), Papilionaceae, is an example. This is a perennial which is thought to have arisen by autotetraploidy from *Lotus tenuis* (2n = 12); a species which is nearly always positive for the cyanogenic property, though two negative plants have been recorded and their identification confirmed by growing them in cultivation.

The dimorphism of *L. corniculatus* is controlled by alternative alleles determining the presence of the glucoside, which is dominant to its absence (Dawson, 1941). The character is, however, not completely penetrant in British material, though it seems to be so in Swedish experimental stocks, consequently some positive plants are capable of liberating relatively large, and others relatively small, quantities of HCN.

Colonies in which the majority of the plants contain the glucoside greatly predominate (thus two on Ranmore Common, Surrey, comprised 145 positives to 8 negatives and 138 positives to 8 negatives, respectively). But many are known in which there is a closer approximation to equality: for example, one of those on Studland Heath was composed of 77 positives and 56 negatives (Dawson, 1941). D. A. Jones, who has scored many populations in southern England and Wales, has indeed discovered a few instances in which the negatives are in excess.

A remarkable aspect of this subject is that a second and ancillary genetic situation is involved. That is to say, another gene, which Jones found to be polymorphic also, is responsible for an associated enzyme facilitating the production of HCN from the cyanogenic glucosides. Yet it is not essential to that process. The gas is nearly always generated upon injury by positive plants, while the enzyme increases the rate of the reaction.

Jones (1962) has obtained information on some of the ecological aspects of this polymorphism. He found that in natural conditions the cyanogenic type of *Lotus corniculatus* is relatively free from damage by invertebrates feeding on the leaves: 27 plants which had been partly eaten all proved to be acyanogenic. Taking into consideration the frequencies of the two phases in the colonies concerned, the chances of obtaining such a result fortuitously were less than 1 in 1,000. Three of the animals responsible for the damage were caught: one was a slug, *Agriolimax reticulatus*, while two were larvae of the moth *Zygaena filipendulae*. The immunity of the latter form to HCN is indeed to be expected since its crushed tissues release that substance in all stages of its life history (up to 200 μgm from a single imago and up to 150 μgm from 50 crushed eggs). The poison is manufactured, doubtless as a protection, by this insect and all other Zygaenidae tested so far, and is not derived from the food (Jones, Parsons, and Rothschild, 1962).

In view of this, Jones carried out a series of laboratory tests in which a total of 200 specimens of the slug *Arion ater* were offered two strongly and two weakly cyanogenic plants, the latter sometimes replaced by the acyanogenic type. Those containing the least glucoside were always the most eaten. In addition, he also obtained evidence that the vole *Microtus agrestis*, two snails, *Arianta arbustorum* and *Helix aspersa*, and the slug *Agriolimax reticulatus* also feed selectively, in a similar way, upon *Lotus corniculatus*. Moreover, he has obtained extensive confirmation of these findings (Jones, 1966, 1967). The liberation of HCN seems then to act as a defensive mechanism reducing the amount of injury inflicted by certain, though not all, predators. Thus Lane (1962) has shown that larvae of the butterfly *Polyommatus icarus* show no preference for acyanogenic compared with cyanogenic plants of *Lotus corniculatus*. He makes a very reasonable suggestion that these insects render the cyanide harmless by converting it into thiocyanate through the action of the enzyme rhodanese (Jones, Parsons, and Rothschild, 1962).

The opposed advantage of the acyanogenic plants, maintaining the polymorphism, is less clear. However Daday (1962) working on the corresponding situation in *Trifolium repens*, which may well apply to *Lotus corniculatus* also, has shown that this phase is favoured in cold conditions, which activate the enzyme. This splits the glucoside, freeing the hydrocyanic acid which inhibits the respiration of the leaves, so retarding growth. It seems also that this cyanogenic form flowers earlier than the acyanogenic: a situation likely to have balanced advantages.

As just mentioned, a very similar mechanism exists in the diploid *Trifolium repens* (Daday, 1954). The plants may possess cyanogenic glucoside (80 per cent lotaustralin and 20 per cent linamarin). They may also contain the enzyme, linamarase, which hydrolyses the glucoside with the production of HCN, to an extent that can cause disease in Ruminants. The presence of the glucoside, which is dominant to its absence, is determined by a pair of alleles, the contrasting members of which are *Ac* and *ac*. Similarly, the presence of the enzyme is also dominant to its absence, the controlling genes being *Li* and *li*. They assort independently of the *Ac* locus, segregation of the two alleles giving ordinary di-hydbrid ratios.

Cyanogenic plants, then, require the presence both of *Ac* and *Li*, and they pass as a polymorph-ratio cline from the Mediterranean Region, where they may comprise 100 per cent of the population, to north-eastern Europe, where in Russia that phenotype is practically absent. About 70 to 90 per cent of the plants belong to the cyanogenic type along the coast of western Europe and in Britain.

The occurrence of *Ac* and *Li* in *Trifolium repens* is closely correlated with the January isotherms, but not with those of July or with annual temperature or rainfall. A decrease of 1 °F. in the January mean temperature is associated with a reduction of 4·23 per cent in the frequency of the gene responsible for the glucoside and of 3·16 per cent in that for the enzyme. It seems, then, that mean winter temperature has played an important part in the evolutionary adjustment of *Trifolium repens*. Cyanogenic plants of this species do not seem detectably protected against invertebrate predation, and they probably produce less hydrogen cyanide than does *Lotus corniculatus* (Bishop and Korn, 1969).

The super-gene

Thus both in *Lotus corniculatus* and *Trifolium repens* two inter-related polymorphisms are established, controlled by pairs of alleles at separate loci which, so far as is known, are not linked. It is easy to see that there could be an advantage in associating certain of these phenotypes: perhaps the combination of the glucoside and enzyme on the one hand and their joint absence on the other. There appears to be no mechanism for ensuring this in either species; but that is precisely what is achieved by bringing together the loci of co-adapted genes so as to form a super-gene. The evolution of this device has already been illustrated in the series *Acridium arenosum, Apotettix eurycephalus, Paratettix texanus* (pp. 111–12).

As explained in Chapter Six, if we find that the switch-control of a polymorphism takes the form of a series of multiple alleles, we have a strong indication, the first likely to attract the attention of the geneticist, that the genetic units concerned have been accumulated to form a super-gene holding together, in the different phases, a set of co-adapted characters. This situation has indeed been encountered in two instances already considered in this chapter, relating to *Drosophila polymorpha* and to the Copepod *Tisbe reticulata*. These, however, were cited for a different purpose; to exemplify another fundamental phenomenon, that of heterozygous advantage.

Turner (1968*a*) studied the ecological genetics of the Tortricid moth, *Acleris comariana*. The work was conducted on an extensive scale and on modern lines. This species is a pest of cultivated strawberries and feeds also on the Marsh Cinquefoil, *Potentila palustris*. The populations on these two food plants are therefore very distinct in their ecology.

The imagines of *A. comariana* are polymorphic for colour-pattern, the variation involving 7 main forms. There is normally a roughly triangular blotch on the outer half of the fore-wings, while the remaining ground colour is subject also to variation. The genetics of these two components had been established by Fryer (1928), but their analysis has been carried much further by Turner.

The costal blotch can be brown (c) or black (C) due to a pair of alleles with full dominance. The three main ground colours are grey, marbled, and brown; controlled by three alleles or closely linked genes, B^g, B^M, and B^B. Dominance ascends in the order given, except that the heterozygote $B^B B^M$ produces an intermediate (mottled brown) effect. The B and C loci form a super-gene with at any rate less than 1 per cent of crossing-over between them. In addition, there exists a dark grey form with no costal blotch, possibly an industrial melanic (Chapter Fourteen). This is a recessive, probably due to another member (b) of the ground colour series, though the data do not quite exclude the possibility that it is at least a separate locus. There is also a pale form of the moth, which also lacks the costal blotch. It is found at a low frequency in north-western England but it has not been possible to establish its genetics.

The colour-pattern genes in this species vary in frequency from one part of its range to another. Turner has examined populations of it in a number of English localities: in the Wisbech district of the East Anglian Fens, the Vale of Evesham, and in the Cheshire, Lancashire, Westmorland area. He finds that the $B^B C$ and $B^M c$ super-genes are predominantly southern; while the melanic form is restricted to Lancashire and Cheshire, where the marbled type is absent in most populations that have been examined. As might be expected, there is, at least in certain localities, a distinction between the genetics of this insect when feeding upon strawberry and upon *Potentilla*. That is to say, the strawberry populations in the Liverpool neighbourhood are more like the East

Anglian, lacking the 'melanics' and including the marbled form, though the moths reared from larvae feeding on *Potentilla* in the same region are characteristically north-western in type.

Acleris comariana could provide valuable material for further study now that Turner has laid down the general pattern of its ecolocial genetics. It would be of much interest to determine what evolutionary adjustments have enabled it to become an agricultural pest, withstanding not only conditions very distinct from those of the marshes which must have been its original home but also such hazards as the burning and spraying of the strawberry fields in cultivation. We have here an interesting contrast with two allied species, *A. latifasciana* and *A. aspersana*, which also feed upon *Potentilla palustris* but have not taken those steps that adjust them to living upon a cultivated plant. Such information as Turner has been able to collect suggests that no marked changes in gene-frequency have occurred during the forty years since Fryer carried out his survey of *A. comariana*; though unfortunately his localities were not recorded with sufficient precision to make this interesting comparison in any detail.

A further example of super-gene formation can usefully be examined at this point. A number of closely related species of *Sphaeroma*, a genus of marine Isopoda, are polymorphic for colour-pattern. Bocquet *et al.* (1951), and Hoestlandt (1955), show that several of the forms are unlinked in *S. serratum*. West (1965), however, studied *S. rugicauda* widely round Britain and concluded that a super-gene is responsible for its colour morphs. Lejuez (1966) has compared the genetics of this species with those of *S. monodi*, *S. bocqueti*, and *S. hookeri*. He has shown that their extensive polymorphisms are in each due to a super-gene within which he has demonstrated rare cross-overs. Furthermore, several of the alleles appear to be common to these species, so it would seem to be the same super-gene in all. Two of the forms determined by it are also linked in *serratum* which is the only *Sphaeroma* so far investigated in which several unlinked genes are involved in this particular type of variation. Furthermore, while the form 'ornatum' is not linked to 'signatum' in *S. serratum*, an allele producing a similar phenotype falls within the same super-gene in *S. bocqueti*.

Bishop (1969) has identified some components of the selection operating upon the colour-patterns in *S. rugicauda*. He shows that the gene for the yellow phase is lethal when homozygous and that during the winter it approximately doubles its frequency in the current season individuals, but not in their parents. It becomes rarer again in the summer. Indeed, Bishop showed experimentally that the yellow individuals survive better than grey at a low temperature. We have here a parallel with the red and black polymorphism of the beetle, *Adalia bipunctata* (p. 325). In *Sphaeroma rugicauda* another morph, that for red colouring, almost trebled its frequency during the 1967–8 winter, but not in other years.

As P. M. Sheppard has pointed out to me, a conclusion of general interest emerges from the genetic studies of these *Sphaeroma* species. In *S. serratum*, in which the genes for the polymorphism are mainly unlinked, the combinations of characters are determined by epistatic interactions. In contrast, the other species, in which the relevant supergene has evolved, have almost no epistatic interactions; for in them the combinations are controlled by the linkage. Indeed, we may reach the deduction, and it is one of general application, that the expression of disadvantageous combinations of characters may be checked by epistatic interactions unless and until the appropriate super-genic control has evolved.

This species reaches its northern limit along the south and west coasts of England where, accordingly, there are striking changes in frequency adjusted to local conditions. It will be shown indeed that, in general, organisms need special adaptations to maintain themselves at the edge of their range (Chapter Fifteen).

The beetle *Harmonia axyridis*, a species of eastern Asia, has a complex polymorphism of elytra-pattern (Komai, *et al.*, 1950; Komai, 1956). This is controlled by a series of multiple alleles which Komai (1956, p. 128) conjectures, no doubt correctly, to represent a group of very closely linked major genes; for he points out that marked differences in the same character are produced, while the heterozygous combinations show 'mosaic dominance'. Moreover, Tan (1946) has obtained evidence of a cross-over within the super-gene so formed.

Dobzhansky (1951, pp. 142–3), in a brief but valuable review of one aspect of the subject, points out that this species includes an approximately monomorphic western race and a polymorphic eastern one. For in west-central Siberia (Altai and Yeniseisk Province) the population is nearly uniform for the *axyridis* pattern (black with, basically, six pale spots on each elytron: these are yellowish or brownish in colour). Further eastwards (Irkutsk, Manchuria, northern China, Korea, Japan) the species becomes polymorphic for a number of additional phases, some of them predominantly pale. These are controlled by four principal major genes the dominance-order of which (from the bottom recessive) is h (*succinea*, the palest form), h^x (*axyridis*), h^s (*spectabilis*, black with two pale spots on each elytron), and h^c (*conspicua*, black with one large spot on each elytron). The effects of these genes are modified by temperature influencing, in particular, spot-number in the paler forms.

On the east coast of Asia, h is much the commonest gene (80 per cent) while h^c is rare (7 per cent). In northern Japan these frequencies are somewhat similar to the neighbouring continental ones (in Hokkaido h over 60 per cent; h^c 20 per cent). Passing southwards down the island-chain there is, however, a gradual increase in h^c (to 60 per cent) and decrease in h (to 20 per cent).

Komai (1956) is of opinion that this cline reflects the path by which *H. axyridis* colonized Japan; reaching the islands in the north, where it most resembles the mainland form, and making its way southwards adjusting itself to the climate as it went. We can be sure that such climatic adaptation has taken place, but the deduction that the gene-frequency also indicates the path taken by the immigrants is surely false. Admitting the adaptation to the climate of the Japanese islands, strung out as they are from north to south, the concurrent assumption that the gene-frequency represents the colonization path seems excluded; even if we did not appreciate the powerful selection operating in such a polymorphism and the opportunities for rapid and accurate adaptation which is presents.

We seem in this species to have distinct evidence that the apparent multiple alleles controlling its polymorphism represent a super-gene. That situation will be studied in the ensuing chapters.

Protein Polymorphism

Reference has already been made to the polymorphism of the human blood groups. In more general terms, studies of protein variation have now been carried out by means of starch-gel and acrylamide-gel electrophoresis in animals (pp. 28–9, 104–9, 341–2). Much of this has also proved to be polymorphic, and already in certain instances there is evidence to show that the phases are at least partly maintained by means of heterozygous advantage (e.g. Fujino and Kang, 1968).

Bearing in mind that the genes have multiple effects (pp. 16–17, 140, 145), we may expect that such cryptic diversity will frequently be correlated with overt characteristics and that these may contribute to its maintenance. Thus a useful study can now be made of distinct phases of morphology, colour-pattern, or habit in such a way as to detect their association, if any, with protein variation, especially in wild populations.

This combined technique is at an early stage, but the success which has attended it so far should encourage its widespread application. As a preliminary example, the investigations in ecological genetics carried out by Milne and Robertson (1965) on the Eider Duck, *Somateria mollissima*, may be cited. The colony of this bird at Forvie, Aberdeenshire, comprises about 1,000 breeding pairs. Approximately two-thirds migrate in winter to the estuary of the Tay and, with some overlap, they nest on their return in a different area from the permanently resident individuals.

Milne and Robertson find a polymorphism in the egg albumen of this species, recognizable by differential mobility as seen in starch-gel electrophoresis. A pair of alleles a, and b each give rise to a single band in different positions when homozygous and, though staining less intensively, both bands are present in the heterozygotes. The gene-frequency of b is the smaller (0·14 and 0·27 in the sedentary and migratory groups respectively). A much rarer situation seen only three times in 258 birds, is also found, in which a further band occupies a third position beyond that produced by b. This is presumably due to an additional gene, c. As Milne and Robertson point out, it will be valuable to look for this in other colonies in which it may be commoner, so as to determine its genetics, which are as yet unknown: a super-gene may have evolved here.

As already indicated, this polymorphism proves, in addition, to be related to the habits of the birds. There is a significant excess of gene b ($P < 0·05$ and $> 0·02$) in the migrant individuals, the numbers of which may well be so adjusted as to reduce pressure upon the feeding grounds in the Forvie and Tay districts though using both of them.

Semeonoff and Robertson (1968) have studied polymorphism for a plasma esterase (E_1) in the Field Vole, *Microtus agrestis*. The condition proves to be widespread in the wild populations in Scotland and is associated with balanced ecological advantages. It is identified by starch-gel electrophoresis and is controlled by four autosomal alleles. These may, in fact, represent a super-gene. One of them, $Es\text{-}1^o$ stops the activity of the enzyme, a recessive condition, while the other three give rise to it: $Es\text{-}1^a$ and $Es\text{-}1^b$ do so to an equal degree, while $Es\text{-}1^c$ doubles the effect. $Es\text{-}1^a$ and $Es\text{-}1^b$ produce different degrees of electrophoretic mobility, though $Es\text{-}1^c$ and $Es\text{-}1^a$ are the same in this respect.

The wild population increases in numbers in the summer, and during that time selection favours those animals which lack the enzyme activity; however, they survive the winter less well than the others. That is to say, we have here a selective balance which must, at least in part, be responsible for maintaining the polymorphism (p. 226).

This mouse is subject to cycles of abundance and rarity, occurring every few years. In the localities where the investigation has been carried out, mean density is about 200 voles per acre; but 600 may be present per acre at the maximum reached before one of these spectacular reductions, which take place in a single season. Such fluctuations must be associated with genetic changes on the lines already discussed (pp. 12–13). As Semeonoff and Robertson remark, it seems improbable that segregation at the *Es-1* locus plays a major part in controlling such cycles of abundance and rarity. Yet as the E1 negative genotype is subject to relative increase at high population density, it certainly seems that the respective frequencies of the alleles must be associated with the striking alterations in population-density which take place in nature.

The fish *Catostomus clarki* is polymorphic for a serum esterase controlled by a pair of alleles (Koehn and Rasmussen, 1967); all three phenotypes are distinguishable by starch-gel electrophoresis. The two homozygotes *Es-1ᵃ* and *Es-1ᵇ* each give rise to single bands, *Es-1ᵃ* representing the slower, anodally-migrating, enzyme. Both bands are present in the heterozygotes *Es-1ᵃ/Es-1ᵇ*.

C. clarki is a large fresh-water species which Koehn and Rasmussen have studied at twenty sites on disconnected and connected tributaries of the Lower Colorado River. The gene-frequency of *Es-1ᵃ* ranges from 0·18 in the head waters of the Pluvial White River in Nevada to 1·00 in those of the San Pedro River, 525 miles away in southern Arizona, where a sample of fifty-eight was monomorphic. Without exception, this allele becomes relatively commoner in more southerly latitudes, indicating adaptation to some environmental condition, perhaps temperature.

As the authors remark, a cline in gene-frequency maintained over such a distance must be due to selection in favour of the heterozygotes, modified to varying degrees. Yet on the available numbers (none exceeding 103 and the majority under 50), the proportions of the three genotypes do not depart significantly from expectation, assuming equal viability, save in one locality. It may be that an excess of heterozygotes is present in the young and disappears with age, as in the transferrin polymorphism of the Tuna (see below). Alternatively, we must suppose that selection for the heterozygote is not strong at present, though it may well have been so and the genetic situation has become partly repaired. Evidence for adjustment in this respect comes from allied species: *Catostomus insignis*, *C. latipinnis*, *C. bernardine*, *C. tahoensis*, *C. macrocheilus*, and *C. ardens*. For all of these have achieved uniformity for the double-banded type, which is heterozygous in *C. clarki*. This has then been attained by distinct genetic means in these related fish, 100 per cent of which are of what appears to be the optimum form.

The power of selection in favour of the heterozygotes in *C. clarki* is shown in two ways additional to its cline in gene-frequency. The population in the Pluvial White River, Nevada, must have been isolated from the Colorado River 100 miles away for about ten thousand years. Yet it takes its place in the cline. So also does the Shoal Creek population of *C. platyrhynchus*, originally monomorphic for $Es\text{-}1^b$, into which the $Es\text{-}1^a$ allele has been introduced by hybridization with *C. clarki* since 1938.

We may now turn to the use of starch-gel electrophoresis for detecting transferrin polymorphisms, choosing an instance in which the necessary work required to demonstrate the selective forces involved in maintaining it has already been carried out. That investigation is due to Fujino and Kang (1968) using three species of Tuna from the Atlantic and the Pacific. In *Katsuwonus pelamis*, the evidence indicates the operation of three co-dominant autosomal alleles at a locus known as *Tsj*, though one of the predictable single banded phenotypes was not encountered among 4,366 specimens. One pair of the alleles was identified in *Tunnus maccoyi* and is responsible for the development of a single banded and a double banded position in the gels: none of the second form of the single banded state that is theoretically possible was found. Five transferrin phenotype phases controlled by three of the *Tsj* genes were also detected in *T. albacares*. Again, though a sixth was to be expected, it was not encountered.

The adaptive importance of these polymorphisms was established by Fujino and Kang in several ways, of which two in particular may be mentioned. (i) An excess of heterozygotes, and deficiency of the two homozygotes was observed in the samples from all four collecting grounds. It was significant in two of them on the numbers obtained. (ii) It is clear that in all three species the superior advantage of the heterozygotes declines with age and is annulled in the oldest specimens. This is well shown by a correlation in which observed divided by expected phenotype frequencies of two homozygotes and the heterozygotes are plotted against time. Three regression lines, well separated in the youngest fish, meet in the oldest. Fujino and Kang also develop an argument demonstrating that the polymorphism observed in these Tunas is balanced. They very reasonably suggest that we have here a situation in which the genes controlling transferrin polymorphism give rise to unequal fertility resulting in heterozygous advantage. Also that the bias thus obtained is gradually eliminated by opposing differential viability during the life of the animals. Thus the gene-frequencies remain unchanged in spite of selection-pressures on the phenotypes.

Though work on protein variation to be detected by means of starch-gel electrophoresis has been prosecuted only in recent years, a great number of instances in which the diversity is polymorphic has already been disclosed by this means. The four examples briefly described here illustrate different aspects of that situation and mechanisms for its maintenance. A bibliography for polymorphism in transferrins is provided by Fujino and Kang (1968) and for esterases by Semeonoff and Robertson (1968). In considering this matter in general, the latter authors (p. 225) remark that variation of this kind

'offers valuable material for the study of population structure in species whose ecology is well enough known, especially where it is possible to follow gene changes over successive generations. The importance of a sound ecological basis for such studies can hardly be overstressed. If this is lacking, the mere accumulation of further records of polymorphism in different species would appear by now to be of limited value, unless the biochemistry of the variants is the main target of inquiry.'

With this conclusion I am in strong agreement, except that it is still necessary to obtain further estimates of general heterozygosity, and the relative frequency of polymorphic loci, in diverse groups of organisms.

CHAPTER NINE

Polymorphism and the Super-gene in Snails

The genetics of *Cepaea nemoralis*

The control of hydrogen cyanide formation in *Lotus corniculatus* and *Trifolium repens* suggested that the switch-mechanism in multiple polymorphism might usefully evolve into a super-gene. Some indication of that occurrence was obtained from those species in which various phases are apparently determined by a series of alleles (p. 112), while clear evidence for it can be derived from the examples to be discussed in this and the four succeeding chapters.

A. J. Cain and P. M. Sheppard, working upon the colour and banding of snails, have been chiefly responsible for one of the most thoroughly analysed studies in the field of ecological genetics. The shells of *Cepaea nemoralis*, the species which they have principally used, may be either yellow (greenish when the animal is within) or relatively dark: pink or brown (Plate 5(2)). These, which represent the three main shell colours (excluding the bands, when present), are controlled by a series of multiple alleles. Brown is dominant both to pink and to yellow, and the latter is recessive to the other two (Cain, King, and Sheppard, 1960). Some of the minor variations to which these shades are subject are also determined by genes which are members of the same allelic series. Thus the pinks may be either dark or pale and so may the yellow. The full allelic series as known today is brown, dark pink, light pink, very pale pink, dark yellow, light yellow. Each colour is dominant to those succeeding it, in the order stated; so that brown is the 'top dominant' and light yellow the 'bottom (or universal) recessive'.

Upon this coloured surface bands may be superimposed. There may be any number of them up to five (six have been recorded as a rarity) or they may be absent (Plate 5(2)). They are recorded on a numerical system, the uppermost on each whorl being scored as number 1. To ensure that the situation is clear, the absence of a band must always be indicated (by a 0). Thus the condition with five bands is represented as

12345, that with one only, placed centrally, as 00300, and the bandless as 00000.

The unbanded type is dominant to the banded of any kind, the various forms of which are controlled by modifiers of banding. Thus the 00300 and the 00345 forms (the latter, of course, being that in which the two uppermost bands are missing) are each simple dominants, but the genes responsible for them work only on the banded genotype. Narrow, and often interrupted, bands are dominant to those of normal width; while a flush of extra banding pigment over the bands, the so-called 'spread band', is a simple dominant also.

Two or more bands may be fused, when they are bracketed in the banding formulae: 123(45) or (12)3(45) for instance. This fusion, being polygenic, is very variable in its expression and may begin at almost any time in the growth of the shell; thus it is sometimes visible as a mere trace, just before the lip. In order to consider primarily the visible characters, Cain and Sheppard (1950) have adopted the convention that fusion shall not be recorded unless it affects bands at or beyond a line drawn across the whorls at right angles to the lower lip of the mouth (that is one quarter of the way round the circumference). A reduction in width of bands 1 and 2, which may affect them in any degree almost to complete absence, is also polygenically controlled.

In describing snails, the colour always refers to the main area of the shell, upon which the bands, if any, are placed; not to the bands themselves unless specifically stated. These, however, also vary in tint. They and the lip are normally blackish, but are occasionally orange or colourless, the latter being known as hyalozonate. Individuals of that type are also referred to in the literature as 'albinos'; but this is misleading, since the rest of their shell is normally coloured. Both they and the type with orange bands are simple recessives to the normal form.

The alleles governing the general shell-colour, from brown to light yellow, and the pair determining whether or not the shells are banded, constitute a super-gene. The linkage between its two components is so close that until recently it has appeared absolute. Now, however, Cain, King, and Sheppard (1960) have obtained results of critical importance in this matter. They report two families in which the cross-over value between these two loci is significantly increased being, in fact, about 2·25 per cent. That the effect is genetic is indicated, were such indication needed in the circumstances, by the fact that in both matings the parent in which crossing-over occurred came from the same brood.

We here have evidence of genetic variation in cross-over frequency between the loci constituting a super-gene: a demonstration, in the very material under study, that they could have been, and doubtless were, brought together by selection favouring increased linkage (pp. 110–12). Fisher and Diver (1934) had indeed reported crossing-over, differing from one family to another, when investigating these same characters. However, Lamotte (1954) has pointed out that their results were invalid because they were using adult snails which could have been fertilized previously. For in this species, living sperm can remain in the spermatheca up to at least three years, giving rise to mixed progeny after a subsequent mating at any period throughout that time.

Though the specific evidence demonstrating the extremely close nature of the linkage involved has been obtained only for the loci controlling the snails' general shell-colour on the one hand and the presence or absence of banding on the other, it is possible that this super-gene consists of a number of units. We know that it is linked with the gene for a 'spread', compared with the normal bands. Thus appropriate alleles could be held together in contrasted and balanced groups (p. 112), yet allowing interchanges at a low frequency within them, enabling the snail populations to adjust themselves to alterations in their habits or to their gene-complex. Polymorphism for shell-colour and band-number is extremely widespread; that for band-colour and for the 'spread band' certainly occurs, but in a few localities only.

The genes for the 00300 and 00345 banding types are not included within this super-gene. They assort independently of it and Cain, King, and Sheppard (1960), who have made some penetrating observations on the subject, have shown that this is in fact to be expected. They point out that as the 12345 form is extremely closely linked with colour, the genetic modifications of it, since they can operate only upon five-banded specimens, must behave very much as if I linked with colour too. Consequently, any tendency to bring such modifiers into the super-gene would be relatively ineffective unless the selection responsible for doing so were very powerful: as indeed it may be for the coloration of the bands, considering that these snails may be subject to heavy visual predation (pp. 181–91). The apparent linkage relationship of the spread band is certainly puzzling: it is just possible that the gene controlling it chanced to be on the same chromosome as the super-gene in question, for we do not know if the cross-over value between the two is in fact very small.

The body-colour of *Cepaea nemoralis* is subject to continuous variation from nearly white through shades of grey to black (one major-gene making it red is known, Murray, 1962). Cain and Sheppard (1952) have shown that it is independent of food and of physiological control by the environment but is genetic and determined on a polygenic basis. In a group of colonies subject to strong visual predation there is close correlation between the frequency of pale and dark body-colours and of pale and dark shell-colours from colony to colony but hardly any in individuals considered separately. That is to say, the polygenes affecting body-colour have little effect upon shell-colour, but the selection which favours light or dark shells favours also light or dark bodies in this snail.

Natural selection in populations of *C. nemoralis*

The proportions of the various colour and banding-types differ greatly from one colony to another. Cain and Sheppard (1950, 1954) find that they are strictly related to the ecology of the site (except in special circumstances, pp. 196–8): the commonest phenotypes in any population being those which are the least conspicuous against the prevailing background, both to the human eye and, as we now know, to that of predators also (pp. 182–91). Thus the most advantageous shell-colours are yellow (greenish when the animal is within) in green areas, pinks on leaf litter, and reds and browns in beech woods with their red litter and numerous exposures of blackish soil. Also the least obvious patterns are the banded ones in a diversified habitat (Plate 6(1)), a mixed hedgerow for instance, and the unbanded in relatively uniform conditions such as those provided by dense woodlands. This is well illustrated by the correlation diagram shown in Figure 10. It indicates how closely the various populations are adapted to the ecology of their habitats.

Cain and Sheppard (1950) studied the frequency of colour and banding types on different backgrounds. Selecting among their localities the five with the most green at ground-level and the five with the most brown, they found that:

The lowest percentage of yellow shells on a *green* background was 41.
The highest percentage of yellow shells on a *brown* background was 17.

Considering also their five most uniform habitats and the five most varied, they found that:

The lowest percentage of unbanded shells on a *uniform* background was 59.
The highest percentage of unbanded shells on a *variegated* background was 22.

Thus the association between colour-pattern and background is a striking one.

In the majority of instances, snails are seen from above, and from that position bands 1 and 2 are fully visible with 3 partially so, but 4 and 5 are hidden by the curvature of the shell. It is only when examined from the side or from below that 3,4, and 5 make any considerable contribution to the diversity of the pattern. Consequently the 00000 type and all others in which numbers 1 and 2 are missing can be grouped together as 'effectively unbanded'.

A detailed survey such as that carried out by Cain and Sheppard (1954, see p. 183) in the Oxford district, shows that colonies with the most diversified proportions of shell-colours and banding may be interspersed within an area of a few square miles where patches both of woodland and of rough herbage are scattered across a countryside traversed by hedgerows. Here the phenotypes are adjusted to the different background of their habitats along the lines indicated in Map 5 overleaf.

It is clear that in such circumstances the striking similarity between colonies inhabiting distinct areas with the same type of background cannot be due to widely acting environmental causes such as climate. Neither can it be the result of migration, since correspondence of phenotype from one population to another is strictly correlated with the nature of the habitat, not at all with proximity. Moreover, the extent to which these snails can wander has been a subject of careful examination. Thus Lamotte (1951) finds that no colony of more than 30 metres radius nor, when linear, of more than 53 metres extent, can be considered as an interbreeding unit. Indeed, Cain and Sheppard (1954) have shown that hedgerow and woodland communities can approach within a few feet of one another without losing their highly distinctive qualities.

The widespread correlation between the appearance of the *C. nemoralis* colonies and their background is a proof that the frequencies of the colour and banding classes, as well as the degree of body pigmentation, are determined by predators hunting by sight (Sheppard, 1952*b*). These may include mammals, such as rabbits, for, although most of the species are colour-blind, selection by tone may also be effective (Cain, 1953); though birds, which can distinguish colours, are doubtless among the chief enemies of these snails. The species which seems to take the largest toll of them in England is the Song Thrush (*Turdus philomelos = T. ericetorum*). This picks up all but the small specimens, which no doubt it swallows, and carries them to convenient stones ('thrush anvils') in order to break open the shells (Plate 6(2)). Here the fragments accumulate, so that it is possible to decide whether or not the birds are collecting a random sample of their prey. Extensive studies on various types of background have established the fact that they capture them selectively, destroying an unduly large proportion of those whose colour and pattern match their habitat the least well. Here, indeed, we have a demonstra-

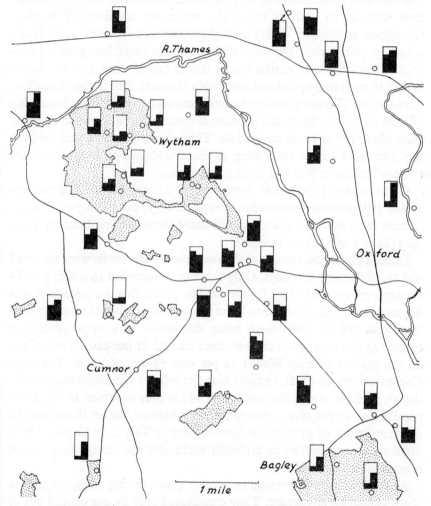

MAP 5. *The Oxford district, showing sites at which A. J. Cain and P. M. Sheppard have collected the snail* Cepaea nemoralis *and the percentages of the yellow shells (left-hand columns) and of the banded shells (right-hand columns). Woodlands are stippled: all colonies outside them are in hedgerows or rough herbage.* Genetics
(*1954*), 39, *101.*

tion of natural selection in action. Before pursuing its implications further, it will be useful to describe an instance in which it has been studied in a way that illustrates very well certain techniques of ecological genetics.

Sheppard (1951*b*) carried out detailed observations on two colonies of *C. nemoralis* occupying clearings of about one and a half acres each, and one and a half miles apart in a mixed deciduous wood (Wytham) near Oxford. In both of them, the vegetation consisted of Bracken (*Pteridium aquilinum*), Willowherb (*Epilobium angustifolium*), and Dog's Mercury (*Mercurialis perennis*), with grass and other low-growing herbage. They were very similar except that in one of them, that in Marley Wood, there is more grass and less Dog's Mercury than in the other, Ten Acre Copse. The work was carried out from early April till the beginning of June 1950 and, due to the advance of spring, the background changed from brownish to green meanwhile. This was rather more noticeable at the Ten Acre Copse site owing to the abundance of Dog's Mercury there which came into leaf during that period.

Collections of living snails were made at Marley Wood. The proportion of yellow shells amounted to 27·9 per cent (57 out of 204) on 26 May. Moreover, it had not altered significantly there, having been 24·2 per cent (80 out of 330) on 14 April.

In both places the broken shells from the thrush anvils were removed and studied every few days. Only those were recorded in which a sufficiently large amount of the lip survived to ensure that the same shell was not counted twice. It was found that in early April when the background was dark, yellow shells were being eliminated to a disproportionate extent: 43 per cent (3 out of 7) on the 11th and 41 per cent (7 out of 17) on the 23rd at Marley Wood; 42 per cent (5 out of 12) at Ten Acre Copse from 19–26 April. In early May the yellows were appearing on the thrush anvils in about the proportion which they occupied in the living population: 27 per cent (9 out of 25) on 7 May at Marley Wood and 22 per cent (2 out of 9) from 28 April–2 May at Ten Acre Copse. These numbers are, of course, individually small, but the change in selection is established by their trend.

As the background became noticeably green during May, the yellow shells gained an advantage. They constituted only 15 per cent (6 out of 40) of the total predated from 19–26 May at Marley Wood and 9 per cent (3 out of 33) from 11 May to 5 June at Ten Acre Copse. The results indicate that they are proportionately less subject to destruction by thrushes as the background becomes greener and, to the human eye at least, they match it the better. One possible, though rather unlikely, source of error to which that conclusion is subject was easily excluded, for the yellow shells had not become rarer among the living population: indeed, as already indicated, at Marley Wood there was a non-significant change in the opposite direction. Alternatively, and apparently more probably, the thrushes might be changing their feeding-grounds during the course of the observations and bringing the snails from a different area, where yellows were relatively uncommon, to break them upon their anvils. But it would certainly be surprising if they were doing this to an approximately equal extent in both localities, as they would have to do to account for the facts.

This possibility was envisaged and skilfully tested by Sheppard (1951b). He liberated at the Ten Acre Copse site an experimental population of 1,358 *Cepaea nemoralis*, 747 yellows to 611 pinks and browns, brought from other habitats. These were identified with a dot of cellulose paint near the umbilicus, a region of the shell which cannot be seen by predators. Provided the periostracum be first removed over the area to be painted, such marks remain intact for many months. It was found that thrush stones thirty yards outside the area where the experimental animals were released never had marked individuals carried to them. Furthermore, additional snails distinguished by a different colour were placed in the vicinity of certain other anvils, and although they were subsequently found broken upon them, they were never brought to those used in the main experiment. The results show that the thrushes were not carrying snails more than twenty yards, and consequently that the observed reduction in the yellow shells destroyed could not be accounted for by the birds moving to new feeding-grounds.

The selective elimination of the colour-classes changed in the experimental population, which was intermingled with the wild, in the same way as it did in the wild one. When released, 55 per cent of the marked individuals were yellows. Yet that class comprised 76 per cent (16 out of 21) of those predated between 28 April and 8 May: while from 11 May onwards (up to 5 June), when the site had become greener, the proportion had fallen to 44 per cent (11 out of 25).

Thus there was an apparent relative decline in the number of yellow shells broken by thrushes as the spring advanced in these two woodland clearings, and it is of decisive importance to determine if it were significant. We are actually dealing with three populations, since at Ten Acre Copse there were both the snails occurring naturally and those that had been brought there from elsewhere and marked. When the percentage of yellows, appropriately transformed into angular measure, is plotted against dates in three graphs, it shows a reduction throughout April and May in each of them, as indeed is indicated by the facts already quoted.

We are concerned to know whether the regressions are significant and Sheppard (*l.c.*) undertook a logit analysis to this end. Since it can be shown that there was no evidence of departure from parallel regression for the data on the indigenous and the introduced snails at Ten Acre Copse ($P > 0.5$) these were combined; and for the regression on the successive totals $P < 0.01$ while in the Marley Wood collections $P < 0.02$. Moreover the χ^2 for the parallelism of the decline in the two habitats indicates no divergence, $P > 0.02$. There is also evidence of regression in the marked population when treated separately, for which $P < 0.05$.

This situation may be summarized by saying that the Marley Wood and Ten Acre Copse sites are clearings, more than a mile apart, in mixed deciduous woodland. They therefore provide habitats in which the background changes from predominantly brown to predominantly green during the onset of spring. In both of them the proportion of yellow- compared with pink- or brown-shelled *Cepaea nemoralis* destroyed by thrushes showed a decline throughout this period. The birds were not exploiting new feeding-grounds, where the yellow type might have been rarer; consequently, hunting as they do by sight, they must have acted as selective predators withdrawing from the snail populations an excess of those individuals which at any given time matched their habitat least well.

It is principally in mixed deciduous woods that the colour at ground level is subject to a marked seasonal change owing to the predominance there of broad-leaved herbaceous plants. These die back in winter leaving dead brown stems and exposures of earth subsequently replaced or covered by a growth of green foliage. In the majority of habitats no such marked transformation takes place. Thus red and brown shells will always possess a cryptic advantage in a dense beech-wood and yellow ones will always do so along hedgerows which, owing to the large proportion of grass present, seldom become predominantly brown or blackish: the latter situation is true of certain downland localities also. Moreover, the advantage of the banded shells is a persistent one in a diversified habitat such as that provided by rough herbage.

In such places as these, Cain and Sheppard (1954) have established a significant correlation between the nature of the habitat and the colour-pattern of *C. nemoralis* (Figure 10). Moreover, it has been shown that a disproportionately large number of the less well-concealed individuals are destroyed by thrushes, their shells accumulating in excess at the 'anvil stones'. Thus Murray (1962, p. 143) working in an area of rough herbage on chalk at Wittenham Clumps, Berkshire, obtained the figures shown in Table 11.

TABLE 11

	Yellows	Non-yellows	Total
Predated by thrushes	5	21	26
Not predated by thrushes	37	43	80
	42	64	106

The frequencies of yellow and non-yellow (pink and brown) shells of *Capaea nemoralis* predated by thrushes and not so predated at Wittenham Clumps, Berkshire. All the shells had previously been marked in the area which includes the thrush stones.

The excess of non-yellows among the shells predated by the birds is significant ($\chi^2_{(1)} = 4.91$, $P < 0.05$). Though the numbers are small in this collection, it has been chosen because all the snails included in it had previously been found in the neighbourhood of the anvils and marked underneath, in a position invisible to predators. Consequently we here have complete evidence that the thrushes were not bringing the shells from a distance, where the frequencies of the types might be different, in order to break them upon convenient stones.

We may also take an instance which involves pattern. Cain and Sheppard (1954) working at Marley Bog, Wytham Woods, a habitat with on the whole a uniform background, found that 47·1 per cent (264 out of 560) of the living snails were effectively banded, but 56·3 per cent (486 out of 863) of those destroyed by thrushes belonged to that form. The significance of the difference is measured by $P < 0.02$, so that there is good evidence that it is due to selective elimination.

It is essential to decide whether or not these birds are important predators of this snail for, even if strongly selective, it is obvious enough that their predation might be of little evolutionary consequence if they are removing but a negligible fraction of the population. Fortunately we have at least some estimate of their effect in a few English localities. Thus of the 1,358 *C. nemoralis* marked and released by Sheppard on 26 April 1950, 46 were found at thrush stones between 28 April and 5 June. That is to say, about 3 per cent of the population were predated in less than six weeks. Moreover, Cain and Sheppard (1954) successfully applied the technique of marking, release, and recapture to the snails at Marley Bog and found that the colony there amounted to about 10,000. The work already described, which showed that the un-banded shells were selectively removed there, suggested an amount of elimination greater than that at Ten Acre Copse. The observations continued from 4–20 July 1951 and during that time about 8·6 per cent (being 863 out of the estimated 10,000) individuals were carried by thrushes and broken on their anvils. It should be noticed that a study of the shells which accumulate around these stones may much underestimate the destruction if the birds swallow the small specimens whole, and of these no record can be obtained.

It is clear therefore that at such sites certain colour-patterns are at a disadvantage and are constantly subject to heavy elimination by predators, notably by thrushes, yet the populations in these places do not become uniform. They respond to the powerful selection-pressure to which they are exposed by a reduction in the proportion of certain colour or banding types, not by their elimination. This impressive fact demonstrates that the super-genes controlling the polymorphic phenotypes have evolved heterozygous advantage, as we should expect them to do in view of the concepts developed in Chapter Six, though *direct* evidence for this is lacking in *Cepaea*: but see Wolda, 1969. Thus, the polymorphism of *C. nemoralis* is *maintained* on a physiological basis; one which, favouring the heterozygotes, promotes diversity: in these localities it is *adjusted* to different frequencies by selective predation.

Thus the polymorphism of *Cepaea nemoralis* is of the normal balanced type. We have, moreover, evidence from shells preserved in Pleistocene deposits that it has persisted since that period (Diver, 1929).

Attention has been drawn by Cain and Sheppard (1954) to a subsidiary mechanism for adjusting the frequencies of the phases. That is to say, de Ruiter (1952) has shown that some birds which prey upon protectively coloured insects tend to hunt for specimens resembling one which they have recently found, even to the exclusion of others that seem more obvious: making use, therefore, of the experience they have just gained. This aspect of predation tends, as Sheppard (1958, p. 147) notes, to make many cryptic animals extremely variable if they be common; it will not have this effect upon rare species. Of this, the immensely variable and abundant moth *Hydriomena furcata* and the rare, relatively constant, yet nearly related *H. ruberata* provide good examples. It is evidently a mechanism which could maintain polymorphism by favouring the rarer morph, whichever that might be: a situation named 'apostatic polymorphism' by B. Clarke (1962).

A similar effect must operate upon polymorphic forms, including the colour-patterns of snails, and so puts the commonest phase at a disadvantage whichever that may be. Thus it contributes to produce, or at least to maintain, polymorphism. We have, however, good reasons for thinking that this is not the principal way in which predators affect *C. nemoralis*. For the experiments conducted by Sheppard both on natural and introduced populations at two localities in the Wytham Woods (pp. 155–8) demonstrated that first one phase and then another was favoured as the spring advanced while their frequencies did not alter in the populations as a whole. In these circumstances, it is evident that the thrushes were not concentrating upon the commonest form. Furthermore, we may recall, in this connection, the results illustrated by Figure 10 and those obtained by Cain and Sheppard in the Oxford district (Map 5) showing a close association between both colour and banding on the one hand and type of habitat interspersed over the countryside on the other (pp. 154–5). A correlation of this kind is not in accord with a situation in which the commonest forms are those subject to the most destruction. It must be remembered, however, that forms which stand out from their background will appear to a predator to be commoner than they really are.

It may be mentioned here that non-random mating provides another mechanism by which polymorphism can be maintained, as discussed in regard to *P. dominula* in which this actually occurs (pp. 139–40. However, Lamotte (1951) and others have undertaken extensive sampling of copulating pairs of *C. nemoralis* in the field and have found no indication of any such tendency in that species.

This snail takes about two years to reach maturity and in this and in other respects it is not very easily bred for genetic work. So far it has not been possible to demonstrate an excess of heterozygotes in back-cross or F_2 families in broods involving the super-gene for colour and banding. Doubtless this will become apparent when more work devoted to detecting it has been undertaken, for of its existence there can be no question (p. 188). Indeed we have evidence that the genes concerned influence the physiology of the organism in addition to their effects upon colour-pattern, for Sedlmair (1956) and Lamotte (1959) report differences in survival value between various of the phenotypes under adverse conditions both in this snail and in the closely related *Cepaea hortensis*.

The super-gene control in *Cepaea nemoralis* saves the population from a high proportion of undesirable recombinants. If the unit be composed of two loci only, two phenotypes instead of four are segregating (remembering that dominance is complete). Unlike the situation in mimicry (Chapters Twelve and Thirteen), only one will be at an advantage on account of its visual qualities. In the majority of habitats this will at all times be the same though, as we have seen, there are certain sites in which the effects of selection will shift from one colour-pattern to another as the season advances. Even there the necessary plasticity could generally be maintained by a single segregating pair of alleles since it will usually be true that an adjustment in colour alone is all that can be needed to accord with the onset of spring, as at Ten Acre Copse (pp. 184-6). One could, however, visualize a site which at that period changed, for instance, from predominantly brown and uniform to predominantly green and diversified. In such circumstances, the super-gene control would be useful even in respect of visual qualities. This must, however, have been evolved on account of its physiological effects. Selection can then operate in favour of a single type of heterozygote, on the lines discussed on pp. 100-110: for it is likely that the physiological response of the various members of each allelic series, which will be closely co-ordinated with the working of the body, will have been adjusted to be more uniform than their visible effects.

It is an interesting fact, constantly encountered, that a rare mutant may have some advantage and become a polymorphic form in one or a very few localities in a widespread species. The *medionigra* gene of *Panaxia dominula* provides an instance of this kind (pp. 103-9) and so does the recessive orange-banded condition in *Cepaea nemoralis* (p. 179). So far as is known, this is merely an exceptional variety everywhere save in one hedgerow community near Oxford where it amounts to 15 per cent of the banded snails. Since the population is a very large one, it does not look as if its occurrence there were due to random drift, unless there has been a great numerical increase in the locality, while that of the *medionigra* form of *Panaxia dominula* at Cothill has been proved not to be so (p. 137). The haylozonate form provides a very similar situation, but it has been found as a polymorphic condition rather more often: Cain and Sheppard (1954) report this in four colonies. Presumably such genes have some physiological advantage in the special conditions provided by these habitats. However, the hyalozonate condition can also have a direct visual advantage in certain places by suppressing the expression of colour, both of the bands and of the rest of the shell, so making the snail paler and more uniform.

Summarizing the situation in *Cepaea nemoralis*, it may be said that the super-gene control of polymorphism in this snail is an expression of the fact that in this species the physiological and cryptic diversity is preserved by heterozygous advantage. Visual selection by predators does not maintain the co-existence of the phases but is responsible for adjusting their frequencies, within certain limits, in a way suited to the ecology of each habitat.

Selection and genetic drift in *Cepaea nemoralis* and *Cepaea hortensis* in England and France

Whenever attention is drawn to a striking instance of evolution in wild populations, attempts are made to prove that it is not due to natural selection. Examples of this kind will be recalled in the earlier chapters of this book, and others are to come. The alternative explanations most favoured today, now that the Lamarckian theory is so discredited, are some form of random genetic drift or else mutation-pressure though, to the latter, selection is sometimes added as a minor adjunct. It should, however, be noticed that in fact Lamarckism involves the implication that evolution is controlled by mutation.

It has been held (e.g. by Taylor, 1907–14) that the colours of *C. nemoralis* depend upon its food-plants. Since they are now known to be genetic, no further reference need be made to that suggestion. Indeed, the yellow, pink, and brown types when all supplied with the same food continue to add to the growing shell segments of the original shade.

Diver (1940) made the important observation that colonies of *C. nemoralis*, and the closely related *C. hortensis*, some of them small and including less than one hundred individuals, may differ greatly in their average colour and pattern. Yet he believed these characters to be of no selective importance, assuming that random genetic drift is responsible for the widely distinct proportions in which they may occur from one colony to another. It is surprising that such a study should have been made without reference to the ecology of the more diverse types of habitat in which these species live. Had such colonies been related to their backgrounds, a correlation of the type demonstrated in Figure 10 should have become evident. It is hard to believe that the colours and bandings of these snails could then have been judged neutral in their survival value: and this quite apart from the fact that the shells broken at thrush stones do not represent a random sample of the living individuals in the neighbourhood. It is possible that Diver was much influenced in his views by sand dune and chalk down populations in which, owing to the rigorous and arid conditions obtaining there, selection for physiological characters is much more important compared with visible ones (p. 196) than it is elsewhere. Moreover, at the date when he wrote he

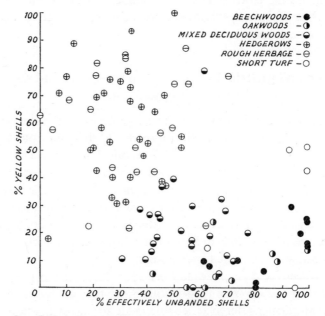

FIG. 10. *Correlation diagram, for percentage of yellow shells, and percentage of effectively unbanded shells, in the snail* Cepaea nemoralis *in different types of habitat.* Genetics *(1954),* 39, 99.

might, very reasonably, not have recognized the importance of physiological selection in these snails. It should indeed be noticed that Diver was carrying out highly original work as a pioneer in a difficult field of inquiry.

There is, however, one criticism of the views so far developed in this chapter which requires analysis since it is based upon an immense survey of *Cepaea nemoralis* and the related *C. hortensis* conducted in France for many years by Lamotte (1951). Briefly, he believed that mutation and migration are the chief agents responsible for maintaining the polymorphism of these species and that the differences between the colonies are due to random genetic drift, with selection playing a very subordinate part. He was thus led to postulate mutation-rates of the order of 1 in 10^4 which, he considered, would be sufficient to maintain the poly-morphism. This is a frequency much in excess of that encountered in other organisms, in which its upper limit tends to be the same in the widest diversity of plants and animals. For this there is a fundamental reason: that if mutation be common enough to provide the diversity required in evolution, it is incompatible with the requisite stability of the genes controlling desirable qualities. It is on this account that a low mutation-rate, with a normal maximum five to ten times lower than that postulated by Lamotte, is (or should be) one of the fundamental proposi-tions of Mendelism. As far as random genetic drift is concerned, the difficulties which it evokes have been discussed in Chapter Three. Furthermore, it should be noticed that if chance survival were important, the variance of the characters concerned should be approximately equal in the different populations of the same size, which it is not.

Lamotte (1959) has now abandoned the hypothesis of an exceedingly high mutation-rate in these snails. Yet he holds that selection is of negligible importance in controlling their polymorphism for the follow-ing reasons.

(1) His data indicated that the differences between small colonies are greater than those between large ones.

(2) The Song Thrush is only a winter visitor in France.

(3) Taking the frequency of the gene for the bandless condition in Aquitaine, he was able to fit it to a curve based on Sewall Wright's formulae, indicating that mutation and migration are important and selection insignificant.

(4) He could find no obvious correlation between the average colour-patterns of the snails and their habitats.

(5) He also found no correlation between variation in *C. nemoralis* and *C. hortensis* when living together on the same background.

Cain and Sheppard (1954) point out that these objections appear to be invalid for the following reasons.

(1) Lamotte (1951) claims that the effects of selection are independent of population-size. Consequently the fact that the small colonies differ from one another more than the large ones must, so he believed, be due to the action of random drift. This is precisely the same fallacy as that already exposed in regard to the greater variability of spotting in *Maniola jurtina* on the small compared with the large islands in Scilly. It is not a result of genetic drift at all but, as explained on pp. 58–73, a phenomenon due to selection. Since, especially in the snail instances, backgrounds

differ much over short distances, large populations will be adjusted to the average of more diverse habitats while small ones can be matched to the particular conditions characteristic of the place in which they occur. Since averages tend to be alike, the larger colonies should therefore resemble one another more nearly than the smaller.

Moreover, it is far from certain that selection-pressure is independent of population-size. For as Cain and Sheppard (*l.c.*) point out, the thrushes generally take snails when other food is scarce or insufficient and, as they always seem able to find them when needed, the intensity of their predation is likely to be more acute in small than in large colonies. Indeed when Lamotte, in his 1959 article, at length studied the relation between shell-colour and background, he did in fact obtain a correlation between them.

(2) In Britain the Song Thrush is by no means the only bird which eats land snails: game-birds do so; so also do some mammals, rabbits for example. Thus the absence of the Song Thrush during the summer in France by no means indicates that *Cepaea* is free from predators there.

Lamotte was in certain instances able to obtain a sufficient number of broken shells to compare them with the living snails. In some areas he obtained evidence of selection but not in others. The latter may represent non-selective winter predation by thrushes, but with the absence of this bird during the summer in France most of the individuals destroyed there will be eaten by other predators in a manner difficult to trace. Moreover, Lamotte does not distinguish between bird- and rodent-predation which, depending respectively upon colour and tone, may work in opposite directions and produce effects which cancel one another.

(3) Lamotte has studied the distribution of the gene for the bandless condition in Aquitaine and, taking its frequency and the values he has derived for population-size and mutation, he found an agreement between these functions on the one hand and a curve obtained from Sewall-Wright's equations on the other. He therefore believed that these could be used to determine selective effects and mutation-rates in his snail populations and deduced that the selection-pressure to which they were exposed was very small. Consequently he concludes that random drift and mutation are chiefly responsible for the proportions of the phases. His deductions are valid only if selection be very feeble and, as indicated here, in this matter he is misled. Moreover, his approach to the problem raises a point of general application. That is to say, it is a dangerous and misleading practice to take theoretically derived curves or types of

distribution-change and attempt to draw deductions by fitting obser-vational data to them without determining the orders of magnitude of the agents concerned, pp. 323–4. In the first place, the error involved is such that curves calculated on other premises could also be fitted to the same field observations; for, to obtain the correspondence between fact and theory, extremely unreliable estimates of certain parameters have generally to be included. Secondly, an effectively similar type of curve to that obtained (for example, on the assumption of very little selection) might also be produced on an entirely different basis (one, for instance, in which selection is very powerful).

The weakness of this device of fitting observational data to theoretically derived distributions is only too apparent in the present instance. The correspondence which Lamotte obtains is produced by utilizing fre-quencies (of the gene for the bandless form) in populations scattered over a considerable area. Yet the proportion of such shells can differ greatly from one to another type of micro-habitat. Thus Cain and Sheppard (1954) obtained very distinct histograms for the percentage of the bandless phase in woodlands and among rough herbage respectively (Map 5). Nevertheless they were able to show that when combined, the result was strikingly like that used by Lamotte to compare with the theoretical curve based upon Sewall-Wright's equations.

(4) The failure of Lamotte to obtain a correspondence between en-vironment and variation in colour-pattern is due partly to the fact that, as he himself points out, numbers of his collections were made in places recently disturbed, where such a correspondence is not to be expected and no direct evidence on whether selection is or is not acting can be obtained merely from the frequencies. Also many of the habitats were of a mixed character. They were described as 'dry places', 'valley bot-toms', and the like: terms that are too broad to have a precise ecological meaning. It is probable, moreover, that the situation in England is very different from that in France. As Cain, King, and Sheppard (1960) point out, since the latter country is the drier and sunnier, especially in the south, accordingly it produces on the whole paler backgrounds and so favours the yellow shells. These indeed are relatively much commoner than in England. In fact, the French environments tend to resemble the the habitats found on English chalk downs (see below). Moreover, in extreme conditions in both countries it is likely that physiological selec-tion is even more important than it is elsewhere and may produce situa-tions which appear anomalous from the point of view of visual selection when the snails are scored for the proportions of their shell characters.

(5) Lamotte holds that selection must be of negligible importance in the evolution of *C. nemoralis* and *C. hortensis* populations since he finds no correlation between the variation of these two species when living together on the same background. This difficulty was raised before their ecological genetics had been sufficiently compared. B. Clarke (1960) has now shown that in each habitat these snails both adjust their colour-patterns in relation to the prevailing background but that they do so by very different means.

Thus the dark colour needed in dark situations such as dense woods or mats of ivy is produced in *C. nemoralis* by selection for red and brown unbanded shells. In *C. hortensis* a corresponding effect is reached by increasing the number of yellow shells with fused bands. The genetics of the various phases appear to be the same in the two species, but the evidence suggests that the genes for the pink and brown background colours are at a physiological disadvantage in the gene-complex of *C. hortensis*, in which therefore a dark coloration is produced by band-fusion, retaining the more advantageous gene for yellow shells. Consequently it is clear why Lamotte detected no correlation between phenotype frequencies in these two snails when found together in mixed colonies. But from this he concluded that they are not subject to a significant amount of selection, a deduction which is evidently invalid.

In general, it seems likely that visual selection, for colour-pattern, is relatively unimportant compared with physiological selection in the greater part of France. This is true also in those regions of England, especially chalk downs, which tend to resemble in important particulars the situation found in southern regions; that is to say, in climatic and certain other environmental features (Cain and Currey, 1963). Such conditions often produce populations fairly uniform as to colour or banding over widely diversified country and give rise to what are called *area effects* by Cain and Currey (1953), who first detected them. These are not the result of the chemical constitution of the calcareous soil itself; for they are absent from several places on the chalk, being superseded there by visual selection.

The area effects are much too homogeneous over large tracts of country, containing great numbers of the snail, to be due to genetic drift or, indeed, to any random process. One such piece of downland, for example, is a triangle with a base of 6·5 km and a height of 5·5 km, which is to be compared with a panmictic unit in these animals of about 40 metres in diameter. Moreover, Cain and Currey (1968a) made a study of *C. nemoralis* in a region of area effects, comprising about 36,500 square feet, and showed by a means of marking–release–recapture that the total population there amounted to about 3,000.

We can further exclude the founder principle (Chapter 3) as being of any importance in this matter. The Marlborough Downs, where much of Cain and Currey's work has been done are well documented as to their history, and evidence exists to show that it is extremely unlikely that former conditions had reduced the snails to minute populations subject to later expansion. In one locality the shells obtained in an archaeological excavation demonstrated that these area effects were already acting in Neolithic times (Cain and Currey, 1968*b*). It was possible to show that there has been a decrease in unbanded shells in this area coinciding with a change in climate.

The general result of these processes is to produce unusually high frequencies of one or two alleles throughout a region containing very different habitats where predation may indeed be taking place. This would normally adjust the frequency of the phases in differing and appropriate ways, as already described, for example, in the Oxford district.

Since area effects produce relatively homogeneous types over extensive districts where large populations of these snails exist, their polymorphism, such as it is, must be due to some systematic process. Its nature here is as yet unknown and has so far eluded experiment. There can be no doubt that it must in some way be associated with local topography and micro-climate and be dependent upon selection (Cain and Currey, 1963; Carter, 1968).

Arnold (1968), working in the Pyrenees, has secured good evidence on the nature of a micro-climate that can produce area effects, for he has detected changes in *Cepaea nemoralis* associated with altitude. That is to say, the banded form increases at intermediate heights, where the climate is more equable, being neither so hot as in the valleys nor so exposed as at higher elevations. For this he subsequently obtained corroboration when working in a different environment, south of the Pyrenees (Arnold, 1969). Here he showed that banded shells are at a fairly high frequency in the bed of a stream but that the proportion of unbandeds rises steeply on higher and drier ground. This happens on both sides of the water course, and often over a distance of no more than 50 metres.

That result is in accord with the experimental finding that non-banded shells survive heating under a 140 W lamp bulb better than the banded type (Lamotte, 1966). It looks therefore as if the excess of non-banded morphs in a very arid locality is adaptive, as a direct climatic effect acting upon the pattern of the shells.

Cepaea nemoralis and *C. hortensis* form a pair of species closely allied morphologically, genetically, and, to a considerable degree, ecologically. They can be difficult to distinguish and their polymorphism is similar, except for the different methods by which they attain dark coloration, as already described. The genetics of *C. nemoralis*, discussed on pp. 178–80, are more fully known than those of *C. hortensis* though enough information on the latter species is now available to demonstrate their close similarity in this respect (Murray 1962; Cook and Murray, 1967). As in *nemoralis*, so in *hortensis*, brown and pink shell colours are dominant to yellow and absence of banding is dominant to its presence; while these two loci constitute a super-gene within which crossing-over is very rare. However, Murray (1963) shows that the genetics of orange banding is probably dissimilar in the two species.

Day and Dowdeswell (1963) working on an ancient field-system on Portland Bill, England, detected two identical clines in *C. hortensis*, each passing from about 20 to 55 per cent of banded shells over a distance of 400 yards. They extend along two lynchets which are about 7 yards apart and vary from 2 to 6 feet in height. The snail populations upon them are subject to predation both by birds and mammals, but Day and Dowdeswell proved that this is insufficient to account for the establishment and maintenance of the clines. They also show that these cannot be the result of immigration. The progressive changes in frequency must presumably be due to the influence of the banding gene on the physiology of the snails. It seems that this enables them to adjust to an environmental gradient along the lynchets which, in fact, slope gently upwards. We have here, in miniature, a situation comparable with the 'area effects' found in *C. nemoralis* (pp. 196–7).

There seems to be a basic similarity in the genetics of many snails, though with interesting differences in detail. General shell colour and banding tend to be controlled by what appear to be two major genes which have been brought together to form a single super-gene. The two units composing it may control a series of background colours and of banding types behaving, respectively, as multiple alleles. These characters may be further modified from unlinked loci. The whole situation is typified by that already described.

Information on this matter has been carried further by a study of the polymorphic *Arianta arbustorum* (Cook and King, 1966), which in

England is the snail most closely resembling the two *Cepaea* species in ecology, size, and appearance. It also is predated selectively by Thrushes, and its remains accumulate with the other two species upon Thrush anvils.

The genetics of *Arianta arbustorum*, as far as they are known, are on similar lines to those of *C. nemoralis* and *C. hortensis*, except for a reversal of dominance by which the presence of banding is dominant to its absence. In *A. arbustorum* the gene involved and the one controlling background colour, with brown dominant to yellow, again forms a super-gene within which no more than 1 per cent of crossing-over takes place. Cain, King, and Sheppard (1960), however, obtained certain homogeneous backcrosses giving a high recombination, a little under 20 per cent, between the units that are normally super-genic; the difference from the normal situation being clearly significant $P < 0.001$. We here have evidence of genetic variation in crossing-over of the type upon which selection could operate to produce the super-gene.

The major loci for banding and colour are not linked with the gene for hyalozonate bands, recessive as in *Cepaea*, nor for dominant mottling: a character absent from *C. nemoralis* and *C. hortensis*. These two genes are in separate chromosomes. In *Arianta* the colours are less sharp and distinct than in *Cepaea*, which presents certain difficulties in scoring.

The ecological genetics of the Gastropods *Limicolaria flammulata* and *L. aurora* have been studied in Nigeria by Barker (1968). These are polymorphic for shell-colour and banding. Both species are predated by birds in the wild and laboratory tests have indicated that the choice is selective. Unfortunately the birds acting as predators in nature do not seem to have been identified. A difficulty is introduced in this matter by the fact that the snails are swallowed whole, as are immature *Cepaea*.

In his studies on a related species, *L. martensiana* in Uganda, Owen (1965) finds that the rarer super-gene for pattern and shell colour is at a higher frequency when the population is large than when it is small, suggesting that the polymorphism is maintained apostatically (p. 189). For predators are more likely to build up a specific searching image in respect of an abundant than of a scarce species.

The dark shell backgrounds of *Limicolaria*, brown and pink, are dominant to grey, which takes the place occupied by yellow in the species previously mentioned. Unlike *Cepaea* but as in *Arianta*, banded shells are dominant to unbanded. The genes for these two characters are again built into a super-gene. Certainly the linkage between them is close: crossing-over of about 3 per cent is indicated. It is interesting to notice that the chromosome frequencies depart significantly from expectation in the direction of an advantage to the heterozygote. Most curiously, it is in particular the coupling arrangement which is deficient.

The genetics of two of the *Partula* species *taeniata* and *suturalis*, found in the Society Islands, have been investigated by Murray and Clarke (1966). The situation proved to be rather complicated and in some instances genic interaction, and in others polygene systems, are involved. However, it may be said that in both these snails the darker forms are in general dominant to the lighter. Also, several banded types segregate among themselves, but all are dominant to the unbanded condition. Here colour and banding have not been built into a supergene, nor indeed do they appear to be linked. There is some evidence in *P. taeniata* that the individuals homozygous for banding are lethal or semi-lethal.

We see in *Partula* the type of arrangement from which the supergenic control of colour and banding found in other genera could have evolved. Moreover, the genetic variation in crossing-over encountered in *Arianta* (p. 199) suggests the way in which close linkage could be achieved once the respective genes had been brought together on to the same chromosome.

Partula and *Achatinella* on Pacific islands

Finally a brief reference must be made to the work of Crampton (1916, 1925, 1932) on the genus *Partula* in the Society Islands, Tahiti and Moorea, and on the Marianas, Guam, and Saipan; also to that of Welch (1938) who collected *Achatinella mustelina* in the Waianae Mountains of Oahu, Hawaii. These studies on snails are widely quoted as giving evidence of random drift and of the 'founder principle', and this they fail completely to provide.

A number of species belonging to the genus *Partula* may be widespread and generally constant on a particular island, but they may develop a series of very diverse populations in one or two of the valleys. It has been suggested that this situation demonstrates different rates of evolution. Alternatively, it has been said that the unchanged communities represent the most recent arrivals (though, as pointed out to me by Dr Bryan Clarke who has examined these snails in their own habitats, only one of the species, *P. attenuata*, occurs on more than one island). No such assumptions as these are warranted, nor are they needed to explain the facts. The natural view, for which there is no contrary evidence, seems to be that several of the species have generally responded to different environments by physiological adaptations only, but that the conditions in certain areas and habitats have necessitated changes in visible characters also.

Much of the work on *Partula* was carried out on Moorea. Here neighbouring narrow valleys are separated by high rocky ridges which have generally been described as difficult for the snails to cross; they certainly do reduce gene-flow, but only in a few places, when dry and fern-covered, do they provide anything like effective barriers. Crampton's data indicate that these habitats, which are largely isolated from one another, contain very distinct populations of the same *Partula* species. It has frequently been maintained (e.g. Huxley, 1942, pp. 232–4) that this situation illustrates 'initial sampling effects acting upon a highly variable population'. By this it is implied that a few individuals, randomized relative to their visible features, have succeeded in crossing into a new valley and established a community there, the characteristics of which were determined by the chance genotypes of the 'founders', not by selection.

Now the methods by which Crampton obtained his famous collection were merely those of a pioneer looking for genetic variation, and this fact needs to be recognized. He drew his samples not from particular types of background but from whole valleys or large sections of a valley. Yet work on such snails, with their marked polymorphisms and differences in visible characters, has a meaning only if related to individual microhabitats. The operation of random drift and of the founder principle has been deduced from the supposedly distinct characteristics of the colonies in these isolated valleys, between which occasional migration can probably take place. But since the various collections may have been taken from different backgrounds, or from mixed backgrounds in different proportions, such conclusions are without foundation. Moreover, the valleys in question, running as they do radically seawards from the centre of the island, are affected by different climatic factors to which the snails have been selected. This is most noticeable in a comparison of slopes facing north-west and south-east. The latter are exposed to the south-eastern trade winds, and the extra rainfall which such areas receive enables them to support a more luxuriant vegetation.

I have had the benefit of discussing the situation on Moorea with Dr and Mrs J. J. Murray and Dr and Mrs Bryan Clarke who have been working on *Partula* there. They kindly allow me to say that they are in agreement with the views expressed in this chapter. The publication of their results, when they have had time to analyse them, will be awaited with much interest.

It has frequently been stressed that the *Partula* work is of special importance since it demonstrates evolution in progress. Crampton (1932) found that samples taken in certain valleys in 1909 differed greatly from others obtained when he revisited these places fourteen years later. But his collecting areas were too extensive, and not ecological entities at all. Indeed Murray and Clarke inform me that they observed significant differences in *Partula* populations in less than fifty yards. Consequently there is no evidence that on subsequent occasions Crampton was studying effectively the same populations as before: a criticism wisely made by Mayr (1947, p. 220). Even if they were strictly comparable, of which we are wholly without evidence, there is no reason to believe that the observed changes were not selective. Thus Cain and Sheppard (1950) point out that they appear to be too large to attribute to random drift.

Moreover, it is said that in some instances *Partula* species had greatly extended their range since Garrett was making observations on them from 1861–88. So too had certain of their forms, and some of these were characterized by marked size-differences almost certainly too great to be of neutral survival-value. A further, and instructive, instance is provided by *P. suturalis* on Moorea. This was reported as wholly dextral last century and it remains so in the area it originally occupied, but sinistrals were common in the regions colonized subsequently. The alternate types of Gastropod torsion seemed at one time unlikely to influence the viability of the snails. More recent investigations, on another genus, suggest that such a view is most incautious. Gause and Smaragdova (1940) working upon *Fruticicola lanzi* showed that sinistrals survive starvation much less well than do dextrals: an observation surely indicating that the two types are associated with important physiological effects. In view of such a discovery, it would be rash indeed to attribute selective neutrality to dextrals and sinistrals in *Partula*.

Crampton had used the shell-dimensions of *Partula taeniata* to calculate 'phylogenetic distances' between the varieties and races on Moorea; that is to say, the extent to which they differ genetically. His data have been re-examined by Bailey (1956) who found that the distinctions between adjacent forms become greater where they meet: a fact which suggests that there are strong tendencies for reproductive isolation to develop between them. This indicates the operation of powerful selection. It is entirely opposed to the concept of random drift so often postulated to explain the evolution of *Partula*. It seems indeed that the different forms must interbreed to some extent in the regions where they come into contact so giving an opportunity for selection to eliminate intermediate types; for these will include both the necessarily ill-adapted hybrids and those individuals that are less closely adjusted to their habitats. In this way, the qualities of any two races would be accentuated at their interface both by the selective process itself and because mechanisms tending to limit further cross-breeding would be favoured there.

The parallel with the reverse-cline effect in *Maniola jurtina* is indeed striking (pp. 83, 85–6). In the butterfly, however, counter-selection is powerful enough to operate upon the individuals so as to eliminate in a single generation an undue proportion of those that are most inter-mediate in spotting between the Southern English and East Cornish types.

It is widely held that the observations of Welch (1938) on the arborial *Achatinella mustelina* in Hawaii provide a parallel to those on *Partula*, and so indeed they do. Here too it seems that many endemic forms are found in neighbouring but largely isolated valleys, some of the populations occupying only a few acres each. Thus the work provides a further illustration of the danger inherent in failing to treat the micro-habitats of such snails separately. As Cain and Sheppard (1950) point out, Welch gives no indication of the predators of this snail or of the backgrounds on which his samples were obtained, highly desirable as it would be to have this information. Consequently we remain in ignorance of the essential features upon which to base an analysis of its evolution.

The techniques of ecological genetics have been used by Murray and Clarke in their work on *Partula* (*l.c.*), and their results show that the diversity of the different populations of this species in the Society Islands is selective. These methods were not employed by Crampton in his well-known studies of that snail, nor by Welch when investigating *Achatinella*. Consequently they came to the conclusions so constantly cited in textbooks on evolution which appear, incorrectly, to provide evidence for genetic drift or the founder principle.

CHAPTER TEN

The Heterostyle-Homostyle System

The heterostyle polymorphism found widely among plants was made famous by Darwin who gave a detailed account of it in 1877. He took the cowslip, *Primula veris*, and the primrose, *P. vulgaris*, as his principal examples and it is these species also that have generally been employed in the more complete analysis of the condition undertaken in recent years using the methods of ecological genetics. Furthermore, it must be emphasized at the outset that heterostyly opens up the possibility of a transition to homostyly and back again: a far-reaching adjustment since it reconstructs the mating system, involving the step from relative out-breeding with diversity to relative inbreeding with uniformity, or the reverse.

The most usual form of heterostyly is the distylic type found, among many other groups, in all the British Primulaceae, including *Hottonia palustris*, with the sole exception of the invariably homostyled *Primula scotica*. Such species as these give rise to flowers of two kinds, thrum-eyed and pin-eyed, each plant producing one or the other but not both.

In the thrum-eyed form the style is so short that the stigma is halfway down the corolla-tube, at the top of which are placed the anthers. The opposite situation is found in the pin-eyed flowers, for in them the stigma, borne at the end of a long style, occupies the mouth of the corolla-tube with the anthers halfway down it. Thus, though the positions of the male and female parts are reversed in the two phases, they are widely separated in both (Figure 11 and Plate 3(1)).

Darwin recognized that we have here a system which promotes out-crossing. These plants are pollinated by insects with a long proboscis: some pollen adheres round its base when pushed down a thrum corolla-tube to reach the nectaries and may be deposited subsequently on the high stigma of a pin flower, which has anthers at a low level. From these, pollen-grains are picked up near the proboscis-tip of a visiting bumble-bee or moth, and may later be transferred to the low-level thrum type of stigma. Consequently this mechanism favours fertilization between pins and thrums and so promotes cross-breeding. Darwin, however, considered it but moderately efficient. He pointed out that the withdrawal of the proboscis from a pin-eyed flower will take some of the pollen up to its own stigma, though pollen can far more readily be carried from the high anthers of a thrum corolla-tube down to the stigma below. Moreover, selfing may be brought about by Thrips and other small insects which often inhabit the flowers and scatter the pollen within them. On the other hand, there is now evidence that the stigma is receptive to pollen before the anthers have dehisced and that this situation exists after the bud has opened sufficiently for the flower to be accessible to pollinating insects: a condition which clearly favours outcrossing and to a considerable extent checks those tendencies to selfing which have just been described.

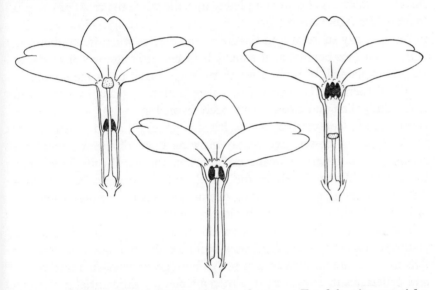

FIG. II. *Primose flowers showing the sexual organs. Top left, pin; top right, thrum; the (long) homostyle type is below and between them (to be compared with the photographs on Plate 3(1)).*

The characters controlled by the heterostyle mechanism are not limited to the position of the sexual organs. The thrum stigma is depressed in the centre and bears short papillae while that of the pin type has a rounded surface with papillae about five times longer. In addition, pin pollen is relatively small, having a diameter only two-thirds that of thrum. The latter fact makes it possible to prove that the mechanical outcrossing device already described operates with at least a fair degree of success. If we examine the proboscis of a moth after visiting a number of primroses, predominantly large-grained (thrum) pollen, with which some small grains are mixed, will be found adhering round its base and predominantly small-grained (pin) pollen, together with some large grains, will be found near its tip.

The most important property of thrum and pin flowers is, however, not morphological at all but physiological. If thrum pollen be placed on a thrum stigma or pin pollen upon a pin one, the reproductive cells prove incompatible and, in general, relatively little seed is set. Thus in the cowslip, thrum × thrum produces one-fifteenth the number of seeds which germinate from thrum × pin, while pin × pin gives one-third the number of seeds produced by pin × thrum (Crosby, 1949).* Less detailed information is available for the primrose but it is clear that thrum × thrum is almost completely infertile while the cross pin × pin though less handicapped has greatly reduced fertility. However, like pollen is rarely successful in fertilizing the ovaries when in competition with unlike (see below), as it usually is in nature; moreover it tends, as already explained, to be in a minority when mixed upon the stigma.

It will therefore be noticed that fertilization is generally successful when the pollen and ovules have been formed at the same level in the corolla-tube ('compatible' or 'legitimate' mating), as they are when unlike types of flower, necessarily growing upon different plants, are crossed. It is generally unsuccessful when the germ-cells have been formed at different levels in the corolla-tube ('incompatible' or 'illegitimate' mating), as in the selfing of a distylic flower or when the pollen has been brought from another plant of the same type.

The barrier to fertility is of two kinds. Both in the cowslip and the primrose, thrum pollen germinates well on a thrum stigma but usually fails to penetrate its surface. Pin pollen both germinates on a pin stigma and penetrates it. However, it is then affected by differential growth in competition with legitimate pollen; that from pin anthers grows well in a pin style but not as fast as thrum pollen. The latter therefore usually reaches the ovary first and so brings about cross-fertilization. It will be apparent therefore that in nature outcrossing generally takes place but that when this fails pins self-fertilize though thrums rarely do so, thus giving some superiority to the pin type. W. F. Bodmer points out to me that a reason for the evolution of greater self-sterility in thrum compared with pin is that thrum pollen will naturally fall down on to the thrum stigma, giving greater opportunities for self-fertility if thrums were self-compatible.

* In all crosses described in this chapter the seed parent is mentioned first.

Heterostyly is controlled by a super-gene the members of which are responsible for anther-height, style-length, and the legitimacy mechanisms (penetration and rate of pollen-tube growth) together with pollen-size and length of papillae on the stigma (Ernst, 1933). These are very closely linked and the block which determines the thrum characters generally behaves as if it were a single gene, S, dominant to that for pin, s. The outcrossing devices, mechanical and physiological, ensure that thrums will normally be heterozygous, Ss, while the recessive pins must of course be ss. Thus the usual reproduction of distylic plants consists in a series of back-crosses of the type $Ss \times ss$ or $ss \times Ss$, which should lead to the segregation of the two phases, in equal numbers. That equality is, however, not accurately attained in nature for, though an excess of thrums has been detected in some populations (Bodmer, 1960), wild colonies generally support a slight excess of pins, probably owing to the advantage of that form already mentioned. As an example of this, and of the occurrence of occasional homostyles in natural conditions (to be discussed later, pp. 211–17), Mr and Mrs S. Beaufoy have kindly allowed me to quote the results of one of their extensive primrose counts in south-east Suffolk, that for 1956. Their total score, from a number of homogeneous populations, amounted to: thrums 1,553, pins 1,827,[*] homostyles 20; total 3,400.

The illegitimacy mechanism is associated with a maternal effect. Pin pollen, which is genetically 100 per cent s, grows more rapidly down a thrum than a pin style. Conversely thrum pollen penetrates and grows the better down a pin than a thrum style. But since thrums are necessarily heterozygous for the controlling super-gene their pollen, unlike that of pins, is not uniform. Half of it indeed carries S but half carries s and, relative to its own genetic outfit, should therefore behave as pin pollen, which it does not do. Thus the growth of the thrum gametophyte must be conditioned by the genetic environment provided by the sporophyte upon which it was formed.

Other comparable instances of the kind are known. A good example is provided by the genetics of dextrality and sinistrality in the snail L. *Limnaea peregra* (Diver *et al.*, 1925). This is unifactorial with dextrality dominant, but the direction of torsion in each individual is determined not by its own alleles but by those of its female parent. A good deal is now known about the physiology of this phenomenon. It seems to depend upon the time when the messenger RNA supplied

---

[*] The departure of thrums and pins from equality is very significant. $\chi^2_{(1)} = 22 \cdot 2$.

by the mother is used up and new messenger RNA is made by the zygote.

A similar situation has been discovered in another snail, *Partula saturnalis* (pp. 200–3) in which, however sinistratity is dominant (Murray and Clarke, 1966). We have here simple Mendelian inheritance the expression of which is constantly delayed one generation. It was long ago suggested (Ford, 1931, pp. 65–6) that this phenomenon may be a widespread and important one controlling the early cleavage of the embryo until its own genes can take charge because its nuclear-cytoplasmic ratio has returned approximately to normal. That possibility, reached by considering the work of Boveri on wide crosses in Echinoderms and of Godlewski on the nuclear-cytoplasmic ratio in that phylum, has I believe never been at all fully explored. This is in part due to the difficulty of obtaining even the initial stages of development in hybrids between species so remote from one another that they are recognizably distinct prior to the earlier stages of gastrulation. However, it should be noticed that Hamburger (1936) obtained evidence of a similar kind in Amphibia. He made crosses between the Salamanders *Triton taeniatus*, *T. palmatus*, and *T. cristatus*. In all of them, including the reciprocals between *T. cristatus* and *T taeniatus*, development was completely maternal up to the limb-bud stage, after which the effects of the paternal nucleus became apparent. When a similar phenomenon occurs in higher plants, Mollusca and Amphibia, it certainly suggests a fundamental situation.

Crossing-over can occur within the heterostyle super-gene but it does so rather rarely (p. 216). It may take place between the two blocks of genes respectively controlling the male and female parts, so giving rise to 'homostyle' flowers. Such genes may be represented as G for the dominant thrum gynaecium (short style and other characters including, of course, the compatibility mechanism), and g for the recessive pin form (typified by the long style); also A, giving rise to the dominant thrum androecium (with anthers placed high and the other associated features) and a to the recessive pin one (typified by anthers low down the tube). Therefore the super-gene responsible for pin, with all its features morphological and physiological, may be represented as $ss = \dfrac{(ga)}{(ga)}$ and that for thrum as $Ss = \dfrac{(GA)}{(ga)}$. Crossing over within the latter results in $\dfrac{(Ga)}{(gA)}$. This produces gametes for short homostyles (Ga) and long homostyles (gA) respectively. Both types are found as rarities in wild populations, but only the long homostyles are known to have established themselves anywhere at high frequencies. Consequently we shall here be concerned solely with that phase so that the qualification 'long' may be omitted, as it normally is in the literature, in considering its occurrence in nature.

The super-gene for long homostyles (gA) is generally designated s' or S^h. However, s^l is a better form as it is necessary to use some symbol distinguishing it from that for short homostyles, which should be indicated by s^s.

The super-gene s^l operates as another allele at the s locus. Such homostyle plants have the dominant thrum androecium, with its high anthers and large pollen-grains, and the recessive pin gynaecium with a long style and rounded stigma bearing long papillae. Thus they behave as recessives to thrums and dominants to pins. Moreover, both their structure and physiology favour inbreeding. The stigma is typically clasped by the anthers, which dehisce directly on to it; but it may be found just above or just below them owing to slight variation in the relative lengths of the style and corolla-tube during the development of the flower. Evidently too, the selfing of homostyles is compatible since it is equivalent to a cross between pins and thrums, the contrasting sexual parts of which are here enclosed within the same flower. For this reason also a cross between two homostyles, similar in structure though they be, is a compatible one. So too are the crosses homostyle × thrum or pin × homostyle, apart from the question of any mechanical difficulties involved in effecting them.

The S super-gene seems to be of a generalized type found in many Primulaceae. Ernst (1933, 1936) worked on it in *Primula viscosa* and *P. hortensis* and has recorded other forms of reconstruction within the units composing it (in addition to that responsible for self-compatible long and short homostyles). These have enabled Dowrick (1956) to analyse and predict its component parts more fully. Thus, when expanded, $S = CGLI^s I^p PA.\star$ There seems fairly good evidence for this order with respect to G, P, A; s is represented by the corresponding recessives, while $s^l = cgli^s I^p PA$.

$\star$ C = thrum area of conducting tissue. G = style length of thrum. L = Papilla length of thrum (this is denoted as S by Dowrick, but the confusion of having an S locus within the S super-gene must be avoided). I^s = thrum style incompatibility. I^p = thrum pollen incompatibility. P = pollen size. A = thrum anther height.

Further light has been thrown on the evolution of this situation by Mather (1950) working on *Primula sinensis*. He finds that additional genes influencing other qualities, colour-pattern and double flowers, could produce the homostyled condition. Thus *a*, for the recessive 'Primrose Queen', shortens the style so that *aa* pin becomes a short homostyle; on the other hand *m*, for recessive double flowers, raises the position of the anthers and consequently *mm* pin is a long homostyle. These further genes are unlinked with the *S* super-gene, and can co-exist with it, but they indicate the type of material from which it has been constructed.

Here then we have a device which controls the mating system. A single cross-over can convert a heterostyled and predominantly out-breeding plant into a homostyled and predominantly selfing one: a result which another cross-over can reverse. Now outbreeding normally, and from a long-term point of view always, has a great advantage over selfing. It provides the heritable variability upon which selection can act, so allowing the organism to evolve and to adapt itself to new or changing conditions. From that benefit self-fertilizing species are excluded to an extent which depends upon the degree to which out-crossing is suppressed in them. Total self-fertilization indeed rapidly reduces genetic variation to mutation-level and so extinguishes the pros-pect of evolution. It is easy to see that this must be so, for it ensures that all alleles already homozygous remain so while one half of the hetero-zygotes became homozygous at each generation.

There is, however, a situation in which selfing has at least a temporary advantage. It is that in which any organism is very well adapted to its en-vironment. For, since genetic variation is at random relative to the needs of the individual, the changes to which it gives rise are themselves most unlikely to improve the harmonious adjustments that have been achieved. The less perfectly a plant or animal may be fitted to the conditions of its life, the greater the chances that from an outburst of variability some-thing worthwhile may be selected. In general terms therefore, it pays an organism to be genetically invariable while the conditions suit it very well, but otherwise to be genetically variable and, moreover, to pass from one to the other situation at need. For environments are not per-manent things: sooner or later they will change however advantageous they may be, and then a species irrevocably committed to selfing will be unable to adjust to the new conditions and must perish: a risk often imposed upon parasites.

It is natural therefore that we should sometimes encounter the means of transition from outbreeding to inbreeding and the reverse and find it attained in diverse ways. It is achieved for example in those Ciliophora which are capable both of binary fission and conjugation and it will especially be noticed that the latter, or sexual mechanism, is resorted to only when the conditions deteriorate. A similar principle is at work in the heterostyle-homostyle device, one which can favour a relatively greater degree of outcrossing or selfing as circumstances dictate.

On general grounds, we should therefore expect heterostyly to be rather widespread in plants and to meet its contrasted phase, that of homostyly, here and there among the species exhibiting it or among their close relatives. This is precisely what we find. Since this book first appeared, Crowe (1964) has published a survey of heterostyly. She finds it occurs in eighteen Orders of Angiosperms.

The homostyle primrose in England

The nature of the advantage conferred by inbreeding due to homostyly may or may not be apparent. There is often strong pressure to isolate populations living at the edge of a species' range in order to preserve their special adaptations (pp. 338–40). Therefore such marginal communities of normally heterostyled plants are frequently homostyled: as with *Primula scotica*, a homostyled 'species' evidently derived from the heterostyled *P. farinosa*, or the distylic *Menyanthes trifoliata* (Gentianaceae) which is homostyled in west Greenland. A similar end, that of isolation from the main body of the species, is also achieved by the polyploidy which often characterizes peripheral plant communities (Darlington, 1956a).

On the other hand, no obvious ecological reason can be deduced for the existence of two predominantly homostyled colonies within the ordinary range of heterostyled primroses (*Primula vulgaris*) in Buckinghamshire and Somerset respectively. Yet it must be remembered that the countryside of southern England not only differs immensely from its primaeval condition but has been much modified by changing forestry practice and by the enclosures during the last two hundred years. It is not surprising therefore that within this region a few areas should be found in which the primrose, with its delicate heterostyle-homostyle adjustment, should be selected for a high degree of inbreeding rather than the normal outbreeding.

The two populations containing homostyles were discovered by Crosby (1940). They are about eighty miles apart and occupy rather similar areas: approximately eleven by twelve miles on the Chilterns and fourteen by sixteen miles or more near Sparkford. They are the only ones known, though the primrose is abundant in suitable places throughout the country. These consist of woods, hedgerows, and roadsides on an acid soil. The plants are able to exist on the chalk of the Chiltern hills only where there are sufficient accumulations of humus. Even so, they are rather sparse and patchy there. It is partly for this reason that the Somerset homostyle area near Sparkford, where primroses are in large numbers, has been so much the more thoroughly studied.

If the viability of the homostyle plants were as good as that of pins and thrums and if, as a superficial study of their structure seems to suggest, they were to a great extent self-pollinated, it is easily seen that they must have an overwhelming short-term advantage over the other two phases. For their complete fertility would constantly be secured by their combined self-compatibility and the juxtaposition of their anthers and stigma, compared with the relative uncertainty and delay of crossing between the two distylic forms. In such circumstances, when a homostyle has arisen by crossing-over within the super-gene, it should, once it has initially gained a footing, sweep through the population and convert the whole primrose community from the heterostyled to the homostyled condition: a statement which would appear to be true for other heterostyled species also.

This view is maintained by Crosby (1949). He holds that now homostyly has established itself in two areas in southern England it must inevitably spread from them through the primroses of Britain. Indeed its reproductive superiority would, according to him, ensure this even if the viability of the $s^l s^l$ phase were only 81·5 per cent that of the heterostyled plants. Thus there should be evidence from both the Chiltern and the Sparkford colony of a decline in homostyle frequency outwards from the point where the condition arose, presumably somewhat towards the centre of each area. That situation is approximately realized on the Chilterns but not in the Sparkford area. However, Crosby (1960), on results obtained by an electronic computing technique, is of opinion that the considerable irregularities in homostyle distribution at the latter site can be accounted for by random fluctuations.

There appear even at the outset to be strong general considerations for rejecting the conclusion that the present spread of homostyles is turning the primrose as a whole into an inbreeding species in Britain. In the first place, it may well be asked why it has remained a distylic plant for so long, and why we are only now seeing the start of such an inevitable process and, incredibly enough, in two small and widely separated areas simultaneously. Moreover, how can it possibly be that more than 120 species of Primulaceae are in fact distylic in nature? They ought to have become homostyled if Crosby's contentions are correct; there are also the many distylic species in other orders whose continued existence might be thought equally mysterious. It is especially to be noticed that homostyles are in fact known in a number of them but are confined to groups or races often, as already indicated, at the edge of their range. We may remark also that it would seem to be in the highest degree likely that a mechanism which alters the reproductive system from outbreeding to selfing is powerfully controlled by selection to adjust the plant to those situations where diversity or relative stability of the genetic material is respectively advantageous.

There are, however, other grounds derived from an analysis of the primrose situation itself for rejecting the idea that this species is now in process of becoming predominantly self-fertilized. Thus there is no tendency for the spread of homostyles to convert the entire population to that type. On the contrary, their frequency rises to about 80 per cent but rarely any higher. At this proportion thrums are excluded, the remaining 20 per cent of plants being pins, partly owing to the advantage of that phase to which attention has already been drawn (p. 206) in situations where the opportunity for cross-fertilization is reduced (but see also p. 215). To produce such a result, Crosby had to assume a 65 per cent viability of homozygous homostyles relative to the other genotypes. There seems to be a relatively rapid change from the normal condition in which homostyles are a rarity to their approximate limit around 80 per cent, so that the homostyle frequencies of the various colonies cluster round these upper and lower values with intermediate ones more sparsely represented.

Two of the localities within the Sparkford area which have been sampled in detail for a number of years by Sir Ronald Fisher and several of his colleagues (including myself) show a significant change in the frequency of the phases, and in both places the proportion of homostyles has actually become smaller. Bodmer (1960) shows that at Sparkford Wood South the reduction (from 82·4 to 73·2 per cent over the sampling period: 1941–8) is significant. Also a linear colony along Laurel Copse hedgerow underwent a significant decline in homostyle frequency, from 34·5 to 20·1 per cent, between 1943 and 1944, the data being homogeneous both before the change (1942, 1943) and after it

(1944, 1948, 1949). These situations are inconsistent with the concept of a steady spread of homostyles.

A number of the populations show very local heterogeneity which indicates, as Crosby (1948) reasonably suggests, that pollen is normally carried from one plant to another over quite short distances only. The heterogeneity observed over a number of years at Sparkford Wood North-West, as well as at other places studied by Crosby (*l.c.*), might well be due therefore to small annual variations in the exact location of the counting, for here the fluctuations are irregular. But minor changes in the precise position of the sampling areas are unlikely to produce a consistent trend, as at Sparkford Wood South. Moreover, Bodmer points out that the Laurel Copse hedgerow is a clearly defined habitat which allows of no such effective variations in the places where the flowers were picked.

As already mentioned, Crosby (1949) reached the conclusion that once homostyles had been established, they must spread unchecked through the primrose population. This was on the assumption that they are always self-fertilizing, a view which has been disproved by Bodmer (1958, 1960). He finds indeed that homostyles possess one of the devices which favour the outcrossing of pins and thrums: that is to say, the stigma accepts pollen before the anthers have dehisced and after the bud has opened sufficiently for the access of pollinating insects. This is approximately at the time when the anthers and stigma can be seen and their positions scored, and the average period between this and dehiscence was 2·75 days. Indeed the flowers are usually fully open before dehiscence occurs.

It is clear therefore that crossing between homostyles is a possibility and Bodmer has shown that it takes place to a very considerable extent, even up to 80 per cent in certain circumstances. These are determined partly by variations in the relative positions of the male and female parts of the flower. Though, typically, the homostyle stigma is surrounded and clasped by the anthers, it is often slightly above them and sometimes even protrudes from the bud. Indeed when it can first be scored the stigma is generally just above the anthers and clear of them and as the corolla grows it is pulled through them until it lies just below.

Evidently the opportunities for homostyle crossing are greatest when the stigma is relatively high and the anthers dehisce late. Both these variables, position of stigma and the time when the pollen is shed, are controlled partly by polygenes but partly by the environment. Thus it comes about that the amount of such cross-fertilization varies from year to year.

Crosby (1959), on the other hand, in examining Bodmer's suggestion of outcrossing in homostyles estimated this event at a much lower level. He does so as the result of growing homostyles and thrums together in cultivation, when he obtained a value suggesting about 7·5 per cent cross-fertilization of homostyles. He also tested 95 thrums for Ss^l instead of Ss, from wild populations containing about 52 per cent homostyles. Two Ss^l plants were obtained instead of 20, calculated on the basis of 80 per cent homostyle outcrossing. This suggests a frequency of 5 to 10 per cent for that occurrence. We need to know, however, whether there be differential viability against Ss^l compared with Ss in nature; the fact that there does not appear to be so in cultivation is hardly relevant. As Bodmer points out to me, it is possible that there may be partial incompatibility of thrum pollen on homostyle stigmas. This would explain the apparent low frequency both of outcrossing in homostyles and of Ss^l heterozygotes. The matter needs further investigation.

It is necessary therefore to take into account some degree of crossing between homostyles, a fact which reduces the postulated reproductive superiority of that phase compared with the other two. For with selfing, all homozygous plants produce homozygous offspring only, while half the progeny of the heterozygotes becomes homozygous in each generation. But with cross-breeding, heterozygous homostyles are maintained, sheltering s which can give rise to pins by segregation. Thus a system in which homostyles are entirely self-fertilized could allow that form to increase unchecked even when associated with marked inviability. On the other hand, if they cross-fertilize to a considerable extent, quite mild inviability would prevent the homostyles from spreading or would reduce their frequency when they are already abundant.

It will be noticed that, where homostyles are favoured, thrums are always open to elimination partly because they are the 'top dominant' of the series so that the gene S is invariably exposed to selection. Pins, on the contrary, are not, because they are the 'bottom recessive' so that s is sheltered in single dose. Here we have a further reason, additional to that mentioned on p. 206, for the rarity and disappearance of thrums but not pins in areas of high homostyle frequency.

The concept that homostyles must inevitably spread through the population is incompatible with their exceedingly restricted occurrence in Britain unless the cross-over which gives rise to them is of immense rarity (and, granting this, it is strange enough that they are established in two small areas independently). It would be fatal to that view if they are in reality widespread at a low frequency in the normal primrose community. Yet this is precisely the situation which we find. If large numbers of primroses be scored anywhere in the country, occasional plants producing homostyled flowers will be seen among them: 0·5 per cent or so is quite usual, p. 207. It is true that Bodmer, Beaufoy, and I have independently grown a number of these and found that many of them are environmental and that the plant reverts to a normal thrum or pin the next season. These rare wild specimens apparently comprise both long and short homostyles, but the latter are often pins with a deformed or crumpled style which has brought the stigma down to the level of the anthers. In such flowers as these the position of the sexual organs alone suggests homostyly, the other characters retaining their normal pin or thrum qualities. But it is now certain that genuine long homostyles occur at a low frequency in normal primrose populations. Thus rather less than 0·25 per cent of the primroses flowering at Appleton Wood, Berkshire, six miles west of Oxford, bear homostyled flowers. Some are environmental, but one of the plants sent to Bodmer at Cambridge and grown there by him proved to be a true long homostyle, verified as $s^{l}s$, while from a capsule on another which had been open-pollinated at Appleton he obtained progeny all of which were undoubted long homostyles. This plant had, therefore, self-fertilized, and of course it is not claimed by Bodmer that all homostyles are cross-pollinated.

In view of the facts mentioned in this chapter, it is clear that the heterostyle-homostyle mechanism provides a means for adjusting the mating system from a relatively great amount of outbreeding to a relatively great amount of inbreeding and that long homostyles do not sweep through the primrose population unchecked when they arise: that is to say, they are not spreading from the Chiltern and Sparkford colonies in such a way as to convert the species to homostyly throughout Britain. There is at present no evidence to show what features favour inbreeding in these two localities. However, as already mentioned, the changes in the countryside consequent upon civilization are certainly sufficient in general terms to account for the existence of occasional areas where it pays the primrose to be relatively invariable. It should certainly be possible to discover not necessarily what are the features at Sparkford and on the Chilterns which promote that situation but to detect certain experimental conditions which do so. For that purpose artificial colonies containing a known proportion of homostyles must be established in isolated open-pollinating plots. This is already being done in laboratory gardens, but the beds should be multiple and far apart, and exposed to widely different conditions, for comparative study. Moreover, the work should be extended to planting homostyles in a number of natural populations, selecting different types of soil and other environmental features.

It seems probable that the primrose, which is known to have produced a high proportion of homostyles in two areas, is particularly well suited to this form of experiment. Indeed it may possibly provide the best material for such tests, whether among distylic plants or those in which the heterostyle mechanism is worked upon some other basis. Species of the latter type have in some instances been subjected to detailed genetic analysis, but as little is known of the ecology and about the relative frequency of their phases in different populations, this is not the place to discuss them in any detail.

Tristylic and 'concealed' heterostyled conditions

A surprisingly complicated elaboration of heterostyly is provided by the tristylic condition found in such genera as *Lythrum*, *Oxalis*, and others. The flowers of *Lythrum salicaria*, the Purple Loosestrife, are of three types, of which each plant produces one only. The sexual organs grow to three lengths: short, intermediate, and long. Any two levels are occupied by the male parts and the third by the female so that forms with short, mid, or long styles are found. Fertile unions are obtained only when the stamens and style are of the same length and since these necessarily grow upon different plants the system ensures outbreeding. For example, a long style can accept pollen from the long stamens of a mid-styled or a short-styled flower, but not from its own mid or short stamens nor from those of any other type of flower.

Tristyly is controlled by two pairs of alleles. One of these decides whether or not the style shall be short; the other has no effect upon short styles but determines whether the remainder shall be mid or long. Long styles constitute the 'double recessive' while short styles are the 'top dominants', so that mid styles are dominant to long but recessive to short. We have here an epistatic system which, in fact, is complicated by polyploidy (Fisher and Mather, 1943). The three phases take different frequencies from one wild colony to another and, indeed, the two pairs of alleles concerned segregate independently.

We notice in *Lythrum*, as in *Primula*, a polymorphism which involves a combination of anatomical and physiological qualities. Yet it is evident that the illegitimacy mechanism could ensure outbreeding without the addition of structural features in which the male and female parts interchange their position from one phase to another, giving higher efficiency because wastage is reduced. This morphological distinction must indeed be secondary to the development of incompatibility, as pointed out by Lewis (1954).

Not only then is there the heteromorphic incompatibility, aspects of which we have been discussing, but there is the fundamental homomorphic type found in primitive plants as well as, widely, in Angiosperms. That situation is one in which the mating groups, which are numerous, are morphologically alike and are controlled by many alleles of the incompatibility genes or super-genes. As Crowe (1964) points out, homomorphic incompatibility falls into two main classes: those in which the mating type of the pollen grains is respectively determined gametophytically or sporophytically; alternatives which illustrate the well-known principle that the time of action of genes is under genetic control, as originally demonstrated by Ford and Huxley (1927). The whole incompatibility situation is responsible for the fact that outbreeding is compulsory in many plant genera. Crowe concludes with evident justice that the first Angiosperms were hermaphrodites with a form of (homomorphic) incompatibility from which the various outbreeding systems now extant have evolved rather than having originated independently.

Primrose pollination

Primroses do not set seed if insects be effectively excluded from them (Woodell, 1960). Their pollinating agents are chiefly small Coleoptera also, to a limited extent, certain Diptera and Lepidoptera and occasionally bees. As far as the primrose in Somerset is concerned, Crosby (1960) believes that *Bombylius*, a genus of bee-like Diptera with very rapid movements, is the chief pollen distributor. Although species of that genus are by no means uncommon parasites of solitary bees, they seem too rare to play more than a minor part in pollinating anything in southern England. I have myself captured moths at primrose flowers which, even at their first appearance, are not too early to be visited by Lepidoptera. Several species of the genus *Orthosia = Taeniocampa* emerge from the pupa at that time and are very common. Moreover, a number of hibernating Agrotidae, belonging to such genera as *Conistra* and *Eupsilia*, which are to be found abundantly on ivy blossom in the autumn, are assiduous in searching out flowers in March and April. However, small Coleoptera, such as *Meligethes picipes*, appear to be the most important agents in pollinating primroses; they are active insects and extremely common in the primrose flowers.

A modern technique rightly deplored by all naturalists may nevertheless have interesting effects upon the heterostyle-homostyle system: that is to say, the widespread use of insecticides and of selective weed-killers when sprayed upon roadside verges and hedgerows. These agents must alter the ecology of many primrose habitats. They must also destroy great numbers of insects, so giving some advantage to self-pollinating devices. It is possible that an increase in homostyles may be detected where such substances are used repeatedly.

CHAPTER ELEVEN

Chromosome Polymorphism

Drosophila

The balanced chromosome systems of *Drosophila* illustrate with exceptional force the importance of the super-gene and of heterozygous advantage in maintaining and controlling polymorphism. These have been
studied by Dobzhansky and his colleagues from 1938 onwards and their
detailed analysis of them, skilfully combining the methods of genetics
and ecology, has thrown much light upon evolutionary processes in wild
populations.

Over thirty species of *Drosophila* are known to be polymorphic for
various types of chromosome inversions (Plate 17). These have been
most fully investigated in D. *pseudoobscura* and the closely allied D.
persimilis, which are widely distributed in the western United States
and Mexico. Both species have five pairs of chromosomes which, however, are not equally subject to such reconstructions; the third being the
most affected. The polymorphism involved is cryptic: all the flies look
alike whatever inversions they may carry, and these can only be identified
cytologically by an examination of polytene nuclei from the larval
salivary glands. Accordingly it was at first thought that they were all
of neutral survival value: a view corrected by Dobzhansky in 1947*b*.

Sixteen gene-arrangements are known in the third chromosome of
Drosophila pseudoobscura and eleven in that of D. *persimilis*, only one of
which ('Standard') is common to them both. The different types are
distinguished by names, generally abbreviated to two letters. They can
be detected in the giant salivary-gland chromosomes both by the formation of inversion-loops in the heterozygotes and, in detail, by sectional
reversals in the known sequence of the cross-striations. These inversions
comprise the independent, included, and over-lapping types and, as
already explained (pp. 110–11), they themselves constitute super-genes
capable of further evolution. Because such reconstructions suppress
crossing-over within the sections involved, even if long enough to
allow chiasmata to form within counter-turned loops, they provide
one of the methods for preventing co-adapted genes from scattering.

This they do without the necessity of bringing these near together on the chromosome or of checking normal chiasma-formation within the section containing them. It will be noticed also that inversions can hold together separate co-adapted groups of genes.

Chromosome polymorphism and selection in *Drosophila*

The selective importance of the gene arrangements in *Drosophila* can in fact be deduced from the occurrence of some of them at high frequencies over great areas. Thus in *D. pseudoobscura* that known as 'Arrowhead' (AR) is established throughout California and extends far into Colorado and New Mexico, 'Standard' (ST) seems to be found in all the Californian populations that have been studied, while Chiricahua (CH) is widespread at a high frequency in Chihuahua and Durengo, Mexico, and exists at smaller percentages from Utah and Arizona westwards to the Pacific coast. Indeed many more types have a range that is almost as extensive both in this and in other species.

However, direct evidence is also available to demonstrate the adaptive significance of the inversion polymorphisms in *Drosophila*. This has been obtained by Dobzhansky and his colleagues from three sources.

In the first place, the frequency of the different gene-arrangements indicates a regular cycle of changes adjusting them to the conditions of the different seasons. Thus at Piñon Flats, Mount San Jacinto, in southern California, ST chromosomes decrease from about 53 per cent in March to about 28 per cent in June. Meanwhile, the CH type increases from 24 to 40 per cent and the proportion of AR fluctuates, though in a regular fashion. In the hot summer period from June to August these numerical changes in the ST and CH inversions are reversed, until the proportions characteristic of the early spring are reached again, for these apparently remain constant during the winter; a sequence which is repeated year after year (Dobzhansky, 1951, pp. 118–19). Such cycles have been detected in many other localities in which, however, they may affect the same gene-arrangements quite differently. For instance, further north at Mather, a locality at the edge of the Yosemite where I have had the privilege of seeing Dobzhansky's work in progress, ST becomes commoner and AR rarer throughout the whole season, while the proportion of CH chromosomes remains fairly constant. Here, in further contrast with Piñon Flats, the proportions return to their spring values during the course of hibernation (Dobzhansky, 1956).

Dobzhansky (1961) states that *Drosophila pseudoobscura* passes through no more than six to eight generations per year at moderate elevations in California, and probably less in most other localities. As he points out, powerful selection-pressures must therefore be operating to bring about the marked adjustments in chromosome-polymorphism which recur during each season.

Secondly, the frequencies of the various inversions change in space as well as in time, being correlated with the habitat of the populations in which they occur. Thus Dubzhansky (1951, p. 138) finds that they are maintained in different proportions from one region to another within the range of each species. Moreover, he has shown (*ibid.*, p. 121) that they are strikingly adjusted to short-range environmental transects of a marked kind. Thus at the season when 46 per cent of *Drosophila pseudoobscura* carry the 'Standard' gene arrangement at 850 feet, the incidence of this type declines to 10 per cent as we climb the Sierra Nevada to 10,000 feet; meanwhile that of 'Arrowhead' is reversed, increasing from 25 to 50 per cent over the same ascent; that is to say, we have here a polymorph-ratio cline.

Also, Stalker and Carson (1947, 1948), using *D. robusta*, made an interesting combined study of inversion-frequency and morphology in the Great Smoky Mountains, Tennessee. They collected at six localities along an 18-mile transect near Gatlingsburg, passing from 1,000 to 4,000 feet. Eleven out of fourteen inversions showed changes in frequency correlated with altitude. These did not closely parallel the differences which in fact exist between more southerly and northerly localities: indeed a few of the characteristically northern gene-arrangements did not alter in proportion, or even increased, in passing downwards to lower elevations though the majority of them became more frequent at the higher levels. This is to be expected since, as Stalker and Carson rightly remark, environments at a considerable height show marked similarities, but are not identical, with those further to the north. Moreover, a given inversion may not have the same adaptive value in different localities, as demonstrated by Dobzhansky when he showed that the seasonal fluctuations of ST and CH are not the same at Piñon Flats as at Mather.

Stalker and Carson combined their work upon inversions in *D. robusta* with a study of five quantitative morphological characters. Four of these (thorax-length, femur-length, wing-width, and wing-length) decreased with reduced altitude, as did three of them (femur-length and the two wing measurements) in passing from north to south. Thorax-length, on the other hand, is greater in more southerly localities. Moreover the morphological changes with altitude occur chiefly between 1,400 and 2,000 feet; not so with the gene-arrangements. Doubtless the polygenes responsible for the morphological variations do not all lie within the inversions; also, they may well have diverse effects in the distinct genetic and external environments to which they will be exposed at different localities.

One further illustration of the selective value of inversion-polymorphism in adjusting a 'natural' population to its environment is of especial relevance here. Following Dobzhansky's discoveries, Dubinin and Tiniakov (1946) demonstrated a strong association between gene-arrangement and urbanization in *Drosophila funebris*. They studied eight inversions in this species. Their occurrence was very heterogeneous in different parts of Moscow, but the proportion of flies heterozygous for one or more of them varied from 17 to 89 per cent of the population within the city. It was approximately 1·5 per cent in each of three rural habitats respectively 60, 115, and 210 km away in different directions. It is unlikely that inversions with an over-all advantage had not yet spread into the countryside owing to isolation; for though *D. funebris* lives in natural habitats it is also a scavenger in human households. Consequently it is rather readily scattered over wide distances. However, Dubinin and Tiniakov rightly recognized the importance of deciding whether or not this striking distinction between the town and country populations is purely a local phenomenon in the Moscow area. They therefore carried out corresponding studies in the city of Ivanov. Here 37 per cent of the flies were inversion-heterozygotes while no inversions at all were found in two rural communities 90 and 150 km away.

The diversity and novelty of environment open to a household scavenger in a large town must be very great, while its habitat must be far more uniform in natural conditions in the country. Thus diverse polymorphisms capable of rapid evolution and adjustment are likely to be highly advantageous to urban communities, especially as they themselves constitute a switch-mechanism by which the ordinary genetic adaptations of the wild population can continue relatively uncontaminated by them. We have here a type of evolution associated with urbanization entirely distinct from the phenomenon of industrial melanism to be described in Chapter Fourteen.

The other type of evidence which demonstrates the adaptive significance of inversion polymorphism in *Drosophila* is derived from laboratory experiments. The flies are kept in population cages of the type devised by l'Héritier and Teissier, one which has been developed and used with outstanding success by Dobzhansky and his colleagues.

Stocks containing inversions at a known initial frequency can by this means be maintained for many generations in controlled, and if necessary optimum, conditions. It has then been found that the various chromosome-types do not fluctuate numerically at random, as they would if selectively neutral, but that they are adjusted progressively to certain frequencies at which they become stabilized. Thus Dobzhansky (1951, pp. 114–16) reports on a population of *Drosophila pseudoobscura* which was begun with 11 per cent ST and 89 per cent CH chromosomes and was maintained at 25 °C for a year which, in these conditions, represented about 14·5 generations. After six months, the proportion of ST had risen to about 52 per cent; it then increased rather more slowly to approximately 70 per cent during the four months following, after which no further change occurred. Evidently in the environment of the population-cage, ST had at first a clear advantage over the CH type. Subsequent events, however, show that the polymorphism was balanced: that is to say, the heterozygotes (ST/CH) were of superior viability to both homozygotes (ST/ST and CH/CH). Consequently neither arrangement could eliminate the other.

So much important work has been conducted along these lines that it is desirable to illustrate it further. Dobzhansky (1961) describes an experimental population of *Drosophila pseudoobscura* which has been kept at 25 °C for more than three years. It is descended from flies obtained at Piñon Flats, southern California, and started with 20 per cent AR and 80 per cent CH chromosomes. The frequency of AR doubled in 1·5 generations and increased to 67·7 per cent in 10·8 generations (that is to say, in 270 days). It then reached an equilibrium, oscillating between 60 and 70 per cent. Since CH was the only other type present, this of course underwent a corresponding decline and became stabilized between 30 and 40 per cent. There is evidence to show that the balanced condition thus obtained arises as a result of heterozygous advantage (pp. 226–30), and Dobzhansky points out that in this instance the result is consistent with a relative inferiority of the homozygotes equivalent to AR/AR = 0·71 and CH/CH = 0·43 when the survival-value of the heterozygote AR/CH is taken as unity.

It is essential to notice that these viabilities are delicately adjusted to the environment, changes in which, even of a small or a subtle kind, may have marked effects upon them. Thus in this instance Dobzhansky (1961, p. 3) states that the relative fitness of the three karyotypes is affected by temperature: it becomes nearly equal at 17°C. In addition, it is influenced by changes in the food. For instance, at 25°C ST/CH heterozygotes from Piñon Flats stock are superior to the homozygotes when reared on a medium containing *Kloeckera* (and some other yeasts). In these conditions, lowering the temperature to 21°C caused no marked change, but when *Zygosaccharomyces* has been added to the diet the heterosis disappears (Dobzhansky and Spassky, 1954).

Furthermore, the balance of inversion polymorphism in *Drosophila* is readjusted in relation to various forms of competition. That aspect of adaptation has also been detected by the use of population-cages and indeed it is difficult to see how it could have been recognized by other means.

Thus Levene, Pavlovsky, and Dobzhansky (1954) have shown that the relative fitness of certain karyotypes in *Drosophila pseudoobscura* is modified by the presence in the same culture medium of larvae with inversions of a different kind. At 25°C, as already mentioned, experimental populations carrying ST and CH reach equilibrium with their chromosomes at 60 to 70 per cent and 30 to 40 per cent respectively; while those dimorphic for the ST and AR gene-arrangements do so at about the same values. From this one would conclude that the relative fitness of the ST/ST and CH/CH homozygotes is approximately the same as that of ST/ST and AR/AR. But if so, a population dimorphic for AR and CH chromosomes should attain stability with the two types in approximate equality, which it does not do. On the contrary, the AR type is then the commoner in the ratio of 7:4 (p. 224). Other instances of the kind have been detected in this species as well as in *D. persimilis* (Spiess, 1957).

An important part of the environment to which any animal is adjusted is contributed by the other animal species present in its habitat: a fact more obvious in the corresponding botanical situation involving the interaction of plants. But, in addition, the ecology of an animal population is affected both by its own density and the nature not only of the inter- but of the intra-specific competition to which it is exposed: that is to say, the local-gene pool of the organism, with the switch-control of its different polymorphic phases, constitutes a part of the environment to which it must be adapted. That proposition, it is fair to say, has never been so clearly demonstrated as by the competing inversion-polymorphisms of *Drosophila* in experimental cultures.

The effect of population-density has also been examined in laboratory stocks. Dobzhansky (1961) points out that the selection which operates automatically in the population-cages parallels certain aspects of the seasonal variation which occurs in the wild. Thus at Piñon Flats, the frequency of ST chromosomes increases, and that of the CH type diminishes, during the hot summer months while little alteration takes place in the winter: just as the ST gene-arrangement has an advantage over CH at 25 °C in the laboratory while the two have approximately equal adaptive values at 15 °C. Yet in the spring there is a reverse trend in nature, with a decline in the frequency of ST and an increase in that of CH, a situation which has never been detected in population-cages maintained in the normal way whatever environmental modifications are introduced. That anomaly was explained by Birch (1955) who realized that in natural conditions there must be a marked numerical expansion of *Drosophila* in the spring, following elimination during the cool winter period, while, on the other hand, the flies in the population-cages always live at the numerical limit imposed by the amount of larval food available for them. A test was therefore carried out in which a population dimorphic for ST and CH was freed from larval competiton by a super-abundant food supply. In these circumstances, and for the first time in any experimental stock, the CH/CH chromosomes proved superior in viability to the ST/ST type: a result which provided a clear parallel with the situation normally occurring after hibernation. As pointed out in Chapter Two, we have here an illustration of the way in which genetic variation may be affected by fluctuation in numbers (and see p. 175).

Darlington and Dobzhansky (1942) have described the situation in which a gene, 'sex-ratio', interacts with one polymorphism, involving a triple inversion, to influence another: that of sex. Nearly all the progeny of affected males are female. The 'sex-ratio' gene and this inversion are carried together in the right arm of X, which is quadri-partite in most first spermatocytes, due to an inversion in the nucleic acid charge and a hastening of the division-cycle. This condition has been found in *Drosophila pseudoobscura* and several other *Drosophila* species. It can also be produced by a spirochaete.

Heterozygous advantage in chromosome polymorphism

The facts already described demonstrate that the polymorphic inversions of *Drosophila* influence the viability of the organism. It is obvious too that among the various phases neither of two alternative arrangements has an over-all advantage since both occur together and remain at considerable frequencies over large areas. That situation could arise if there were an excess of pairings between unlike types, as in *Panaxia dominula* (pp. 139–40). However, Dobzhansky has repeatedly demonstrated that mating is at random relative to the chromosome arrangements, at any rate in *Drosophila pseudoobscura, D. persimilis,* and in some other species. It might also result if flies carrying alternative inversions were adjusted to different successive or concurrent aspects of the environment. In fact the seasonal changes already quoted suggest, and indeed demonstrate, that condition. Thus at Piñon Flats, southern California, the ST third chromosomes of *Drosophila pseudoobscura* are at an advantage compared with the alternative CH form in the late summer, while the reverse is true in the spring: a situation in itself capable of maintaining both of them in the population if selectively nearly equal. Yet it is not alone responsible for doing so, since the heterozygote ST/CH is more viable than is either homozygote. Indeed it has already been pointed out in Chapter Six that the genetic switch-control of polymorphism will automatically tend to evoke heterozygous advantage, even if there is also an ecological basis for diversity, as in this instance or in Batesian mimicry (Chapter Thirteen).

The existence of such 'inversion heterosis' in *Drosophila* has been established both by ecological studies in the field and by experimental analysis in the laboratory. Moreover, it has been demonstrated in a number of species.

In the first place, several wild populations have been detected in which more than 50 per cent of the flies are heterozygous for certain inversions. Thus Pavan *et al.* (1957) found that near Recife, Brazil, *Drosophila willistoni* carried 61 per cent of heterozygotes for an inversion (J) in the third chromosome, while 80 per cent of *D. paulistorum* were heterozygous for an inversion in a large sample collected at Urubamba, Peru. These are extreme instances, but the heterozygotes have repeatedly been found to exceed expectation in wild populations of *D. pseudoobscura* and *D. persimilis* and other species (e.g. Dobzhansky and Levene, 1948).

The advantage of inversion heterozygotes is often very clear in experimental studies. Indeed an instance of the kind has already been described in *D. pseudoobscura*; for it was pointed out that in a stock from Piñon Flats kept at 25 °C the relative viabilities of AR/AR, AR/CH, and CH/CH were 0·71 : 1 : 0·43. Dobzhansky (1947a) has also shown that in experimental cultures from any one locality the heterozygotes for the various polymorphic gene-arrangements consistently exceed expectation, calculated from the numbers of the homozygous classes.

This superiority is manifested in a number of ways, particularly by resistance to unfavourable conditions of all kinds, including over-crowding. These are qualities which automatically produce differential mortality which, however, is not a necessary result of certain other aspects of heterosis, such as longer life or increased sexual activity. Further, and concrete, examples of this latter type of effect are provided by the studies of Spiess (1958) on *Drosophila persimilis*. He showed that heterozygotes for the third chromosome-inversions known as Whitney and Klamath have greater fecundity and develop faster than their corresponding homozygotes.

Thus inversion heterozygotes of *D. persimilis* and other *Drosophila* species are generally better adapted than either homozygote, provided that the chromosome-pairs are derived from the same population. If, however, they are of different geographical origin the heterozygotes show no such superiority. For instance, Dobzhansky has repeatedly found that they have a decided advantage over the other two genotypes in all three comparisons involving ST, CH, and AR, both in stocks from Piñon Flats and in those from Mather. Dobzhansky (1950) has also studied this situation in crosses between flies from these two localities, which are three hundred miles apart. He then found that the heterozygotes were no longer superior but intermediate in viability between their corresponding homozygotes when the members of the chromosome-pairs were derived from these two sources. He also made a geographically wider cross, involving the same inversions, between strains from Chihuahua, Mexico, and Piñon Flats: a distance of approximately seven hundred miles. The localities are, moreover, isolated by extensive desert areas. In the broods segregating for ST and CH, heterozygotes were again of intermediate viability, consistent with values of $1.26:1:0.87$; but in another test, involving CH and AR, the heterozygous class was actually the least viable, to a degree which can be enumerated as $1.53:1:1.16$. (In both these calculations, the fitness of the heterozygotes is taken at unity.)

We have here clear evidence that this type of heterosis is not a necessary outcome of the heterozygous state, but that it has *evolved*. In the same way, the effect of harmful recessives tends to be obliterated by their dominant alleles in wide outcrosses. Each inversion, acting as a super-gene (pp. 110–11), has doubtless undergone selective adjustment of its effects so as to make its disadvantageous qualities recessive and its advantageous ones dominant, from which heterozygous advantage must accrue. But the selection mainly depends upon the individual adaptation of each local gene-complex (as well as substituting different alleles of major genes), consequently its effects are largely lost on outcrossing; a procedure which gives us an insight into the condition before such genetic evolution had taken place. It will be noticed that this situation, detected by means of laboratory experiments, will arise in nature if two isolated populations, each adjusted to its own environment, extend their ranges and meet (pp. 334–8); it may arise also in certain instances of mass migration.

It is clear, therefore, that evolution has adjusted the effects of inversion in the past. Dobzhansky (1958) has obtained evidence that it is doing so rapidly at the present time. Briefly the facts are these. The Pike's Peak (PP) gene-arrangement, always common in Texas, was a great rarity in California prior to 1946 (four instances of it had been encountered in 20,000 chromosomes studied). Since then it has increased over a vast area, representing the greater part of the State, where about 8 per cent of all chromosomes now carry it. Similarly there has been a corresponding widespread decline of the CH type. ST has also become rather commoner and AR rather rarer. The change cannot be ascribed merely to mass migration of flies from Texas since PP has not increased in Arizona. It does seem just conceivable, however, that a widespread movement from Texas may have taken place, the PP gene being at a strong disadvantage in the climate of Arizona but favoured on reaching California. In this way little trace might be left along the migration route. Such an event could not, however, in itself have been responsible for what has occurred. PP did exist, though at an exceedingly low density, west of the Sierra Nevada previously. Some chromosome-reconstruction of the type suggested by Dobzhansky (see below) must surely have operated also. If the possibility of such migration can indeed be excluded, and the fact that PP does not seem to have increased in Texas is against it, two types of explanation remain.

In the first place, the Californian environment may have altered by natural or human agencies in a way favourable to PP and unfavourable to CH. It is difficult to prove a negative of this kind, but no recent widespread change has been detected in that region. The State is indeed exceedingly diverse in its climate and ecology, yet the increase of PP seems to have over-ridden all local conditions. Nor is it limited to, or particularly associated with, industrial areas such as the Los Angeles or San Francisco districts, nor to regions where insecticides are widely used: indeed it is just as striking in extremely remote areas. It is true that there is a testing ground for atomic weapons near Charleston Peak, but Dobzhansky points out that the prevailing winds tend to carry the fall-out away from California.

It appears probable therefore that these singular events are the result of a genetic change. As Dobzhansky says, a mutation or a cross-over in the right place might provide a chromosome that reacted more favourably with some inversions, especially PP, and less so with CH. Here we face a difficulty which a climatic modification does not involve: the immense area over which these alterations in gene-arrangement have taken place in a few years. The extent to which the flies normally scatter has indeed been determined experimentally in this species: it amounts to a radius of 1·76 km ten months after release, and is therefore on an entirely wrong time-scale to account for these events. Remington has drawn attention to the way in which moths larger than *D. pseudoobscura* are not uncommonly drawn up to considerable heights by air currents so that they can be spread widely by this means. Such occurrences might certainly facilitate the changes observed, though it does seem difficult to suppose that they provide an amount of movement needed to spread an inversion throughout California in so short a time. However that may have been achieved, it is clear that very powerful selection-pressures must have been operating to bring about evolution on this scale.

It may here be remarked that with the great force of selection now known to be acting, accurately balanced genetic systems are in fact likely to respond to subtle environmental changes and may in fact provide the first means of detecting them. Kettlewell has shown (Chapter Fourteen) that the spread of industrial melanism outside manufacturing districts has provided information on the extent to which the apparently normal countryside is in reality polluted. The widespread change which affected the spot-frequencies of *Maniola jurtina* in southern England about 1956 (pp. 56, 81) must surely have been climatic in origin: it followed an exceptional spring, and the population largely returned to normal subsequently. That modification must therefore be of a different kind from the spread of the PP inversion in California. It nevertheless serves to show that the genetic adjustment of organisms may act as a sensitive indicator of ecological events otherwise difficult to detect.

Other organisms, and conclusion

Chromosome-polymorphism in other organisms may take a form quite different from that just described in *Drosophila*: one involving number.

In *Nicandra* there is a polymorphism in germination-time controlled numerically by the iso-chromosomes present; plants with two germinate faster than those with one (Darlington and Janaki-Ammal, 1945). Indeed Darlington has made the important suggestion that variation in the frequency of B-chromosomes occurring in many plants may provide a switch-control for polymorphism in such qualities as germination-date, drought-resistance, and other features (see Huxley, 1955).

Staiger (1954) has studied polymorphism in chromosome number, and the adaptations concerned, in an animal: the marine Gastropod, *Purpura lapillus*, in the neighbourhood of Roscoff, Brittany. In the form in which n = 13, there are 5 metacentric chromosomes (M) which correspond with 10 acrocentric chromosomes (A) in that with n = 18. Each A chromosome represents part of one arm of an M, while the other 8 chromosomes do not undergo fragmentation. The heterozygous condition A/M involves 5 different and distinguishable trivalents corresponding to the 5 different A types and the combinations A/A, A/MM, M/M are formed freely for each of them, so producing all intermediates between the n = 13 and the n = 18 conditions.

The n = 13 form is found in places with a high degree of exposure to waves while n = 18 is restricted to sheltered areas. Intermediate localities are colonized by populations intermediate in their chromosome complement. The adaptive values and particular forms of the A/A, A/MM, and M/M karyotypes depend on the conditions of each locality, while in such places the heterozygotes, A/MM, are the most advantageous, so maintaining the polymorphism.

Polymorphism due to pericentric inversions has not yet been reported frequently in animals, probably because it has not been sufficiently looked for. An instance has been discovered in the mouse, *Peromyscus maniculatus* (Ohno *et al.*, 1966) and there are others.

An interesting example of combined chromosomal and other types of polymorphism has been studied in the North American White-tailed Sparrow, *Zonotrichia albicollis*. Lowther (1961) has shown that there are two polymorphic phases in this species, they are recognizable principally by the colour of the median crown-stripe. This may be either white or tan, a distinction association with other, but minor, plumage characters. White-striped males seem to be slightly the rarer in Newfoundland (43 per cent) and the commoner in western Canada (63 per cent) while the reverse is true of the females (50:22 per cent), though the difference is not fully significant. Lowther also shows that there is a strong tendency for the unlike types to pair P < 0·001.

The chromosomes in this bird are of the usual avian type: some large and easily recognized, while there is a cloud of small ones so difficult to count that it is still doubtful if n = 41 or 42. Thorneycroft (1966) shows that three of the macrochromosomes are involved in a polymorphism. They are of equal size. That labelled number 2 may be present once, twice, or three times; number 3 may be absent or present once or twice; while another, 'M', may be present as one chromosome or absent. The arrangement is so adjusted that the total numbers of the three types always amount to four. The differences do not affect sex; a distinct pair of sex-chromosomes has, indeed, been recognized. It is clear that some relationship exists between the chromosome constitution and the colour dimorphism. Its nature is not fully established, but chromosome M is present in all the tan-striped birds, and absent from all the white-striped ones. As Thorneycroft remarks, the colour variation, assortative mating, and chromosome polymorphism seem to combine in an interlocking system to promote heterozygosity. The advantages of hybrid vigour are thus gained in this species while the proportions of the less viable homozygotes are reduced.

Inversion-polymorphisms resembling those of *Drosophila* occur very widely also. They have been recognized in various other Diptera, in which they can similarly be studied with special facility in the giant salivary gland chromosomes characteristic of that Order. They are common in several species of *Chironomus*, Nematocera (Acton, 1957). In that genus they are, of course, associated with male (as well as female) crossing-over, which has been demonstrated within the heterozygous inverted segments (p. 125). The selective importance of these inversions is indicated by the fact that, in *C. tentans* for example, they occur in distinct and characteristic frequencies in different European localities, sometimes no more than a mile apart. Moreover, the species inhabits Canada also, where the banding-patterns differ much more from the types found in Europe than they do from one population to another in either continent (Acton, 1959).

Outside the Diptera, inversions must be detected cytologically by a study of the first meiotic prophase when, if they are long enough, their presence gives rise to loop-formation owing to an attraction between corresponding loci. An included chiasma then results in a bicentric chromatid, forming a bridge, together with an acentric fragment and, consequently, in the destruction of the cell concerned. By means of these prophase configurations, relatively long inversions have been recognized in a considerable number of plants and animals, in which they must produce the beneficial heterozygous effects noted in *Drosophila*. Unfortunately, however, the investigations necessary to demonstrate these have rarely been carried out, nor have such conditions often been related to their gene-structure by adequate genetic studies.

It will be observed that, as Darlington (1956*b*) has stressed, an inversion or other structural change in a chromosome establishes isolation at the super-genic level; for it normally divides a population into two groups within each of which recombination can occur, though it cannot do so between them. The value of this situation lies both in the possible selective adaptation of the phases to different conditions and in the heterozygous advantage which will evolve from it. These two features may be separable, as shown by Darlington (*l.c.*) in his brilliant studies on the relative frequency of interchange heterozygotes in *Campanula persicifolia* when subjected to different breeding systems. Here the situation, clearly an exceptional one, enabled him respectively to promote and to reduce heterozygosity by inbreeding on the one hand and by outbreeding on the other. With crossing, the basic homozygotes proved viable, but the interchange homozygotes were lost owing to the action of homozygous genes within the interchanged segments. With selfing, only the (ring forming) heterozygotes survived for, in addition, the basic homozygotes were eliminated by recessive lethals within their interchange region.

The most extreme example of chromosome-polymorphism is certainly provided by various species of *Oenothera*. These are famous for maintaining themselves as permanent interchange-heterozygotes, with elimination of the corresponding homozygotes, so securing universal heterosis even with self-fertilization. This represents, as it were, the end-product of a special type of chromosome-polymorphism in which most of the populations, though entirely heterozygous, are phenotypically monomorphic.

Interchange-polymorphisms have very seldom been detected in animals, though they are known in a few species. For instance, Lewis and John (1957) have found them in populations of the American Cockroach, *Periplaneta americana*, isolated in South Wales coalmines. The condition is there associated with the change from outbreeding to inbreeding, but it needs further ecological study.

It is obvious enough that desirable qualities must generally be produced by the interaction of groups of genes, so that a mechanism must exist for holding these together. Something is needed, that is to say, to oppose crossing-over. This can be achieved by localization of chiasmata and by means of inversions so as to maintain numerous small blocks of loci relatively intact. These of course may include major-genes, polygenes, or the two together. Mather and Harrison (1949*a* and *b*) have shown that some units comprising a polygene-system are, at any rate in *Drosophila melanogaster*, carried on different chromosomes. However, in their quantitative studies on abdominal chaetae they demonstrated that selection could build up linked polygene combinations, each having a considerable effect, and that these could act virtually as distinct units. It is evident that there may, in certain circumstances, be advantages in holding each of them together as a super-gene.

A condition in some ways comparable is provided by the work of Lewontin and White (1960) on the Australian Grasshopper *Moraba scurra*. They have shown that two pericentric inversions, carried on different chromosomes, for which the species is polymorphic, interact to influence viability; for they are not combined at random in adult males in certain natural populations. Lewontin and White reasonably suggest that their observed frequencies depend upon their interaction, relative to viability, in the fluctuating environment of the natural habitat in which this species lives. When inversions are too short to form loops they can, outside the Diptera, be recognized only by the existence of small unpaired segments at meiosis. But since a chiasma cannot occur within them, they do not reduce fertility. There is no reason therefore why, with the advantage they confer, they should not be frequent: and so indeed they seem to be. In *Drosophila* an examination of the polytene nuclei has demonstrated numbers of these short inverted segments, though few of them have as yet been detected genetically. It seems that the genes situated in them have not so far been studied, probably because many of them are cryptic; the polymorphism being limited to physiological characters, in which its true value resides. One thinks of the purely physiological homomorphic 'heterostyly', compared with that of *Primula* in which a structural difference is added to the 'illegitimacy mechanism' common to them both (p. 218), and of the protein variation to be detected only by electrophoresis (pp. 28–9, 104–6).

Except when salivary gland chromosomes are available, these short inversions are almost indetectable and their widespread occurrence is only beginning to be appreciated. They represent not only a genetic necessity, holding together co-adapted groups of genes, but they explain a number of obvious features evidently difficult to interpret in terms of simple Mendelism: the transmission, for instance, of marked family resemblances for many generations in certain human pedigrees. The existence in all organisms of short inversions acting as super-genes must indeed be one of the fundamental features of particulate inheritance. The longer type, demonstrated cytologically in Man (Koller, 1937), allows loop-formation and constitutes a slightly different problem. It gives an opportunity for forming a considerable co-adapted region of the chromosome; a benefit generally to be balanced against some reduction in fertility, the disadvantage of which has, however, been greatly exaggerated (p. 125). Moreover, it is possible that selection both against the formation of included chiasmata, and for 'short' compared with 'long' inversions, may have been effective in some species.

CHAPTER TWELVE

Mimicry

General survey

Mimicry is the resemblance of one species to another for protective, or aggressive, purposes. It is not dependent upon affinity and involves only those characteristics which can deceive the senses of a predator. Thus the markings on the abdomen of a wasp may be copied upon the closed elytra of a beetle. Also the similar colours of two mimetic species may be of very different chemical constitutions or one may be pigmentary and the other structural. A mimetic resemblance can involve any of the features by which one animal may judge the identity of another: structure, pattern, coloration, behaviour, scent, and sound-protection. Wickler (1968) gives a good popular account of the subject in which its genetic and evolutionary aspects are treated at a superficial level, but the illustrations, both in colour and in black-and-white, are excellent.

Mimicry generally falls into either of two main categories, the Batesian or Müllerian types. The Batesian involves an association in which the members play two very different parts: (1) The *model*, which is a species possessing some form of inherent protection such as a nauseous flavour, poisonous qualities, or a sting. (2) Its *mimics*, of which there may be one or more; these gain a potential advantage by resembling it, for they are relatively lacking in such protective devices.

Müllerian mimicry, on the other hand, is dependent upon no such unequal partnership but includes a number of species all of which are protected and have come to resemble one another. In some instances they do so closely, in others sufficiently only to recall an unpleasant experience by means of some striking feature which they possess in common. Their similarity reduces the inroads made upon them by predators when learning their harmful qualities.

Many unpalatable or dangerous forms, whether models or forming part of a Müllerian association, or indeed if not involved in mimicry at all, exhibit *warning coloration* and specialized warning behaviour in order to advertise their means of defence. Thus they possess colour-patterns of a conspicuous and easily recognized type, often composed of red or yellow and black, such as stand out clearly from the background. Moreover, they tend to move slowly, attracting attention, so that potential enemies are given time in which to identify them. These features are copied by their mimics which, being relatively vulnerable, would otherwise adopt the opposite plan of *cryptic coloration* and often of agile movement. It is, however, noteworthy that, in insects at least, many models have developed one asset not found in their mimics: that is to say, they are tough and leathery and resistant to injury, so that they are not necessarily killed by an inexperienced or mistaken predator before it has discovered their nauseous or dangerous properties. Such a device would be meaningless in the mimics, for their resources are a sham and they have no concealed weapons to disclose. Thus one can inflict injuries on Danaine butterflies without doing them serious harm, whereas a similar degree of pinching, crushing, or wounding would cause death to a mimetic *Papilio* or Nymphaline.

Instances of mimicry are to be found in very diverse groups of animals, in some of which they fall outside the scope of the classical Batesian or Müllerian types. This indeed is generally true of plant mimicry, two instances of which will suffice.

Certain weeds have come to resemble very closely the crops with which they grow. They have done so owing to human elimination of those plants and their seeds which are the most easily recognized. In this way, flax is frequently impoverished by very distinct species which greatly resemble it but are unknown elsewhere, though it is possible to determine their wild progenitors (Berg, 1926). Thus *Camelina linicola*, with its thin unbranched stalk and small pale leaves, is easily mistaken for fibre-flax among which it is found. Its fruits also correspond fairly closely in shape and weight, and are much heavier than those of *C. gabrata* from which the 'mimic' is almost certainly derived. Consequently they become included with the flax-seed sown for the ensuing crop, and in this way the mimic is perpetuated and spread. Professor C. D. Darlington informs me that rye came into cultivation in a similar manner, through its resemblance to wheat.

The Algerian Orchid *Ophrys speculum* and two allied species are pollinated by the males of burrowing Aculeate Hymenoptera. These insects attempt to copulate with the flowers (Godfrey, 1925) which much resemble them in appearance and, probably, in scent; (see p. 260). In Europe, including England, the Fly Orchid, *Ophrys insectifera*, is pollinated by a burrowing wasp, *Gorytes mystaceus*, and other instances of the kind are known.

Among animals, mimicry has been widely reported and that of cuckoos' eggs will be briefly discussed on pp. 255–60. It is, however, in insects that the phenomenon has been most fully studied. It occurs very commonly with Hymenoptera (bees, wasps, and ants) as the models. Their mimics are generally monomorphic, while in butterflies many instances of polymorphic mimicry are known. This is perhaps due to the fact that in the latter group several suitable though widely differing models are often available together, while wasps, for example, are not only very powerfully protected but much resemble one another, being Müllerian mimics.

Butterfly mimicry is largely, but by no means entirely, a feature of the tropics and sub-tropics with their intense competition in which every available ecological niche is occupied to the full. It is also particularly associated with forests, with their extreme contrasts of light and shade which make accurate identification difficult and give some advantage to the first stages in the evolution of a mimetic resemblance. Moreover, I have always found it difficult to judge the size of a butterfly in these circumstances. The eyes have not time to adjust for distance and it is by no means easy to determine whether one catches a glimpse of a larger individual further away or a smaller one nearer at hand. It is interesting therefore to note that there may be surprisingly little correspondence in size between the species involved in mimicry. Moreover, the Hon. Miriam Rothschild (1963) has pointed out that selection may actually favour such disparity. She draws attention to the fact that, other things being equal, birds generally select the largest insect-prey available (Morton Jones, 1932, 1934). Indeed during the summer of 1962 she made tests in which caged birds were given insect species of which they had no previous experience. These were offered in pairs, one large and one small, and the birds almost invariably chose the larger (with a 1 per cent error).

Consequently it is frequently obvious that Batesian mimics are smaller than their models, rarely the reverse which, curiously enough, is consistently found in relation to *Papilio dardanus* and its models (Chapter Thirteen). Thus *Mimacraea marshalli*, 60 mm across the expanded wings, has achieved a remarkable resemblance to the much larger *Danaus chrysippus* (80 mm), and numerous other instances of the kind could be quoted. It is to be noticed also that such a difference in size would delay selection for distastefulness in the mimic. As Rothschild indicates, one would expect similarity, not divergence, in size where the mimicry is of the Müllerian type, as indeed seems usually to be true.

It is noteworthy that the Batesian mimicry of butterflies is generally restricted to the females, never to the males, though there are a few species in which both sexes are involved, as in *Hypolimnas dubius* and *Pseudacraea eurytus*. This is precisely the same situation as that characterizing the non-mimetic polymorphism of butterflies (but not moths, in which polymorphism may be limited to the male, as in *Parasemia plantaginis*, pp. 158–9), and is to be explained in the same way. The genetic basis of this sexual difference is discussed on pp. 165–6, but the need for it is probably in part determined by courtship stimuli. Since in butterflies these are partly visual, we may expect selection in favour of maintaining a constant male pattern capable of eliciting the final female response. This colouring is generally the 'pre-mimetic' one, somewhat modified, since it so frequently resembles that found in related species: for instance, the males of *Hypolimnas misippus* and *H. bolina* display the typical '*Hypolimnas* pattern'. In *Papilio dardanus* the males, and the females when non-mimetic, differ basically but little from *P. nobilis*, and many other instances could be quoted. It is evident then, that the general association of butterfly mimicry with the female is not connected with the fact that this is the heterogametic sex in the Lepidoptera. Since in most butterflies a single male can fertilize several females while the latter sex tends to be exposed and easily captured when egg-laying, the need for mimicry is much greater in the female than the male. In this connection, Rothschild (1970) points out that when, as is usual, the models are sexually monomorphic, a species in which Batesian mimicry is restricted to the female obtains approximately twice the protection gained when both sexes are mimetic. It is possible that the limitation of polymorphism to the male in some moths may be associated with the sexual stimulus also, as this is very different from that of butterflies.

Rothschild's suggestion that a large size attracts predators is relevant here. Since females are often larger than males, there will be a tendency for purely female mimicry or crypsis to develop. Indeed in her tests with caged birds she found that the females were chosen when the sexes were offered together and the male was the smaller (Lane, 1956). Thus a difference in size may contribute to restricting polymorphism to the female sex in so many butterflies. It is unlikely, however, that this is the sole cause of that condition since it occurs in many instances where neither mimicry nor crypsis seems operative and the sexes do not differ in wing-expanse, as in the genus *Colias* (pp. 161–4).

As there must be all degrees from extreme edibility to extreme distastefulness, it might be thought that there can be no clear-cut distinction between Batesian and Müllerian mimicry. This in one sense is true, in that a non-mimetic phase of a Batesian mimic may act as model for a more palatable species. On the other hand, the two concepts have effects which do, on the whole, discriminate them into one or the other type. The Müllerian situation evidently tends to produce uniformity, since the presence of any one species in such an association, whether at a low, high, or changing density, does not threaten the security of the others. It is only required that the warning colour-pattern of the groups should become similar and there appears to be no theoretical limit to the number of individuals or species taking part, at least when insect life is abundant.

Yet there are a few curious instances of polymorphic Müllerian mimics and they are still something of a mystery. The tropical South American butterflies *Heliconius melpomene* and *H. erato* provide the most striking of them (Turner, 1965). They have an immense range and Brower *et al.* (1963) have already shown that birds find both species distasteful and that they confuse them. Though monomorphic over large areas, each has about thirty phases, many involving mutual mimicry. Some of the pairs are, however, allopatric and some forms are non-mimetic. Their genetics have been studied extensively (Sheppard, 1963; Turner and Crane, 1962) and of the genes controlling them, some are unlinked and some loosely linked while two constitute a super-gene.

Turner (1968*b*) shows that the red and yellow mimetic forms of *H. melpomene* and *H. doris* result not from failure to diverge from a common ancestral colour-pattern but from convergent evolution. He suggests that much of the polymorphism arises from the hybridization of species once isolated by climatic cycles in the Pleistocene, and there is considerable evidence for this. Other, perhaps additional, possibilities exist. A species may be distasteful enough to protect several distinct forms within it and so gain heterozygous advantage without the necessity of attempting to obliterate the visible effect of the alleles responsible for that asset. It is possible, also, that the situation may be partly apostatic (p. 189).

The polymorphism of one Müllerian mimic has already been cited in *Zygaena ephialtes* (pp. 154–7). This, however, is rather a special instance, the contending advantages of which are now fairly evident. On the other hand, that of the Old World butterfly, *D. chrysippus* in its east-African habitat involves unresolved problems of a type more like those posed by the *Heliconius* species just mentioned.

Batesian mimicry provides a contrast in which the safety of the model is threatened by the existence of the mimics. As a Batesian mimic becomes relatively commoner in any given area, so the advantage it gains from its mimicry, and the safety of its model, wanes, vanishes, and is converted into a disadvantage. The latter situation occurs when a particular colour-pattern, necessarily conspicuous since evolved for warning, becomes associated by predators with something edible rather than inedible. Evidently, therefore, the frequency of the mimic will be stabilized at the point where the advantages and disadvantages of mimicry are equal, so that the mimetic form is of selective neutrality.

We here touch upon a subject of difficulty and violent controversy: the significance of selective advantage. Selection will favour the increasing perfection of Batesian mimicry since, other things being equal, those individuals which best resemble their model will contribute most to posterity; but it will also ensure that the quality selected for will cease to be an asset. For any increase in the relative numbers of a mimic compared with its model, whether due to more perfect deception or other cause, can only be carried up to the point of equipoise, that at which the value of the resemblance is balanced by the danger of making a species conspicuous when it is in fact, relatively palatable and defenceless.

This situation must be considered with reference to a further aspect of mimicry. It has been said, and it is obvious enough, that the Müllerian type favours uniformity; not so the Batesian. If a genetic variation chances to give a palatable but already mimetic species some degree of resemblance to another 'protected' one, the new form will increase up to the point of selective neutrality as before; its relative numbers when this occurs will become greater as the deception involved is improved. Consequently, in Batesian, unlike Müllerian, mimicry there is no barrier to mimetic polymorphism, which is in fact frequent. The different phases must all be maintained in equilibrium at their point of selective neutrality, when they confer neither advantage nor disadvantage upon their respective forms however imperfect or perfect they may be. Nevertheless selection will favour their improvement; a situation which throws further light upon the theory of polymorphism. Indeed in Batesian mimicry polymorphism is very frequent because it confers a potential, though in a sense unreal, advantage. It does so in the following way. The number of suitable models in any area is limited and usually quite small, consequently they may be copied by several distinct species: *Danaus chrysippus* in parts of Africa is a good instance of the kind. As the relative numbers of mimics sheltering under the guise of a distasteful or dangerous form rises, so the benefit they derive from doing so declines in terms of the total required to annul the advantage of that particular mimicry. Therefore selection will favour the tendency for a Batesian mimic to spread its repertoire of deception over a range of models. It must not, however, be thought that it gains greater protection thereby; that is obtained while the relative numbers of a mimetic phase are rising to the point of stabilization (relative to those of the species it copies), not when they have reached it.

The situation just discussed is in a sense the idealized one; it will certainly limit the numbers of a Batesian mimic if and when it obtains. But in general it does not do so, in the sense that the mimics are rarely common enough to annul the advantage of the corresponding colour-pattern in the model. Thus it is difficult to suppose that *Danaus chrysippus* can be threatened in that way in Africa by the combined existence of its various, on the whole uncommon, mimics. It is quite impossible to think that the powerful protection enjoyed by the large bees in England can be effectively diminished by the two relatively quite rare species of Lepidoptera (the Bee-hawks, *Hemaris fuciformis*, *H. tityus*) which copy them.

But the nummerical increase of Batesian mimics must usually be checked not by the extent to which they constitute a danger to their models, but by aspects of their ecology not directly related to their mimicry. So too must increase in the population-size of the models which, as far as their protective qualities and warning coloration go, have everything to gain from augmenting their numbers.

Yet, since in Batesian mimicry the individuals which enjoy the best mimetic protection will be at a selective advantage, this will on the whole improve the deception which they practice. Thus, as will be shown (pp. 250, 268–9), Batesian mimics may survive quite well beyond the range of their models, but their mimicry tends to be imperfect in such regions.

It has in the past seemed mysterious to many students of mimicry that the non-mimetic forms of a species and some or all of its mimetic ones should often persist together in the same locality. There are numerous well-known instances of the kind: *Papilo polytes* in the Oriental Region and *P. dardanus* in Abyssinia provide examples of them. The reasons for this should now be clear, for the mimics will reach their equilibrium frequencies and, as long as they are maintained at them, will have no superiority compared with the non-mimics. Where these latter survive in Batesian mimicry, they will have a potential value of their own, generally due to the evolution of heterozygous advantage, such as maintains polymorphism in those species which are not mimetic at all. That conclusion is confirmed by the fact that a non-mimetic phase may co-exist with several mimetic ones, as in the Abyssinian *Papilio dardanus* (pp. 270, 280–1).

In 1945 Goldschmidt published a long article on butterfly mimicry which showed not only that he had misunderstood the theory of the subject but that he was unacquainted with many of the important facts relating to it. During the course of his remarks he says (p. 163): 'In some instances there are found non-mimetic and mimetic females.* One would expect the non-mimetic females to disappear.' Of course not: all the polymorphic phases, whether mimetic or non-mimetic, must be balanced at such frequencies that each receives equal protection.

Experimental proofs of mimicry

The general principles of ecology, as of genetics, are assumed without explanation in this book. Thus the evidence which demonstrates the reality of mimicry need not be summarized here except for work which leads on to new developments in that subject.

Jane van Z. Brower fed North American mimetic butterflies and their models to caged Florida Scrub Jays, *Cyanocitta coerulescens*, which acted as predators. In the first series of experiments (Brower, 1958*a*), these birds ate *Limenitis archippus* (Nymphalidae) in every test. They were then supplied with its very similar model, *Danaus plexippus* (Danaidae); this they constantly refused after initial tasting, which provoked violent aversion. Following more than fifty trials, that species was replaced by the mimic, which was then consistently avoided.

Brower proceeded to analyse other well-known instances of mimicry in a similar way. In one of these (1958*b*), the model was the blackish-green *Battus philenor*, which belongs to the distasteful tribe Troidini of the Papilioninae. Three of its mimics were used: *Papilio troilus*, *P. polyxenes*, and the black female form of *P. glaucus* (pp. 253–5) and they all proved acceptable to the Jays which, however, invariably rejected *B. philenor* after once tasting it. Two of the birds then refused to eat *P. troilus* and *P. polyxenes* when substituted for the model, though a third was not deceived and accepted both species as edible. The black female form of *P. glaucus* was apparently recognized as distinct from *B. philenor* by two birds but not by a third; however, the available numbers of this species were too small for significant results to be obtained. Indeed Brower's work provided many instances which show that individual birds vary in their ability to distinguish models from mimics.

A further example of mimicry to be analysed in this way (Brower,

* In butterflies, Batesian mimicry is often, but not always, limited to the female sex (p. 239).

1958c) involved *Danaus gilippus berenice*, a reddish-brown species with black and white markings, and its mimic *Limenitis archippus floridensis*. In this work the former control birds, which were never given the un-palatable species, were used in the experiments and those previously employed in the tests became the controls. Although the Danaine model was clearly distasteful, the birds failed to remember this, perhaps owing to their long experience of relatively palatable butterflies. Three of the controls refused the mimics on sight alone, probably because of their previous experience with the distasteful *D. plexippus* (in the first series of experiments) which this form of *Limenitis archippus* much resembles. Yet the latter species is certainly palatable, for a new bird with no laboratory experience of models ate it consistently. As Brower remarks, this tendency for birds to mistake a mimic for a model belonging to a related mimetic-complex indicates the selective value even of incipient mimicry.

Brower (1960), working at Oxford, also undertook important studies on mimicry in which artificial models were used. For this purpose caged Starlings, *Sturnus vulgaris*, were employed as predators and meal-worms, *Tenebrio mollitor*, which they eat voraciously, formed the prey. Muhlmann (1934) had been the first to colour these larvae and make them distasteful. Brower manufactured 'models' by dipping them into a 66 per cent solution of quinine dihydrochloride; she distinguished these visually with a band of green cellulose paint. Others were merely dipped into distilled water and therefore remained palatable; these were either painted green like the 'models', so as to produce 'mimics', or orange. The latter group was used to discover whether the birds could learn to associate what is normally a warning colour with something edible. Green, on the other hand, is generally found in cryptic species unprotected by a repellent flavour. Consequently it was chosen for marking the models since the starlings were unlikely to have a previously acquired aversion to it. Control birds were fed with 'mimics' only, and their ready acceptance of them showed that the green paint was not itself unpalatable.

Two groups of starlings, one comprising four and the other five birds, were used. These were given 160 trials, generally ten per day, consisting of two meal-worms each: one of which was orange-banded 'edible' and one was green-banded. Six of the birds received these latter in the proportions of 10, 30, and 90 per cent mimics respectively, the remainder being models. One bird (in the second group) received 60 per cent mimics (to 40 per cent models), while two birds (one in each group) formed the controls. Eight of the nine starlings used ate all the non-mimetic edibles in all trials. So did the ninth bird, which was receiving 30 per cent mimics, until the sixty-second trial when it would no longer eat meal-worms.

After initial tasting and violent rejection, the models were generally recognized by their appearance and avoided and, in consequence, their mimics proved to be protected also. In fact about 80 per cent of them escaped predation when they comprised 10, 30, or 60 per cent of the green-banded meal-worms. Indeed the two starlings which received only 10 per cent of models left 17 per cent of the mimics untouched.

There was some variation in the ability of the individual birds both to learn and to memorize the association of colour with edibility. Brower notes, however, that most of the starlings made occasional mistakes, by pecking, killing, or even eating models after learning to avoid them. As she says, this type of reaction is doubtless an important one for, as suggested by Fisher (1930*a*), it allows a predator to detect changes in the relative frequency of models and mimics.

Brower's experimental proofs of mimicry (see also pp. 246–9) mark a great advance in the study of protective coloration. So too does her demonstration that, if distasteful enough, a model may give effective protection even when considerably out-numbered by its mimics provided that their resemblance to it be an accurate one. Her methods were new and original; they have made it possible to approach the subject from a quantitative aspect.

More recently, Morrell and Turner (1970) have extended this type of investigation to studying the response of wild birds to artificial prey. This consisted of cylindrical pieces of pastry (2·2 × 1·0 cm) mounted on triangular cards (height 3·5 cm) and coloured red (models and mimics) yellow (imperfect mimics) and green (controls). A less perfect mimic was also manufactured by painting a black bar (0·5 cm wide) across a triangle. The dummies which acted as models were made distasteful by soaking the pastry in a 75 per cent solution of quinine hydrochloride, after which they were left to dry.

The results corroborate the behaviour of caged predators. Thus the starlings, which were the species chiefly involved, began by attacking all the 'prey' equally. However, they learned to continue attacking the pastry on the green cards and largely to avoid the red models and mimics. Moreover, they attacked poor mimics more often than good ones. Thus it is extremely probable on this as on other evidence that Batesian mimicry evolves, a fact which it is important to confirm from as many aspects as possible.

A further step had been taken by Brower, Cook, and their colleagues when they investigated mimicry using live prey, and wild predators in their natural habitat. For this purpose they worked in Trinidad treating as a mimic the males of the palatable North American moth *Hyalophora promethea* which is unknown in the Island. This insect is day-flying and they painted it to resemble the local unpalatable butterfly *Atrophaneura anchises*. Their results indicate that the experimental mimics survive better in the natural environment than the non-mimetic controls (*H. promethea* painted black). They also demonstrate that this advantage is lost as the relative frequency of the model becomes too low (Brower, Cook and Croze, 1967; Cook, Brower and Alcock, 1969).

Some attention must be given to the protective devices of warningly coloured models. The qualities which enable these to escape or survive predation must necessarily be diverse and to some extent interlocking. They must be effective and not a sham, as in the mimic. Unpleasant scent, if widely applicable, may well be the most useful of them all: and this for two reasons. Firstly the predator will not have to damage or kill the insects producing it in order to learn that they are to be avoided. Secondly the force of their advantage will be but little diluted by mimicry. It may be that their colour-pattern, acting as a signal to be observed from afar, can usefully be employed by a relatively unprotected species, but a disgusting scent saves the model from molestation (Haase, 1893). We are quite ignorant how often warningly coloured forms escape attack by their smell. For it is clear enough from studies of courtship that insects frequently produce scents imperceptible to man. A further pointer in this direction is provided by the fact that an odour judged to be powerful by some people (that of freesia blossoms and of ferrets for example) may be undetectable to others.

Further, and perhaps more usual, forms of protection are less satisfactory because they cannot be descried from afar, except by experience. Distasteful or poisonous qualities, or even a sting, are a last line of defence, for by the time they are called into service an individual may be crippled if not killed: hence the importance of warning colours, and indeed of the rattle to the rattle-snake which seeks not to use up its venom.

Moreover, mimicry involves adaptations at various levels. Rothschild has pointed out that a mimetic butterfly must copy, though to a somewhat less degree, the slow displaying flight of its model; but it should as a last resort be able to give up its disguise and dart away using the characteristic adroit flight naturally acquired by the more palatable group to which it belongs. Also noxious qualities should be associated with a tough leathery cuticle from which the body fluids can exude from bleeding points on the application of pressure and with glands producing a repellent secretion. A Danaine butterfly can survive pecking by a bird which would be severe enough to crack the exoskeleton of a Nymphaline and cause its immediate death.

Such ancillary features must accompany the best known of a model's protective devices in the Insecta, the possession of an unpalatable flavour or of poisonous qualities. Haase, in the course of his book on mimicry (1893) pointed out that edible species of Lepidoptera feed only in very rare instances on poisonous plants. Indeed the possibility that the repellent properties of certain insects are derived from their larval food is one that has often been discussed. It has in recent years been studied in detail in the butterfly *Danaus plexippus*.

As already mentioned (p. 243), feeding experiments conducted by Brower (1958*b*) have shown that this species can be highly distasteful and effective as a model. Brower and Brower (1964) carried out further feeding tests by offering to predators sample butterflies taken from groups normally regarded as palatable and from others that are thought to be distasteful. They confirmed these attributes. The member of the Danaini used was *Danaus plexippus*. In addition, the Browers (*l.c.*) restated the suggestion previously made by various authors that the unpleasant qualities of the imagines were derived from the plants on which their larvae had fed: a deduction not indeed verified chemically but based upon the strong correlation between the occurrence of noxious properties in a series of butterfly species and their larval food plants. For these latter belong to several very distinct botanical groups some of which are widely regarded as poisonous and others not so.

Subsequently, Parsons (1965) demonstrated and partly purified a toxin obtained from *D. plexippus* and showed that it resembles digitalis both pharmacologically and chemically, and he suggested that this species escapes predators owing to the presence of that substance in its tissues. Having pointed out that so far as is known insects are unable to synthesize steroids, he goes on to say, 'if the butterfly toxins are steroids, as seems probable, study of their biochemical origin and their relation to cardenolides of the food plant will provide information relevant to both ecology and evolution.'

The matter was carried further at a Conversazione of the Royal Society held in May 1966 when Rothschild *et al.* gave a highly original demonstration by which they showed that the tissues of *D. plexippus* contain two powerful heart poisons, calotropin and calactin, derived from the larval food (Asclepiadacae) and that it is these substances which protect the species from predators. Their results were made available in the general literature by Rothschild (February 1967) and by v. Euw *et al.* (April 1967).

Brower, Brower, and Corvino (April 1967) achieved an outstanding advance by selecting a strain of *D. plexippus* which will feed on cabbage, though not without heavy mortality. They found the resulting imagines were non-poisonous and fully acceptable to Blue Jays, which had not learnt by previous experience to avoid this insect. Also that palatable specimens of *D. plexippus* can be reared by feeding the larvae on *Asclepias tuberosa* or *Gonolobus rostratus* (Asclepidaceae), for these plants are not poisonous. Since the larvae of *D. plexippus* feed on them to a limited extent in the wild, palatable specimens of the butterfly must exist in the normal population. When studying this matter quantitatively in Massachusetts in the autumn of 1968, Brower (1969) forcibly fed one *D. plexippus* each to fifty Blue Jays and found that twelve of the butterflies were emetic and the remainder (76 per cent) non-emetic. These latter must derive protection from their visual identity with distasteful examples: a situation which Brower *et al.* (1967a) name '*automimicry*'. It must also be one of degree, for Brower *et al.* (1968) find that, in addition to the plants already mentioned, *D. plexippus* will eat other Asclepiadaceae in which the amount of the poison differs. Larvae which feed on *Calotropis procera* give rise to butterflies which are slightly more emetic to Blue Jays even than those reared on *Asclepias curassavica*, though as a larval food plant this produces imagines nearly five times more powerful in their effect upon the birds than does a *Gomphocarpus* species. Yet the latter, unlike *Gonolobus*, is not quite lacking in the poison. Thus there must be a tendency for the less poisonous members of a normally distasteful species to shelter under the protection afforded by the warning colour-pattern of those individuals that are more unpleasant and dangerous. That situation may well constitute a widespread and important aspect of mimicry.

Outstanding success was achieved by Reichstein when he elucidated the chemistry of calactin and calotropin (Reichstein [*et al.*], December 1967). He extracted these substances both from *D. plexippus* and from *Asclepias curassavica* and *Calotropis procera*: see also Reichstein *et al.*, 1968.

The repellant substances which may protect Lepidoptera, or indeed other Orders, from predators are of course not invariably derived from the larval or imaginal food· They may be manufactured by the insects themselves as shown, for instance, by Jones *et al.* (1962) in the Zygaenidae.

It should be noticed that even when a butterfly such as *D. plexippus* is acceptable to some predators because its larvae have fed on non-poisonous food, it may be protected against others owing to a repellent scent or to a poison manufactured by the insect itself. Also The Hon. Miriam Rothschild points out to me that African Danainae of the genus *Amauris* only rarely feed upon *Asclepias* containing calactin and calotropin; yet even when the food plant is *Marsdenia* or *Cyanchum*, these insects are generally safe from attack. That is to say, a repertoire of protective devices may be needed to ward off a series of predators, and because a butterfly is relatively secure from one potentially dangerous species it may not be so from another.

Mimicry, as so often in complex subjects, appears more intricate as one's knowledge of it increases. The old concepts of Batesian and Müllerian mimics have become less clear-cut, but that is not to say that they have become meaningless. We see them now as representing quantitative rather than absolute differences. There are gradations between one type and the other and it may well be that a Batesian mimic in one region can act as a model in another, or form part of a Müllerian association there. That is to say, a butterfly, for instance, may be palatable enough to benefit from resemblance to a better protected form, yet distasteful enough to attract others to copy it.

Brower (1960) showed that mealworms largely escaped predation by starlings if only one in three were made distasteful (pp. 244-5). We need feel no surprise therefore on learning that some specimens of *D. plexippus* are effectively palatable although the species as a whole acts as an efficient model. Nor is any basic difficulty introduced into mimicry theory if a higher proportion of certain insects be unacceptable to predators at one season of the year, or at one stage of their emergence, than at another; and this is likely enough if the imago obtains its noxious qualities from its larval food, in which the toxins may vary in quantity seasonally and from plant to plant (Rothschild *et al.*, 1970a). We are concerned with average advantages and disadvantages. Nothing is more revealing in the general study of mimicry than the discovery that the mimetic forms of *Papilio dardanus* exist perfectly well outside the range of their models, but the accuracy of their mimicry breaks down there (pp. 268-9). That example, applicable as it is in general, involves polymorphism and to that situation we must now turn.

Mimetic polymorphism in butterflies

An example of mimetic polymorphism is provided by the African Nymphaline butterfly *Hypolimnas misippus* (Plate 8, Figures 3–6), which measures approximately 80 mm across the wings. The genus *Hypolimnas* is a large tropical and sub-tropical one widely distributed in the Old World. The basic pattern of the upperside (that of the underside is cryptic) is seen in several of the species including the non-mimetic males of mimics. It is blackish with a circular white patch, surrounded by iridescent purple, in the centre of each wing. It is found in the males of *H. misippus* (Figure 3) in which, however, the females (Figure 4) are extraordinarily modified in mimicry of *Danaus chrysippus* (Figure 1), an abundant and widespread butterfly in the warm parts of the Ethiopian and Oriental Regions. This has reddish-brown wings with a narrow black border and a black area, normally marked with a curving row of confluent white spots, in the apex of the fore-wings. It belongs to the Danaidae, one of the groups that are powerfully protected by distasteful qualities. The female mimicry of *H. misippus* is so successful that inexperienced collectors often catch it in mistake for *D. chrysippus*, while the males of the two butterflies have been seen to follow the wrong females.

In east Africa though not elsewhere, *D. chrysippus* is dimorphic owing to the presence of a second form, *dorippus* (Figure 2), lacking the black area marked with white in the apical region of the fore-wings, the result being an almost completely reddish-brown butterfly. This is mimicked by the *inaria* form of *H. misippus* (Figure 6) which, curiously enough, is found throughout the range of that species: the greater part of the Ethiopian region and western India. The mimetic value of *inaria* is attested by the rarity of intermediates (Figure 5) where its model is present. I am not aware of random assessments of their frequency in east Africa, but Leigh (1904) speaks of them as 'well-known but not very common'. This seems quite different from the situation where *dorippus* is absent, for there the intermediates appear to equal or outnumber true *inaria*. Thus Edmunds (1969) has provided an excellent random collection at Legon, Ghana, in which he scores the frequencies of *misippus*, *inaria* and the intermediates between them as 293, 42, 52. The problems involved in the polymorphism of a distasteful butterfly such as *Danaus chrysippus* have already been indicated on p. 240, while the difficulties posed by mimics existing outside the range of their models, as with the *inaria* form of *H. misippus*, are discussed in respect of the Polytrophus race of *Papilio dardanus* in Chapter Thirteen.

Though in east Africa the hind-wings of *D. chrysippus* are orange, they are always white in Ghana where some of the *H. misippus* females are similarly marked, but with a varying amount of white. These latter become relatively common compared with the normal orange form when *D. chrysippus* is abundant, but relatively scarce when it is rare (Edmunds, *l.c.*). This is the situation to be anticipated if the latter species in fact acts as an effective mimic for *H. misippus*.

The two female forms of *H. misippus* are determined by a pair of auto-somal alleles sex-controlled in their expression. *Inaria* is recessive and *misippus* seems to be fully dominant. Nearly all the bred families so far obtained are derived from females fertilized in nature, and it must be re-membered that even were the cross made in captivity the male genotype could not be determined by inspection. Moreover, the method of hand-pairing, which proves so successful in some of the more primitive groups of butterflies, such as the Papilionidae and Pieridae, is very difficult to apply to the Nymphalidae, in which the external genitalia are much more enclosed within the body. The majority of broods which have been reared have given a back-cross result, but four have been recorded which show F_2 segregation. I have shown (Ford, 1953*b*) that these are homogeneous and that they depart from a $3:1$ ratio in the direction of $2:1$ to an extent which exceeds formal significance (the totals are: 99 *misippus*, 47 *inaria* = 146; $\chi^2_{(1)}=4\cdot0$). This suggests a deficiency of homozygous dominants, as discussed on p. 102. When white is present on the hind-wings of *H. misippus*, it seems to be controlled polygenically.

Another African *Hypolimnas*, *H. dubius*, provides a useful comparison with *H. misippus*, throwing further light upon mimicry. Genetically it is, in a sense, the simpler since both sexes are mimetic, the genes not being sex-controlled; this, however, as already mentioned, is exceptional.

The species is dimorphic and no non-mimetic form is known. On the west coast one phase, *dubius*, copies *Amauris psyttalea democlides*, a black butterfly with white spots on the fore-wings and a white basal area to the hind pair. The other, *anthedon*, is a beautiful mimic of *Amauris niavius niavius*, also a black species but with large white patches on the fore-wings and a white base to the hind ones. The genus *Amauris* belongs to the strongly distasteful Danaidae.

On the east coast of Africa where the models available in the west are not found, the two phases of *H. dubius* have been differently adapted. The form *dubius* becomes *mima*, which has a closely similar though not identical pattern, but the basal area of its hind-wings is yellow in mimicry of *Amauris echeria* Stoll. and *A. albimaculata* Btl. *A. niavius* is still present but in a different race, *dominicanus*, in which the white basal area of the hind-wings is greatly extended, and *anthedon* is correspondingly modified to produce a form closely resembling it, known as *wahlbergi*.

Dubius is a simple dominant to *anthedon* and the same relationship holds for their adjusted east coast phases, *mima* being dominant to *wahlbergi*. Attention should, however, be drawn to a possible source of error in butterfly genetics especially when broods are obtained from females fertilized in nature: that is to say, the occurrence of multiple pairings. Platt (1914) found two *wahlbergi in copula* and raised a family consisting of 104 *mima* and 94 *wahlbergi* from the female. This, taken at its face value, of course indicates that *wahlbergi* cannot be a recessive. But apart from other evidence to the contrary, it will be noticed that this represents a clear 1:1 ratio; not a 3:1, or something between that and 2:1, which would have been obtained were the two insects in reality heterozygotes. Thus it is evident that the female had paired previously, with a heterozygous *mima*, and that the male *wahlbergi* observed to be mating with it had not fertilized its eggs: a type of occurrence reported also in other species.

There is a strong indication, unfortunately not yet fully proved, that the genes controlling the mimetic phases of this butterfly affect its habits as well as its colour-pattern. Leigh (1906) and Platt (1914), who were experienced in the ecology of African butterflies, have both discussed this matter and are agreed on the following conclusions. *Wahlbergi* is sun-loving and has a floating flight, rests upon the uppersides of leaves, and emerges from the pupa in the morning. *Mima* rests in shady places on the underside of leaves, takes shorter and more rapid flights, and emerges from the pupa in the afternoon. In these various particulars each phase resembles its model. It is interesting that when Leigh wrote it had only recently been suggested that *mima* and *wahlbergi* were forms of the same species, a fact now long established. This view he opposed on the ground that having an extensive experience of these insects in the wild, he was so much impressed by their different habits and behaviour. Genetically controlled differences in habit are now well known and instances of them have already been discussed (pp. 139–40, 163). It is obvious that they are likely to be of much importance in mimicry.

In the two species so far mentioned, the mimetic polymorphism is dimorphic only. Its control when multiple phases are segregating will be discussed in the next chapter. However, attention must be drawn to instances in which distinct and clear-cut mimetic phases are the result of exceptional genetic mechanisms.

One might certainly assume, though incorrectly, that the sex-controlled dimorphism of *Papilio glaucus* is genetically of a straightforward type. This is a North American Swallowtail butterfly about 80 mm across the expanded wings. Its larvae feed on *Liriodendron*. The males are invariably monomorphic and non-mimetic: yellow with black markings. The females are dimorphic: either male-like or blackish in mimicry of the blackish-green *Battus philenor*. This, also a member of the Papilionidae, belongs to the strongly distasteful tribe Troidini (as does the genus *Atrophaneura* already mentioned). The larvae eat *Aristolochia*.

In the Chicago neighbourhood about 90 per cent of the *P. glaucus* females are black but thence there is a steep ratio-cline northwards as the model becomes rare and disappears, and in Canada all the females are yellow. Further south they are all black except in Florida where there exists a very distinct race in which the females are generally yellow. The species is not found west of the Rockies and the change to the condition in which all the females are black occurs at lower latitudes along the eastern seaboard; thus that form is rare in New Jersey. Indeed Brower and Brower (1962) have recently sampled 3,387 butterflies of the *Battus philenor* mimicry-complex in four localities in the south-eastern United States. In passing northwards from south-central Florida to the Great Smoky Mountains of Tennessee and North Carolina, they found that the frequency of *P. glaucus* increases relative to its models from nearly 0 to over 85 per cent. Moreover, since a high proportion of the black form of *Papilio glaucus* coincides with areas in which the model is abundant, irrespective of latitude, its occurrence cannot be an adaptation connected with temperature or humidity and must be due to its mimicry.

One would have expected the dimorphism of *P. glaucus* to be under simple genetic control as in so many comparable instances: that of *Hypolimnas misippus*, or *Colias philodice*, for instance. Yet Clarke and Sheppard (1959a) find that in bred families the female offspring are generally uniform in each brood and of the same colour as their female parent irrespective of what male parent they may have had. Thus, in general, all the female offspring of a yellow female will be yellow and those of a black female black even when the same male has been used to fertilize the two female parents. Out of 70 families reared by Clarke and Sheppard 6 showed segregation while, of those which did not, 13 contained 8 or more (some many more) females. Among the segregating families, either female form may be heavily in excess, always in the direction of the maternal type. Thus the two largest reared by Clarke and Sheppard contained respectively 23 yellow: 2 black (the mother being yellow) and 3 yellow: 16 black (the mother being black). The segregants themselves showed no specieal tendency to produce segregating off-spring.

It is clear therefore that the switch-control cannot be an autosomal gene. Nor can it be normally sex-linked. If it were, some families would comprise female offspring none of which resembled the mother, for the female is heterogametic in the Lepidoptera; also segregation would then be frequent, which it is not. Moreover it did not at first appear that the control could be a gene in the non-pairing region of the Y chromosome since segregating families would then be impossible.

Clarke and Sheppard have analysed the situation in *P. glaucus* care-fully and a survey of their views may be given here. They point out that their results make it possible to exclude the action of partial sex-linkage in determining this dimorphism. For instance, since a single male mated with two yellow females gave 2 black and 23 yellow female offspring from one and 16 female offspring all yellow from the other, yellow would have to be dominant to black. Also with partial sex linkage, if black were dominant, the male must have been a heterozygote. But if so, black and yellow offspring should be in equality, which they obviously are not.

It can be shown also that cytoplasmic inheritance is an extremely im-probable alternative. Were it operating, a form unlike the female parent could be due to cytoplasmic factors carried by the sperm. There must, however, be a threshold value working, or intermediates instead of dis-tinct classes would be obtained. This indicates that such cytoplasmic factors would be particulate. A female is known in which one side is yellow and the other black: such a condition would be most unlikely with cytoplasmic inheritance unless the number of particles were very small, but in that event they would often fail to pass into both cells at mitosis. Mosaics would then be common, which they are not.

There is rather good evidence to exclude the possibility of partheno-genesis, with the sperm necessary to stimulate the egg and occasionally fertilizing it, which could account for the situation found in *P. glaucus*. In crosses between that species and *P. rutulus*, as well as with *P. euryme-don*, the progeny were undoubted hybrids while the gene for a recessive mutant which appeared in the *glaucus* stocks bred by Clarke and Shep-pard was transmitted through both sexes.

It is just possible that the external environment might switch the de-velopment so as to pass from one phase to the other: black to yellow or the reverse. Yet, bearing in mind the existence of some segregation, one has to admit that the reaction would be only a semi-stable one so that, with environmental control, mosaics should be common, which they are not.

Haldane and Spurway have suggested an interpretation which at the time seemed to accord with the known facts of this dimorphism. They postulated that in this species the genes for black and yellow colouring are in the pairing regions of the X and Y chromosomes and that they cross over with a low frequency, say 5 per cent. Also that only the gene in Y controls the colour of the butterfly, perhaps because of a position effect. This, however, does not appear likely, since position effects have so far been detected in but few organisms: *Drosophila melanogaster* and *Oenothera lamarckiana*. On the other hand, further data obtained by Clarke and Sheppard (1962) have failed to disclose any more segregation. This suggests that the gene is carried in the non-pairing region of Y, the previous segregating families being due to chromosome abnormality. Further work on this subject is needed. It is at any rate clear that in *P. glaucus* we have an efficient switch-control of mimicry which nevertheless does not work upon simple Mendelian lines.

Mimicry in brood parasites

The brood parasitism of cuckoos provides the most remarkable examples of mimicry controlled by exceptional means. About two hundred cuckoo species, Cuculidae, are known, and many of them lay their eggs in the nests of birds belonging to very different families, to be incubated and their young fed by these foster-parents. Such parasitic habits are found in other families also; for instance, in the Icteridae, containing the genus *Molothrus*, the Cowbirds of North America, and in Weaver Birds, Ploceidae, in which the adaptations involved have been very highly specialized. In these, each species is predominantly adapted to one type of host only; not so in the cuckoos of the sub-family Cuculinae, some of which are polymorphic in respect of their parasitism.

This subject has been obscured by concentrating too much attention upon the European Cuckoo, *Cuculus canorus*, which is the only species that breeds in the British Isles. For though this provides the best-known example of parasitic mimicry it is, in fact, rather a poor one.

In general, the mimetic resemblances involve the colour, pattern, and size of the eggs and, sometimes, the appearance of the nestling. Moreover, many additional adaptations have had to be developed in order to produce successful brood parasitism. These include the formation of a tougher egg-shell, a shorter incubation period, and profound changes in the habits of the adult birds.

The eggs of non-parasitic cuckoos are white or bluish-white but in the parasites they include every colour and type of marking required to produce a close resemblance to those laid by a wide assortment of host species. They are adjusted in size also. Generally they are as large as might be expected in birds of the cuckoo's build but they are tiny when adapted for laying in the nests of warblers and other small species, and there may even be polymorphism in this respect as well as in colour-pattern.

Thus the Indian Koel, *Eudynamis scolopaceus*, mimics the eggs of crows with great exactitude both in size and colouring, while the relatively small blue eggs of the Pied Crested Cuckoo, *Clamator jacobinus*, are extremely difficult to distinguish from those of the host genus *Argya*. Cott (1940) cites a telling example of polymorphism in the Large Hawk-cuckoo, *Hierococcyx sparveroides* of south-eastern Asia. This lays two types of eggs: one small and dark olive-brown in the nests of the Great Spider-hunter, *Arachnothera magna*, the other larger and clear pale blue when parasitizing Laughing Thrushes, Timeliidae. Thus the two forms are each of an appropriate size and colour. Moreover, the mimicry is adjusted to the distribution of the models. The blue eggs are laid throughout the entire range of *H. sparveroides*, from the north-western Himalayas to Assam, but the brown type is restricted to the latter area for it is only there that the Spider-hunter is common enough to act as an effective host.

It is well known that, using the special depression on its back, the newly hatched European Cuckoo ejects the young of its foster-parents from its nest, which is always that of relatively small species, so securing the whole food supply for itself. A further specialization for this purpose is rapid incubation, consequently the parasite has an initial advantage in size compared with the other nestlings. However, this habit of destroying its brood-mates is far from universal among cuckoos, the young of which mimic those of their hosts when they are brought up with them. Two examples of this kind can usefully be selected from several mentioned by Cott (*l.c.*). In India, *Eudynamis scolopaceus* parasitizes *Corvus splendens* and its offspring are black-headed like those of the crows, a corresponding head-colour being sufficient to produce the necessary resemblance when looking into the nest. Once the nesting stage is ended the cuckoo moults and the head becomes brown. In *Clamator jacobinus*, already mentioned, the change is in the reverse direction, from a lighter shade (brown) to the dark head-colour of the adult, so as to accord with the relatively pale nestlings of the genus *Argya* with which it is reared. Furthermore, Rothschild and Clay (1952) point out that in the African Widow-birds, *Vidua* (Ploceidae), mimicry of the foster-parents' young is carried to an extreme degree and includes even the distinctive specific markings to be seen within the gaping mouth.

Rothschild and Clay (*l.c.*) have discussed the evolution of brood parasitism and their reasoned judgement on the matter is convincing. They suggest that an important early stage in the development of this habit may well have been the tendency, manifested in many normal birds, to usurp the nests of other species. As they point out, transitional stages between such behaviour and complete dependence upon foster-parents can be studied in the Cowbirds, *Molothrus*, of the United States. In that genus also, a progressive weakening of the protective territorial instinct of the males can be detected, and to these two attributes the origin of brood parasitism may well be traceable, while a parallel series in the development of that condition is observable among cuckoos.

Unfortunately we have not as much information on the ecology of the more perfectly adapted cuckoo parasites as of the European, and British, *Cuculus canorus*. It is thus necessary to comment briefly upon the less accurate mimicry of that species. In Britain the bird most parasitized is the Meadow Pipit, *Anthus pratensis*, which lays dusky-brown eggs speckled with darker brown, but the Hedge Sparrow, *Prunella modularis*, the eggs of which are clear blue, is often selected. Among others frequently chosen are the Robin, *Erithacus rubecula* (eggs whitish grey, speckled and blotched with light red), the Pied Wagtail, *Motacilla alba yarrellii* (eggs whitish with fine ash-grey speckles), and the Reed Warbler, *Acrocephalus scirpaceus* (the eggs of which are pale dull green blotched with olive). The eggs of all these species are copied with a fair degree of accuracy, but by no means exactly, by *C. canorus* except those of *P. modularis*. Though the European Cuckoo is capable of laying blue spotless eggs, which it does in the nests of the Redstart, *Phoenicurus phoenicurus*, in Continental Europe, they are not often recorded in Britain. Rothschild and Clay (*l.c.*) point out that *P. modularis* is a most complacent fosterer and will accept almost any kind of egg given to it. Consequently there has been no selection for blue eggs among the cuckoos parasitizing it. It seems, as Rothschild and Clay suggest, that few birds are so uncritical as *P. modularis*, the nests of which could easily become over-parasitized; consequently *C. canorus* has not been able to restrict itself to hosts laying eggs that need not be copied (though, doubtless, egg-mimicry is its original habit here).

Southern (1954), who gives a detailed discussion of the brood parasitism of the European Cuckoo, points out that its egg-mimicry is rather better on the Continent than in Britain, where the isolation is not so complete and a larger number of species is victimized in a smaller area. Moreover, change in land-usage seems to have led to a deterioration of accurate mimicry by separate gentes. However, *Cuculus canorus* frequently chooses hosts in other countries that are not generally selected here: the Brambling *Fringilla montifringilla*, in Finland and the Red-backed Shrike, *Lanius collurio*, in Germany.

This is not the place to discuss the habits of the European Cuckoo in any detail. They have often been described, and an excellent account of them is provided by Stresemann (1927–34, see pp. 417–27 and 818–19). It is necessary to mention only those features which pose special problems in ecological genetics and these can be stated quite shortly.

Cuculus canorus is extremely promiscuous in its mating habits, nevertheless it is now certain that each female lays eggs of one colour-type only; a total of about twelve in the season is usual but never more than one in each nest, and this is always chosen so that the mimicry is effective (except when the Hedge Sparrow is the foster-parent). It will be realized therefore that female European Cuckoos are subdivided into groups, called 'gentes' by Newton (1893), each laying a certain type of egg and parasitizing the host suited to it. It is evident that this situation cannot be determined by an ordinary genetic switch-mechanism, otherwise segregation would occur within the clutch laid by a single female. What, then, is the nature of this control, and how is it that the female always selects the correct nest?

The latter problem, which may be taken first, seems to have been solved by an important suggestion due to Rothschild and Clay (1952, see p. 255). As they point out, it is now well known that many birds respond to behaviour-patterns stimulated by sights and sounds encountered very early in life. This, for example, may attach the young to the first individual they see, of whatever kind; generally of course their parent. If, however, it be arranged that a newly hatched gosling first encounters a man, crawling on all fours, it will in future react to him as to its mother. Thus the plumage and song of birds similar to those which reared the female cuckoo, and the appearance of their nest, may evoke in her the stimulus to lay and to parasitize the species which fostered her.

The female of *C. canorus* must often be mated by males belonging to other gentes, for though the male of this species does establish a territory this rarely corresponds with that of any particular female. In spite of that, she invariably lays one form of egg only. As originally suggested by Punnett (1933), this result could be achieved if the necessary genes were carried in the Y chromosome, since birds are heterogametic in the female sex. Selection would then favour translocations of the requisite autosomal genes to the non-pairing region of Y; and in the instances in which the genetics of egg-colour is known in other birds, the inheritance is not maternal. It had been suggested that there is a difficulty here if it is true that birds are generally XO. Professor C. D. Darlington had kindly informed me, however, that this condition does not seem to occur in Vertebrates. But the Y chromosome has now been discovered in *Gallus domesticus* (Fréderic, 1961; J. Owen, 1965); also in a Sparrow, *Zonotrichia albicollis* (Thorneycroft, 1966).

It appears then that the mimicry of the cuckoo may depend upon an exceptional genetic mechanism combined with an appropriate behaviour learnt in the nestling stage. In these circumstances, natural selection would ensure that the choice of foster-parent would be related to egg-type, since those birds for which the two did not correspond would contribute little to posterity, while the young reared in the correct nest would themselves tend to develop a suitable behaviour-pattern so as to parasitize the species which had fostered them. There would thus be a general similarity between the cuckoo's eggs and those of its host, one which could be improved, as to details of colour-pattern, by selection operating on the gene-complex to adjust the effect of the controlling gene in Y. This is likely to be powerful, since the host species frequently destroy the cuckoos' eggs intruded into their nest, so putting a premium upon accurate resemblance to their own.

A general comment seems appropriate at this point: that is to say, ornithologists have in the past apparently relied upon the methods of ecology, not those of ecological genetics, for the analysis of their problems. Those surrounding the egg polymorphism of the cuckoo could presumably be solved by a series of breeding experiments combined with behaviour tests. *Cuculus canorus* must obviously be a difficult bird to rear in captivity but it is encouraging to know that this is now being attempted. A chromosome count does not seem to have been made in this species. Even though it cannot provide easy cytological material, it is certainly extraordinary that this has not been carried out. It is to Rothschild and Clay (1952) that we owe the general analysis of brood parasitism in birds; one which has demonstrated the principles of its evolution.

European Cuckoos return annually from South Africa and southern Asia to countries such as England, Germany, or Finland where the hosts which they parasitize are different and the mimicry of their eggs reaches different degrees of perfection. Thus the migration of this species is a supreme example of the purely genetic determination both of habit and of mimicry.

Aggressive mimicry

Brood parasitism is a special example of aggressive mimicry. The subject has been studied experimentally and well summarized by the Browers and Westcott (1960). They define aggressive mimicry as 'the superficial visual similarity of some stage in the life-history of a predator to its prey or a parasite to its host; as such, the predator or parasite is the aggressive mimic and the prey or host is the model'. Poulton (1890) was originally responsible for this concept. However, he abandoned it in 1904 nor did he revive it even when he later published van Someren's South African field-studies on flies of the family Asilidae which mimic Carpenter Bees, *Xylocopidae* (Poulton, 1924). This resemblance allows the Asilidae to approach and attack their model and lay their eggs in its nest, for they prey upon it in both imaginal and larval stages.

Brower *et al.* (*l.c.*) working in Florida investigated the mimicry of Bumble-bees by Asilids and made the suggestion that this is aggressive but at the same time Batesian relative to vertebrate predators, a view for which they obtained strong confirmation by means of combined laboratory and field-studies. They showed that the flies capture and eat their models in natural conditions and that, in the laboratory, toads rejected the Asilids owing to their resemblance to the Bumble-bees once they had been stung. As Brower *et al.* point out, it is unlikely that these Amphibia are serious natural enemies of the Asilidae, but the observations certainly show that the species studied may well derive protection as well as aggressive advantages from its mimicry.

Sound and scent mimicry

The orchid *Cyptostylis leptochila*, from New South Wales and Victoria, depends upon the Ichneumon *Lissopimpla semipunctata* for its pollination. This is due not to similarity of appearance, as often reported, but of scent. It is so effective that sperms have been found in the pouch of the flower after a visit by the insect (Coleman, 1928).

The existence of scent mimicry has indeed only recently been recognized. They deserve brief mention as they may well be important phenomena, to be studied more fully by the techniques of ecological genetics in the future.

Sound mimicry seems to have been detected by Gaul (1952) in the wasp *Dolichovespula arenaria* (Vespidae) and the fly *Spilomyia hamifera* (Syrphidae). These two species, which are found in the north-eastern United States, are extremely similar in appearance and Gaul had great difficulty in distinguishing them upon the wing. He is in doubt whether the mimicry between them is Müllerian or Batesian and, if the latter, which species is the model. Using appropriate modern equipment, he found the standard wing-beat of *D. arenaria* to be 150, and that of *S. hamifera* 147, strokes per second; and he had previously shown that the frequencies are unlikely to be influenced by temperature in insects of this size. Consequently both species emit an audible tone, between D and D#, when flying; a circumstance which, in this and other comparable instances, must contribute to the efficiency of a mimetic resemblance.

However, the investigations of Brower and Brower (1965) carried this matter further. They had shown that experimental Toads which had been stung by Honey Bees, *Apis mellifera*, ate significantly fewer both of them and of their mimics the Drone Fly, *Etisratis vinetorum*, than the controls which had not been stung. When the wings had been cut off to reduce the buzzing, the experimental Toads rejected fewer both of the models and of the mimics; so demonstrating the protective value of sound in addition to the resemblance in pattern.

The Hon. Miriam Rothschild (1961) has made a study of the repellent scents emitted by warningly-coloured insects. This she has carried out both by means of direct comparison, so far as the human olfactory sense allows, and (with the help of collaborators) by the use of gas chromatography. The scent so analysed is due to substances secreted by various glands, sometimes perhaps by scattered gland cells.

Rothschild finds that one fundamental ingredient of many defensive odours, warning predators of poisonous or unwholesome qualities, resembles the objectionable yellow secretion of Coccinellidae; or at least it produces the same sensory effect as that substance, which may be a quinolene derivative. This gives rise to an acrid and highly characteristic smell, with which there is generally combined one of a different kind akin to that of nettles. The two must be chemically distinct, since they can be separated: if a filter-paper impregnated with lady-bird secretion be passed through a stream of liquid air and then dried in a container heated in an oil-bath, the quinolene scent is driven off while the nettle-like one remains.

This combination of two defensive smells appears to be extremely widespread among aposematic Lepidoptera and other insects. Those possessing it are often members of Müllerian associations and Rothschild makes the important deduction that this similarity of odour must contribute substantially to the success of their mimicry. She also suggests that further research will detect Batesian mimics that have developed the scents as well as the colour-patterns of their models. Also she wisely makes the point that the effectively similar repellent smells produced by protected secies may not all be due to substances that are chemically the same, any more than are their warning red pigments (Ford, 1942b, 1944a and b). We may therefore suspect the existence of polymorphism for scent production. There is some suggestion of this though, curiously enough, in a protected species. For Fritz Müller (1878), whose observations were very reliable, says that the males of the American Swallowtail butterfly *Battus polydamas* emit a very strong odour, but that this is of two quite distinct types produced, respectively, by different individuals. It is noteworthy that so little attention has been paid to the subject of butterfly scents that this observation seems neither to have been confirmed nor extended up to the present day.

CHAPTER THIRTEEN

Papilio dardanus and the Evolution of Mimicry

Few species provide such complex examples of mimicry as *Papilio dardanus*. It is, moreover, particularly well fitted to throw light upon the evolution of these resemblances since they occur in a number of distinct geographical races. Most fortunately therefore this is the very butterfly in which the genetics of polymorphism, both mimetic and non-mimetic, have been most fully investigated. The credit for that work is to an outstanding degree due to C. A. Clarke and P. M. Sheppard. They have overcome great difficulties in obtaining the necessary stocks and in breeding these largely tropical and sub-tropical insects in Liverpool on an extensive scale. Their analysis of the results so obtained has verified the theory of mimicry based upon the concept of polymorphism described in Chapter Six and has provided a secure foundation for it. Consequently it has seemed best to reserve the study of *Papilio dardanus* for treatment in a separate chapter, where it could lead on to a general inquiry into the evolution of mimetic resemblance as a whole.

The races of *Papilio dardanus*

Papilio dardanus feeds upon *Citrus* plants. It is entirely restricted to the Ethiopian Region, where it occurs in eight or nine principal races. The males are invariably tailed, monomorphic, and non-mimetic (Plate 9, Figure 1). Their upperside is yellow, due to pigments that are fluorescent in ultra-violet light. On the fore-wings is a black border widening towards the costa. The pattern of the hind-wings is subject to considerable geographical variation. It consists of black submarginal and marginal markings, the latter extending into the tail, the tip of which is yellow. The fore-wings are rather similar on their two surfaces while the underside pattern of the hind pair is cryptic, for they alone are visible when the insect rests among vegetation. In the race Meriones, confined to Madagascar, the females also are tailed, monomorphic, and male-like (Plate 10,

Figure 2), except for a black curved mark on the fore-wings running along the proximal half of the costa and turning back into the cell. This is present also in the rather similar male-like females which constitute the most abundant form in Abyssinia (the race Antinorii) (Plate 10, Figure 6) where, however, a small proportion of tailed female mimics is also found (p. 270).

MAP 6. *The distribution of* Papilio dardanus, *showing the regions occupied by each race.*

On the whole African mainland from South Africa approximately to Northern Kenya and Uganda in the east and along the entire western region to the limit of the butterfly's distribution a little north of Sierra Leone, the females are tailless. They are, moreover, very different from the males and normally lack their fluorescent pigments even when marked with yellow, though these are present in the male-like females of Madagascar and Abyssinia. Some of the forms are mimetic and others non-mimetic and their relative occurrence and frequencies differ from one part of the continent to the other; so does the size of the butterflies, which varies from 9 to 11 cm across the expanded wings though

264

this does not seem to be associated with size-differences in the models.

Within the region just delimited, the species is subdivided into five races (Map 6). Two of these, Dardanus* and Meseres, comprise a western type. They are larger than the others and the males are less heavily marked with black; moreover, their genital armature differs slightly from that of the eastern and southern forms but not in such a way as to prevent interbreeding with them. This difference is indeed controlled on a single factor basis, subject somewhat to the effect of modifiers: the presence of a long spine on the inner surface of the valve in the eastern type is domin-ant to its absence, which characterizes the western one in which genitalia of the eastern type occur rarely (Turner, Clarke, and Sheppard, 1961). The point is worth noting, since the anatomy of the genitalia is so often treated uncritically by taxonomists as a criterion of specific differences in the Lepidoptera and some other insects. Statements on the structure of these organs are generally made in an unscientific manner: the number of individuals in each species that have been examined in order to establish their characteristics is not recorded nor is any indication given of their variance. Yet such information should always be made available in view of the variability to which all organs are necessarily subject. The need for it is here illustrated by the fact that the spine diagnostic of the male genitalia in the eastern races of *P. dardanus* is found as a rarity in the western ones.

The race Dardanus extends up the whole western coast of Africa from Angola to Sierra Leone and far inland. Here one of the mimetic phases, *hippocoon* (p. 266), so predominates that from Calabar northwards prob-ably 99 per cent of the females belong to it. Southwards from the Camer-oons this value must be slightly less owing to the intrusion there of a small proportion of other forms. Meseres is found in Uganda and Tangan-yika territory east of Lake Victoria, whence to the west it merges with the race Dardanus.

The eastern and southern type is represented by three races: Cenea, Tibullus, and Polytrophus. The Cenea race occupies South Africa and perhaps extends as far as Angola in the west. In the east it reaches Delagoa Bay, where it meets Tibullus. This occurs as far as Mombasa and thence for an unknown distance northwards. These two races are very similar except that the males of the latter are the more heavily

---

* It is unfortunate that this name is used for the species as a whole as well as for one of its races. It will not be italicized in the latter connection, in order to assist the distinction.

marked with black. Their females comprise the same striking series both of mimetic and non-mimetic phases which differ only in their proportions as we pass from one region to the other.

It will be helpful first to give a brief account of the female polymorphism of *P. dardanus* and its genetics in these four races. Polytrophus is so different that it must be considered separately.

Apart from those females which are approximately male-like (and which are absent from the area under consideration), the basic female pattern is that of *hippocoonides* (Plate 11, Figure 3). It is the one which departs least from that of the male, which is also that of the species most closely allied to *P. dardanus*; moreover, this form represents the 'bottom recessive'. In it the yellow has been replaced by white and the black markings are enlarged. On the fore-wings this is due primarily to an extension of the costal stripe of the male-like females to which attention has been drawn. The result is a black-and-white butterfly. Its black fore-wings carry white spots and two white patches, one sub-apical and the other on the inner margin. Its hind-wings are white with a narrow black border bearing white spots. This produces a beautiful mimic of *Amauris niavius dominicanus* (Plate 11, Figure 4), which therefore also much resembles *Hypolimnas dubius wahlbergi* (p. 252). It comprises about 10 per cent of the female population in South Africa but 85 to 90 per cent of it north of Delagoa Bay (the race Tibullus) where *Amauris niavius* is very much commoner.

On the west coast of Africa and in the Central Region occupied by the race Meseres (that is to say, Uganda and Tanganyika east of Lake Victoria), *Amauris niavius* is abundant but in the dark form *niavius* in which the hind-wings are black with a white basal area (Plate 11, Figure 6). Here its *dardanus* mimic, known as *hippocoon* (Plate 11, Figure 5), is correspondingly modified to represent it: just as are the respective forms (*wahlbergi* and *anthedon*) of *Hypolimnas dubius* (pp. 251–2). That is to say, the black border to the hind-wings is broad and spotless, or nearly so; its inner edge is less definite than in *hippocoonides* and narrow black rays run far into the much reduced white patch at the base.

Throughout the vast region of Africa under consideration, some *P. dardanus* females are strikingly transformed to mimic *Danaus chrysippus* (Plate 8, Figure 1), producing a butterfly with the pattern but not the colouring of *hippocoonides*. This is known as *trophonius* (Plate 9, Figure 2). The fore-wings are black with a white patch near the tips and a reddish-brown area extending forwards from the inner margins. The hind-wings are also reddish-brown. They have a narrow black border spotted with white and, unlike *hippocoon(ides)* and its *A. niavius* models, this is no wider in the western than in the eastern and southern races, for *D. chrysippus* is everywhere the same.

Trophonius is an excellent mimic though always rare. Generally about 5 per cent of the females belong to it, but only about 1 per cent do so on the west coast, where it has been given the name *trophonissa*. This is distinguishable only because the thin black rays running inwards from the border of the hind-wings are rather longer than in *trophonius* itself.

It has been mentioned that in the females of *Hypolimnas misippus*, which always copy *D. chrysippus*, a form *inaria* loses the black area marked with white on the fore-wings, in mimicry of the *dorippus* phase of their model (p. 250). A corresponding adjustment in *trophonissa* produces *niobe*. *D. chrysippus dorippus* is absent from its range, but a similarly marked and strongly protected species, *Bematistes tellus*, is found from the Cameroons to Angola and in the Congo Basin, and acts as a model for *niobe*, which comprises several per cent of the *P. dardanus* females in that region. Further north, where *B. tellus* is absent, it exists as a rare variety only. *Salaami* is phenotypically a similar modification of *trophonius* in the eastern races. There, however, it is a rare variety without a model, and indeed it is genetically distinct from *niobe* (p. 271).

The form *cenea** (Plate 9, Figure 3) has diverged more than any other from the original pattern of *P. dardanus*. It is black with small spots on the fore-wings varying from pale yellow to white while the hind-wings have small marginal white spots and a buff-coloured basal area. It is an excellent mimic of *Amauris echeria* (Plate 9, Figure 4) and *A. albimaculata*. It is characteristic of the Cenea race in South Africa, where it comprises 85 per cent of the females. Further north it is unknown on the west coast and much rarer in the east, amounting to about 4 per cent of the females in the race Tibullus.

All the mimetic forms of *Papilio dardanus* so far mentioned copy Danaidae except *niobe*, whose model, *Bematistes tellus*, belongs to a different family, the Acraeidae. So does that of another phase, yet more striking and distinct. This is *planemoides*, which is an excellent mimic of *Bematistes poggei* and of the males of *B. macarista*. It is a black butterfly with a broad orange band running from near the costa to the inner margin of the fore-wings, while the basal half of the hind-wings is white. The

* It is unfortunate that this name is used for a race and also for one of the polymorphic forms of the female. It will be italicized in the latter connection only, in order to assist the distinction.

mimicry so produced is excellent. *Planemoides* is principally associated with the race Meseres, and the eastern part of race Dardanus which occupies the area in which its models are common, and here it comprises about 15 to 30 per cent of the *P. dardanus* females. It spreads westwards for some distance, becoming progressively rarer and disappearing towards the coast.

In addition to the mimics already described, there exists in these four races several non-mimetic but polymorphic forms of this butterfly. One of these, *salaami*, has already been mentioned but several others also are to be found in the south and east and attention must be drawn to two of them here. (i) *Leighi* resembles *cenea* except that the fore-wings have the subapical patch of *hippocoonides*. Moreover, both this and the other spots are of a rich fulvous shade. It is a decidedly variable form and is least rare in the Cenea race where perhaps 0·5 per cent of the females belong to it. It is known also north of Delagoa Bay (the race Tibullus) where it is even more uncommon. (ii) *Natalica*. This is a modification of *hippocoonides* subject to considerable variation, in which the pale areas are ochreous brown instead of white. It approximately reverses the distribution of *leighi*, being found more often in the Tibullus than in the Cenea races. It is, however, infrequent and generally less than 1 per cent of the females belong to it, though as much as 10 per cent may do so in certain small areas.

One striking non-mimetic form, *dionysos*, occurs in the west: indeed it is restricted to the race Dardanus. It is nearest in pattern to *hippocoon* but the amount of white on the fore-wings is much increased and very variable; sometimes, indeed, the black is reduced almost as much as in the male-like females of Madagascar. The pale basal area of the hind-wings is greatly enlarged and of a lemon-yellow shade. It is a very uncommon form.

Although the distribution of the Polytrophus race falls between Meseres and Tibullus, it must be mentioned after the normal forms have been briefly described because of its exceptional features. These result from its ecology. It inhabits the mountains of central Tanganyika and Kenya, east of Lake Victoria, and here the models copied by *P. dardanus* are rare or absent. Consequently its mimicry breaks down. In addition to their imperfect patterns, the female forms so produced are often remarkable in two other ways: they sometimes have a trace of tails and they frequently possess small amounts of the characteristic male pigment (p. 263) recognizable because it is fluorescent in ultra-violet light. The principal female forms occurring in the Polytrophus race are indicated in the following list, in which the imperfect mimics (see below) are distinguished by the prefix '*proto-*' attached to their names.

cenea	47
proto-cenea	18
hippocoonides	44
proto-hippocoonides, with several sub-forms	9
trophonius, which is much rarer than the succeeding modification of it	1
proto-trophonius (= *lamborni*) with several sub-forms	16
proto-salaami, with several sub-forms	15
planemoides	—
proto-planemoides, with several sub-forms	—
	150

The figures in the last column refer to the numbers of each form in a random sample from Nairobi. They therefore give a fair indication of relative frequency; and those in which the mimicry is imperfect, which are so rare elsewhere (p. 270), comprise 38 per cent of them. It is curious that no *planemoides* or its modifications are included, as the latter are not uncommon in other collections from the area.

There is thus in Polytrophus a considerable proportion of females in which mimicry has broken down. It has just been said that this is due to the rarity of their models in the region which this race inhabits; a matter on which we fortunately have fairly accurate information for we are able to compare the normal situation, as found at Entebbe (race Meseres), in which the models greatly outnumber their *dardanus* mimics, with one in which the models are relatively rare as they are in the mountains near Nairobi (race Polytrophus). Very large random collections, in which all butterflies seen, of every species, were caught as far as possible, are available from both places. The numbers of female *P. dardanus* and their models sorted out from the great mass of specimens so obtained are given in Table 12 (Ford, 1953*b*). The results show that at Entebbe the models are seventy-three times commoner relative to their mimics, while imperfect forms of the latter are eight times rarer, than at Nairobi. Nor is this an isolated instance. Carpenter (1913) demonstrated a similar type of association between the variability of mimics and the abundance of their models in his studies of *Pseudacraea eurytus* on Bugalla island in Lake Victoria and on the neighbouring mainland.

Moreover, a precisely similar situation is to be expected, and in fact exists, when we contrast the mimetic phases of *P. dardanus*, which are

K

very constant in regions where the models are abundant, with the rare non-mimetic forms that fly with them, for these are markedly variable; as already indicated in describing *dionysos*, *natalica*, and *leighi*. That distinction indicates the selective importance of mimetic compared with non-mimetic patterns.

TABLE 12

| Locality | Models (totals) | P. dardanus (♀) mimics | |
		Totals	Per cent imperfect
Entebbe	1,949	111	4
Nairobi	32	133	32

Frequency of Imperfect Mimics, Relative to Models, in Female *Papilio dardanus*.

It has already been mentioned that the race Antinorii which inhabits Abyssinia is strikingly different from any other on the mainland of Africa (except Byatti from Somaliland, of which little is known). Here alone the females are tailed and, generally, resemble the males except for the black costal bar, p. 264, as they all do in Madagascar (the race Meriones, pp. 263–4). About 80 per cent of the Antinorii females are of this male-like type (giving, in fact, a gene-frequency of 53·8 ± 4·2 per cent) but with them fly smaller numbers of some extraordinary forms, the tailed mimics. The least infrequent of these is *niavioides*: a modification of *hippocoonides* but a less perfect mimic both on account of its tails and its hind-wing pattern which does not represent that of the model quite so well. Still rarer is *ruspinae*, which looks like a tailed *trophonius*. A tailed *cenea* has also been seen but on two or three occasions only.

The genetics of *Papilio dardanus*

We can now consider the genetics of *P. dardanus*. As already mentioned, this subject has been unravelled by Clarke and Sheppard (1959b, 1960a–d; see also Sheppard, 1961b) working in Liverpool. They arranged for the live butterflies and eggs to be sent to them by air from various parts of the Continent and have even been able to secure living material from Abyssinia and Madagascar. They bred and paired the butterflies in heated greenhouses. At first they fed the larvae upon *Citrus* for which they later found a good substitute in *Choisya ternata*, a Mexican Rutaceous shrub not infrequently planted in English gardens where it grows well. Some of the recombinants produced in their work represents forms only occasionally reported in nature. Not all of these will be listed in this chapter as their interest is rather a specialized one best suited to a monograph on *dardanus*, while we are here concerned with those broader principles which throw light upon its ecological genetics.

The inheritance of the tails is due to a single pair of alleles, autosomal but sex-controlled. In the $T^T T^T$ genotype both sexes are tailed, while in the alternative homozygote $T^N T^N$ the males are tailed but the females tailless. The heterozygous females, $T^T T^N$, have tails of intermediate but rather variable lengths (Plate 10, Figure 4).

The colour-patterns of the females, whether mimetic or non-mimetic, are generally determined by a series of allelic major-genes, or supergenes. These are carried at a single autosomal locus not linked with T and are sex-controlled in effect.

Hippocoon, *hippocoonides*, and *niaviodes* all represent the bottom recessive (hh) and the differences between these three geographical forms are polygenic, while *niavioides*, being in addition tailed, carries $T^T T^T$ in place of $T^N T^N$. The next in the series is *natalica* (H^{Na}) which is dominant to *hippocoon* and recessive to the other forms.[*] *Cenea* (H^C) is dominant to both the foregoing but recessive to the *trophonius* group (H^T) which comprises *trophonius*, *trophonissa*, and *ruspinae*. These three geographical modifications of the same phase differ polygenically from one another, the latter in addition carrying $T^T T^T$, as with the *hippocoon* series.

The sequence of these alleles is continued by H^L, producing *leighi* which is dominant to all those previously mentioned except *trophonius*, with which it interacts to produce *salaami* ($H^L H^T$). Yellow male-like females, H^Y, are recessive to nothing else. They produce recognizable heterozygotes only in racial crosses, not with sympatric forms. Three more genes which continue the allelic series are H^{Ni}, H^{Pl}, and H^{La}. These respectively produce *niobe*, *planemoides*, and *proto-trophonius* (the latter known as *lamborni*). They give rise to detectable heterozygotes with most of the other forms.

It has already been pointed out that *niobe* and *salaami* are phenotypically similar modifications of *trophonissa* (in the west) and *trophonius* (in the east) but that genetically they are produced in two different ways. As just indicated, *niobe*, which is a mimic, is due to a major-gene substitution at the H locus while the non-mimetic *salaami* is generally the heterozygote ($H^L H^T$) between *leighi* and *trophonius* though it can also arise as the result of a 'single gene'. There is, in addition, some evidence that *dionysos* is a heterozygous form variable in expression (Ford, 1936).

[*] Not completely recessive to‾*trophonius*.

271

It is not known whether the gene controlling it is an allele at the *H* locus.

Clarke and Sheppard (1960*a*) have obtained information on the genetics of the imperfect forms of Polytrophus (see also Sheppard, 1961*a* and *b*). *Proto-trophonius* (*lamborni*) is, as already mentioned, controlled by an allele (H^{La}) at the *H* locus. *Proto-salaami* is a phenotypic not a genetic entity. It includes one very distinct form, *poultoni*, produced by another member of the same series (H^P). This is dominant to *hippocoonides* but gives rise to recognizable heterozygotes both with *cenea* and *trophonius*. *Proto-cenea* (H^{PC}) is allelic and dominant both to *hippocoonides* and *cenea*. It is also produced occasionally as the heterozygote $H^Y H^C$, which is sometimes distinguishable. Though H^Y is generally fully dominant in Abyssinia it gives rise to the heterozygote *proto-hippocoonides* in Kenya when combined as $H^Y h$. It will be noticed, therefore, that the gene for the male-like yellow female, characteristic of Antinorii, exists at a low frequency in the mountains of Tanganyika and Kenya: an area which approaches nearer to Abyssinia than any other part of the insect's range in Central and Eastern Africa.

The evolution of mimicry

Mimetic polymorphism confronts us with two facts which at first sight appear contradictory. On the one hand, the resemblances may be extremely accurate and involve a number of different characters such as colour, pattern, shape, and, apparently, habits. Great numbers of forms displaying these are known, and they are often adjusted to the geographical races of their models. On the other hand, the various phases are controlled by single-gene switch-mechanisms the different alleles of which must have arisen suddenly by mutation. Yet it is inconceivable that the complex mimetic resemblances which they evoke can have appeared fortuitously: the chances against such an event would be immense even were we dealing with a single instance only. Recognizing the force of this argument, three distinct theories have been advanced to resolve the difficulties inherent in it.

Punnett (1915) endeavoured to explain these seemingly opposed facts by means of parallel mutation. He suggested that the same genes, responsible for the same sets of effects, had mutated both in mimic and model. That view is so clearly contrary to the evidence that it need not detain us. Indeed it endeavours to account for a situation which does not exist: the occurrence of identical characters in mimetic species and those they copy though belonging to different groups. The facts are far otherwise. It has already been mentioned that one of the most usual features of mimicry is its complete superficiality: model and mimic have nothing in common but their appearance. Their similarities, being such as deceive the eye but no more, are attained in entirely distinct ways. The pigments producing corresponding colours are generally quite different, the various resemblances commonly extend only to those parts that are visible, and markings that produce a deceptive identity very frequently occupy parts of the wings or body that do not correspond. These facts are indeed so fully recognized that this theory has long ceased to be considered seriously.

Surprisingly enough therefore, Goldschmidt put forward a modified form of it so recently as 1945. His account does not always make clear the precise nature of his views, but he holds that the colour-patterns and shapes of Lepidoptera may, owing to the limitations of development, be restricted to certain forms only; consequently the mutation of distinct genes, producing different effects, may bring about an approximately similar appearance even between members of distinct families. He holds also that this can occur often enough to account for mimicry. Thus the resemblance between mimic and model, though involving many features, is according to Goldschmidt due to the mutation of a single, though different, gene in each of the species concerned. This view he describes (Goldschmidt, 1945, p. 215) as 'parallel mutation as seen from the angle of the phenotype'.

I have already discussed this concept rather fully and pointed out that it is quite untenable (Ford 1953*b*), while Carpenter (1946) has dealt with certain non-genetic features of it and drawn attention to the fallacies which they contain. It is only necessary here to summarize briefly certain aspects of these criticisms and to show the way in which their force has been substantiated by recent work.

In the first place, Goldschmidt's knowledge of mimicry was not sufficiently extensive to enable him to analyse the subject correctly. Thus his deductions have frequently been founded upon errors which involve plain mis-statements of fact. For instance, in order to support his views upon 'parallel mutation' he says (p. 157) 'The widely different models of mimetic females of a species are related among themselves; it does not occur that one female form mimics a Danaiine and another a poison-eater *Papilio*.' Consider *Papilio dardanus*: there are no 'poison-eater' *Papilios* (that is, members of the tribe Troidini) in the Ethiopian Region save for a single rare species (*Atrophaneura antenor*) confined to Madagascar. But while the mimetic forms of *dardanus* generally copy Danaidae, the models of two of them, *planemoides* and *niobe*, are members of the Acraeidae; a complete contradiction of Goldschmidt's assertion.

He also states that mimicry 'is confined to a few nearly related members of a few systematic groups'. It is not clear what this remark means, but it is incorrect in any sense. If it suggests that mimics and models are necessarily nearly related, it represents a gross error. Many Coleoptera mimic wasps, many Homoptera mimic ants, and one genus, *Oeda*, the cocoon of a distasteful moth, while Lepidoptera may mimic bees or other Lepidoptera. When butterflies copy one another the facts already given in this and the last chapter show that no close relationship is necessary either within the groups of models and of mimics or between them.

Two further points may be mentioned to illustrate Goldschmidt's failure to appreciate the working of mimicry and the errors which he makes in discussing the processes upon which it depends. First, since he relies upon mutation to supply the similarities involved, he has to provide an explanation for the variation which may affect those phases that are under the control of single genes; yet it will later be shown (pp. 275–9) that such variation constitutes a fundamental aspect of these phenomena when correctly interpreted. He suggests that the occurrence of imperfect though unifactorial mimics may be due to a change in dominance and remarks (p. 219), 'selection for dominance does not have anything to do with selection for pattern modifiers'. It is only necessary to make a brief study of dominance-modification to appreciate that such a statement is incorrect: selection acting upon pattern-modifiers can be directly responsible for changes in dominance. Secondly, attention must again be directed to a point which has already been discussed (p. 243): Goldschmidt's surprise that the non-mimetic females of a species frequently persist flying together with those that are mimetic (p. 163 of his paper). He says 'one would expect the non-mimetic females to disappear'. He would have been wiser to discard his views when he himself found them leading to conclusions contrary to the facts of nature.

The difficulties which seemed inherent in the evolution of mimicry and which, as will be further shown, the theories of Punnett and Goldschmidt have so signally failed to overcome, were resolved by Fisher (1927). His solution of them depends upon the fact that the *effects* of a switch-gene (as of other genes) are subject to genetic variation. They can therefore be modified by selection operating upon segregation taking place within the gene-complex, leaving the main gene itself unchanged. He therefore suggested that when a mutation chances to give an unprotected species some slight resemblance to a protected one, the mutant gene would spread owing to the advantage it confers while its effects would be gradually modified, and the mimicry for which it is responsible progressively improved, by selection acting on the gene-complex. If polymorphism be evolved, the original mutant would then remain as the switch-gene controlling alternative forms. Though it arose suddenly by mutation we are not to suppose that the adaptations for which it is responsible, in all their perfection, did so too; for these would have been attained by gradual evolution within the ambit of the major-gene. This view, both as a generalization and in its special application to mimicry theory, has been further developed by Carpenter and Ford, 1933; Fisher, 1930a; Ford, 1937, 1945, 1953b; also to a notable extent in recent years by the experimental studies and the deductions of Clarke and Sheppard (1959a and b, 1960a, b, and c) and of Sheppard (1958, 1959, and 1961b).

Information on the existence of genetic variation affecting the expression of a controlling major-gene has indeed long been available in mimetic polymorphism as well as in all other relevant situations. Thus Leigh (1904) obtained a rare variety of *Hypolimnas misippus* (p. 209) intermediate between the dominant *misippus* and the recessive *inaria* forms (Plate 8, Figure 5). From it he bred 8 female offspring: 4 *misippus*, 3 *inaria*, and one intermediate. Furthermore, Lambourne (1912) captured a female *Hypolimnas dubius* whose pattern was somewhat transitional towards that of the recessive *anthedon*. It produced 7 offspring: 4 *dubius*, also transitional towards *anthedon*, and 3 normal *anthedon*. The reappearance of the intermediate form, as caught in the wild, when the species was bred in the laboratory demonstrates both in this and the previous instance that the breakdown in the expression of a mimic, unifactorially controlled, is genetic not environmental. Thus it represents precisely the type of variation upon which selection can operate to improve a mimetic resemblance. These and other similar facts were of course available in the literature, though apparently not taken into account, when Punnett was devising his concept of mimetic resemblance through parallel mutation, published in 1915, while that of Goldschmidt appeared thirty years later.

This subject has now been investigated in detail by Clarke and Sheppard (1960a, b, c, and d; also Sheppard, 1959, 1961a and b). Their work is of exceptional importance. It was carried out under even greater difficulties than were their studies upon the genetics of the different forms of *Papilio dardanus* already mentioned, since in order to conduct it they had to secure stocks from various parts of the Ethiopian Region, including Madagascar and Abyssinia whence it is very difficult to obtain material, and maintain them in England available for interbreeding as needed.

When the monomorphic race Meriones from Madagascar, in which all the females are male-like, is crossed with the Cenea race from South Africa the mimicry of the phases *hippocoonides*, *cenea*, and *trophonius* breaks down in the F_1 generation. The black pattern of *hippocoonides* is then imperfect and its pale areas, which should be white, are light fluorescent yellow but paler than in Meriones. When back-crossed to South African stock, the black pattern returned partially to that of *hippocoonides* but the pale areas remained light yellow in the heterozygotes. The effect of the *cenea* gene is completely altered in the F_1 gene-complex produced by crossing Madagascan and South African insects, for it is hardly distinguishable from the abnormal form of *hippocoonides* occurring in a similar F_1 hybrid (Plate 10, Figures 3 and 4). In the same racial cross the mimicry of *trophonius* is also destroyed: yellow is then present on the fore-wings in addition to the orange which clouds the white patch near their tips.

It is clear therefore that though each of these three forms is under the control of a single gene, their adjustments break down when placed in a new gene-complex partly derived from a race in which they do not occur. There can be no doubt therefore that the South African population of *Papilio dardanus* possesses modifiers which materially increase the resemblance between the mimics and the models. That is to say, the mimicry has not arisen as the result of a single mutation the effects of which have thereafter remained unchanged, as required by the theories of Punnett and of Goldschmidt: their views have thus been put to an experimental test, and it has disproved them. On the other hand, the results are in full accord with the improvement of mimicry by gradual selection as originally suggested by Fisher.

Clarke and Sheppard also crossed Antinorii from Abyssinia with the South African race carrying the gene for *trophonius*, which latter, in the hybrid gene-complex of the F_1 generation, produced abnormal specimens in which the mimicry had broken down. These were different from the imperfect *trophonius* which appeared when South African material was crossed with that from Madagascar, and they differed also from *proto-trophonius* as found in the Polytrophus race. Since the Abyssinian *ruspinae* is just like a tailed *trophonius*, and therefore quite a good mimic of *Danaus chrysippus*, we must, as Clarke and Sheppard (1960c, p. 168) remark, suppose either that the gene for *ruspinae* is different from that for *trophonius* though the two have similar effects, or that the *trophonius* gene produces an abnormal colour-pattern giving imperfect mimicry in the hybrid gene-complex arising from the Antinorii × South African cross. The latter is the more likely supposition, providing, as it does, a parallel with the results obtained when Madagascan material supplied the gene for yellow females. If it proves correct, it will be noticed that the same gene has been adjusted by different means to give the same effects (apart from the tails which are independently controlled) in the Abyssinian and South African races, just as the *curtisii* gene of the moth *Triphaena comes* has done in the Hebrides and in the Orkneys (pp. 147-9).

Clarke and Sheppard have also hybridized several of the African races of *P. dardanus* in which the females are never male-like. Thus they have crossed *hippocoonides* females from South Africa with males of the *hippocoon* stock from West Africa (the two forms of female are shown on Plate 11). There was some variation between individual families in F_1, depending on whether the male parent had come from Ghana or Uganda. Both in F_2 (Plate 12), and in the first back-cross to West African material, there was complete intergrading without segregation from *hippocoon* to *hippocoonides*, though the various distinctions between these two forms did segregate: thus on the hind-wings there was independent assortment of the width of the border, the occurrence of spots on it, and the presence or absence of black rays running into the white basal areas. That is to say, *hippocoon* and *hippocoonides* differ from one another multifactorially, not by a pair of alleles. Here therefore we can assess the value of Goldschmidt's own conclusions based upon his theory of mimicry. He says (his 1945 paper, p. 213), 'Thus I confidently expect that future research will reveal that the differences between geographic races of mimics (and models) will be of a multiple allelic character involving the locus controlling the mimetic pattern.' The crosses between *hippocoon* and *hippocoonides* stocks just described provide precisely the information required to test Goldschmidt's 'confident' prediction, and they have proved it wrong.

Similar types of crosses involving other phases of *P. dardanus* have confirmed these results. Basically, this work has consisted in mating several of the female forms, inherited as simple and complete dominants, with males from a different race. It was arranged that each of the latter should carry homozygous *hippocoon(ides)*. Since this is the bottom recessive (p. 271), the effects of the other switch-genes can be studied in F_1 as heterozygotes in abnormal gene-complexes.

Cenea from South Africa, where it occupies 85 per cent of the female population, has been crossed with males from regions where that form is uncommon (7 per cent or less around Lake Victoria), rare (the Poly-trophus race), and unknown (on the west coast; Ghana). In the first two of these tests the F_1 hybrids (*hippocoon/cenea*) are markedly variable and the white spots near the tip of the fore-wings are confluent. In the third of them these spots were replaced by a large white patch like that of *hippocoon* (Plate 11(1)). As Clarke and Sheppard remark, these results indicate that in regions where *cenea* is at a low frequency it is adapted by modifiers to give quite good mimicry but not so perfect as in South Africa, where it is abundant. In the west, however, where it does not occur, the appropriate modifiers are lacking and the resulting mimicry is very poor.

By crossing *planemoides* females from race Meseres with *Polytrophus* males, *planemoides/hippocoonides* heterozygotes were obtained. These resembled normal *planemoides* except that the orange band on the fore-wings was broken into two by a broad black bar much as in *proto-plane-moides* found in the Polytrophus race. That is to say, the modifiers which perfect the mimicry of *planemoides* where the models for that phase are common have not been properly adjusted to do so where they are rare.

Each of these three forms was altered when the single switch-gene con-trolling it was placed in an abnormal gene-complex, half of which was derived from a region where the phase concerned is differently adjusted (*hippocoon*) or else rare (*cenea*; Plate 11, Figure 1) or imperfect (*plane-moides*). A striking contrast with that situation was obtained when *tro-phonius* ($H^T h$) was the dominant tested. South African females of that form have been crossed with males carrying *hh* from Uganda (race Meseres) and from Ghana (race Dardanus). In both these matings the heterozygotes ($H^T h$) produced in F_1 were perfect mimics (Plate 11, Figure 2). That is to say, their adaptations do not break down when the phase under examination is very widespread; indeed it is mimetic and approximately identical in each of the three races that were hybridized with one another. The so-called *trophonissa* (p. 267) differs from *tro-phonius* only in possessing longer rays running inwards from the narrow black border of the hind-wings. This is a trivial feature with hardly any apparent effect upon mimicry and it is the only one which behaves in an unusual way (varying towards a *trophonissa*-like condition) when the *tro-phonius* gene from South Africa is placed in a gene-complex half of which is derived from races of the western type. It is especially to be noticed, however, that when the *trophonius* form is crossed with the Madagascar race in which it is unknown, its pattern becomes imperfect in the F_1 hybrids, just as that of *cenea* or *planemoides*.

The evolution of the tailless condition in the females of *P. dardanus*, studied so thoroughly by Clarke and Sheppard (1960b and c), has thrown much light upon the development of mimicry. In the race Meriones from Madagascar (Plate 10, Figures 1 and 2) the tails of the females are as long as or longer than those of the males (means: males 14·4 mm, females 14·8 mm) and in both sexes their variance is very small. When crossed with the South African race, the female heterozygotes in F_1 (Plate 10, Figures 3 and 4) are variable in tail length, the mean being 4·4 mm. This is again slightly reduced in the tailed specimens ($T^T T^N$) of the first back-cross to South African material suggesting, as Clarke and Sheppard point out, that the Cenea race has modifiers which shorten the tails of the heterozygotes. The situation is different in the race Antinorii, in which the yellow females have significantly shorter tails than the males (means: males 14·2 mm, females 12·5 mm) while the tail-length of the mimetic Antinorii females is even less than that of the male-like ones (9 mm, $P < 0.001$). Moreover, the variance of both forms of female is greater than that of the male and than that of the Madagascan male and female ($P < 0.001$). When Antinorii is crossed with South African stock, the F_1 females have shorter tails than the corresponding Meriones hybrids, the mean length being 1·9 mm. Indeed though they must all carry $T^T T^N$, about half the individuals are tailless; in them therefore that condition is dominant.

These facts clearly illustrate the evolution of dominance by means of selection operating upon 'modifiers'. The absence of tails, even their re-duction in length, increases the effect of mimicry in *P. dardanus*, inhabit-ing as it does a region from which *Atrophaneura*, the only genus contain-ing powerfully protected tailed butterflies, is absent.* On the other hand, the retention of tails by all male *dardanus*, also by the females in Madagas-car, where none is mimetic, and in Abyssinia where only about 20 per cent are mimics, strongly indicates that this feature, or the genetic and physiological background associated with it, is advantageous in the absence of mimicry.

The tailed and tailless females of this butterfly are allopatric. Had they been sympatric, it seems clear that full dominance could have evolved in respect of this pair of characters just as it has in fact done in the common mimetic colour-patterns; for all except the rarer of these show complete

* Except for the rare *Atrophaneura antenor*, restricted to Madagascar, which does not act as a model for *P. dardanus*.

dominance in respect of one another sympatrically but not allopatrically. It is instructive to notice that in the oriental *Papilio memnon* (pp. 283–6) we see sympatric (polymorphic) tailless and tailed female forms and in them dominance has been fully evolved. Indeed Sheppard (1958) has argued, without doubt correctly, that in the sympatric situation disruptive selection automatically promotes dominance. For if we consider a pair of alleles switching, for example, the colour from one to another mimetic form, we have two optimum conditions for a character but three genotypes controlling it. Consequently selection must favour the convergence of the phenotypes produced by the heterozygote and one of the two homozygotes.

The origin and maintenance of the tailed mimics found in Abyssinia is a problem which needs consideration. On the one hand, it is evident that they have not arisen independently from the mimetic forms occurring elsewhere, the resemblance between the two being merely due to convergence through copying the same species; for Clarke and Sheppard have supplied full data which show that the major-genes controlling the different phases in both areas are the same. They are clearly still being maintained in Abyssinia by selection, because the models are present and quite common, while the mimics are remarkably constant (in striking contrast with the Polytrophus race in which, owing to the rarity of the species copied, mimicry is not effective), also their tails are reduced both compared with the males and the non-mimetic females.

In spite of this, I have heard it suggested that the existence of mimetic females in Antinorii is due to immigration at the present time, not to their own mimicry. This is a quite wild hypothesis, and for two reasons additional to those just given. First, in that event the migrants must come from Polytrophus which is the only mimetic race which approaches Abyssinia geographically. Moreover, Meseres and Dardanus, which are next in proximity, have the western-type genitalia while in Polytrophus and Antinorii these are of the eastern type. Tibullus, the closest of the normal eastern races, is completely cut off from Abyssinia by Polytrophus. But the country between the forested highlands of Kenya and Abyssina is now quite well known. Apart from two isolated areas of forest-clad mountain, where colonies of *P. dardanus* with exceptional features linger (p. 281), it is a desert across which that species could not migrate. Secondly, the females are tailless in other mainland races including Polytrophus in which though the gene T^T indeed exists it is very rare. Thus, were the Abyssinian mimics maintained by migration from any source today, approximately half of them would be tailless since nearly 50 per cent of the population carries genes for mimetic forms (p. 270), whereas all are in fact tailed.

Consequently migration at the present time is excluded as the source of the Antinorii mimics. However, its occurrence in the remote past, and from the Polytrophus race, seems clearly to have been responsible for them (Clarke and Sheppard, 1960*d*). A long period of increased rainfall, of which there is evidence during the last Ice Age, would allow the highland forests of Kenya and Abyssinia to become continuous by spreading across the intervening desert country where, even today, populations of *Papilio dardanus* survive in isolated forests on two mountain tops. These relict colonies contain an exceptional female form, *ochracea*, which mimics the peculiar local race of *Amauris echeria*, known as *septentrionalis*, found only in these two places.

There is some evidence of such past hybridization between Antinorii and Polytrophus. Clarke and Sheppard (1960*d*) show that females of the latter race have a fore-wing pattern which much resembles that found in hybrids between Abyssinian and Central African stocks. In Polytrophus, alone in African material south of Abyssinia, the gene T^T is present, though at a very low frequency. Also Polytrophus males have reduced black markings but genitalia of the eastern type, a combination characteristic of no other race but Antinorii. In fact hybridization is the only reasonable explanation of the recessiveness of *niavioides* to the male-like form in Abyssinia.

It is clear that at the period of such hybridization the Kenya females had already lost their tails. Had they not done so we should have to suppose that the ineffective mimicry in that district was yet strong enough to spread T^N almost completely through the population there. On the other hand, the presence of tails in all Abyssinian females is nevertheless consistent with their hybridization with a tailless race. The T and H loci are not linked, so that the genes occupying them will be distributed at random in a race that is polymorphic for both. Even if the absence of tails were strongly favoured in the mimics, a slight disadvantage of T^N in those that are male-like, for which we have evidence (p. 279), would eliminate that gene, since only 20 per cent of the females are mimetic. The reverse process would preserve the tailless condition in Kenya where, on the other hand, the male-like females are rare (only about 6 per cent).

Mimicry is likely to be attained in most instances by the incorporation of a number of distinct features, each giving some advantageous resemblance to a model in colour, different elements of the pattern, and, occasionally, in shape. Each would be selected for in its own right as conferring some benefit by producing mimicry, even though imperfectly. It is improbable that a mutation should result in a single gene tending to influence all these features in appropriate ways which could be improved subsequently by the selection of 'modifiers'. On the contrary, the various items will probably each be the result of separate mutations. Evidently it will be desirable to bring and keep together major-genes which thus co-operate in a useful way. The analysis in Chapter Six will suggest that this can be done by the evolution of close linkage between them so as to form a super-gene. The various genic combinations taking place in it, responsible for the control of distinct phases, will thus behave as if they were multiple alleles. These will act as switch-units, the effects of which are perfected, and adjusted when necessary to the geographical races of their models, by selection acting upon the gene-complex. When we consider the allopatric situation, we must expect to find that the finer adaptations have been achieved by building up a different set of 'modifiers' in each race.

This is precisely what occurs in butterfly mimicry, as exemplified especially in those examples of it involving distinct geographical races which have so far been analysed extensively. Especially in *Papilio dardanus* where Clarke and Sheppard (1960*d*) have shown that the female phases of that species, though strikingly dissimilar, are attained by combining in a number of ways the variation of a few simple characters: these include an extension of the original black costal border on the fore-wings, the coalescence of the primitive black marginal and submarginal markings on the hind pair to form a border, narrow or wide, and differences in the colour of the pale areas. Moreover, it will have been noticed that all the diverse forms are controlled by what appears to be a series of multiple alleles, a most improbable situation unless it had been *evolved*, and that the effects have been improved by selection operating upon the gene-complex; as also in *P. memnon* (pp. 283–6).

These facts provide indeed striking evidence for the control of mimetic polymorphism by means of a super-gene. For precise proof of that contention we have to await the occurrence of a cross-over within the unit itself (such as that which has produced the homostyled primrose by reconstruction of the super-gene controlling the pin and thrum phases). However, this may well have become a very rare event, perhaps down nearly to mutation frequency.

Clarke and Sheppard (1960*d*) point out that two further considerations support the super-gene hypothesis in *P. dardanus*. (1) *Niobe*, a mimic of *Bematistes tellus*, is produced by one of the 'alleles' at the *h* locus. But also it is sometimes formed as the heterozygote between *trophonius* and *planemoides*, which are both sympatric with it. Thus a crossover in this heterozygote could evoke *niobe* and cause it to be inherited as a single unit (the other cross-over class would be *hippocoonides*). Here then a mechanism exists for producing a mimic by crossing-over and, as they remark, if the gene-complex had already been adjusted in the heterozygote, the mimicry resulting from the new '*niobe* gene' would be well adapted from the start. (2) Polytrophus is the only race on the African mainland south of Abyssinia in which the gene H^Y for yellow male-like females exists; it is the only one also in which are to be found imperfect mimics with some of the yellow fluorescent pigment of the male. These have the constitution $H^Y h$ in *proto-hippocoonides*, though the others are not heterozygous for that form. Clarke and Sheppard therefore make the reasonable suggestion that those individuals in which white is replaced by fluorescent yellow result from a cross-over in the super-gene between units controlling respectively the formation of this particular yellow pigment and the elements of the mimetic pattern. Such recombinants would no doubt have been eliminated had their models been common.

Clarke and Sheppard have now extended their genetic work on mimicry from *Papilio dardanus* to an equally thorough study of the oriental butterfly *Papilio memnon* (Clarke, Sheppard, and Thornton, 1968). This also has involved the analysis of many forms from different regions and breeding them in England, though the species is strictly a tropical and subtropical one. Its distribution extends from southern Japan as far south as Java, and from the extreme east of India across to the Philippines.

Although a 'swallowtail' butterfly, *P. memnon*, as well as those members of the genus most closely related to it, is normally tailless in both sexes, except in Palawan where the males as well as the females are always tailed. That form has been described as a separate species, *P. lowi*. Clarke and Sheppard have, however, crossed it with true *P. memnon* and shown that the two are fully interfertile and specifically identical. That is to say, '*P. lowi*' is merely a local race of *P. memnon*.

The situation found in this insect is, in many respects, similar to that in *dardanus*; the males are monomorphic and non-mimetic while the females are polymorphic, the phases differing in frequencies and kind from one part of the range to another. Only in Japan are the females monomorphic, though they differ from the males. Elsewhere some are mimetic and others non-mimetic. The latter are, no doubt, maintained by heterozygous advantage adjusted to certain environments, as with numerous other non-mimetic polymorphisms in the Lepidoptera: the white and yellow females of *Colias* for example (pp. 261–4). Frequency-dependent selection is also possible here, since we are dealing with mimicry and only one common non-mimetic form is found in any given area.

Papilio memnon had previously been bred by Jacobson (1909, 1910) in Java. de Meijere (1910), who analysed his results genetically, concluded correctly that the three forms there are controlled by single genes. They are *laomedon*, the bottom recessive, and *agenor*, both non-mimetic (though in some places *laomedon* is a mimic of *Atrophaneura nox*); while *achates*, which here copies *Atrophaneura coon*, is the top dominant. However, Baur in 1911, wrongly as we now know, substituted the theory that *achates* results from a modifying gene interacting with that for *agenor* but having no effect upon *laomedon*: an interpretation copied in principle by Fryer (1913) when he investigated three forms of *Papilio polytes*, in which it is today apparent that this view is also fallacious.

Clarke *et al.* actually examined the genetics of seventeen of the female forms of *P. memnon*. They showed that these behave as if determined by eleven autosomal alleles, all sex-controlled to act only in the females. There is good evidence that these are included within a super-gene of at least five closely linked loci between which occasional crossing-over occurs: T, t, for the presence or absence of tails; B^Y, b^y, and b for the colour of the abdomen (yellow, yellow tipped, and black); and W controlling the colour-pattern of the hind-wings. This latter exists in about nine distinct alleles or closely linked genes, of which W gives rise to the *agenor* type while W^{al}, and W^d produce the two forms of *achates* adjusted to copy *A. coon* and *A. aristolochiae* respectively. Clarke *et al.*, moreover, supply good evidence to suggest that T and B lie at opposite ends of the super-gene with W between them. Loci F, for forewing colour-pattern, and E for the colour of the 'epaulette' (a spot at the base of the fore-wings, p. 287), are also involved in the linkage, but their position relative to the others has not yet been determined.

The loci responsible for two of the colour differences are not included in the super-gene. They produce, respectively, pale compared with dark fore-wings which, unlike the fore-wing colour pattern in general, is not material to the mimicry; also the yellow suffusion of the hind-wings characteristic of the race in Borneo which, therefore, is not a poly-morphism.

After their previous work on mimicry, Clarke and Sheppard decided to study *P. memnon* because this is a species in which the genetic control of the polymorphic females had already been analysed and interpreted along lines differing from the one which they had established for *P. dardanus*. Their results showed that Baur's deductions as to the genetics of *P. memnon* were wrong and that they are, in fact, of the *P. dardanus* type. It now seems that this is likely to be general in polymorphic mimicry. The system may have wide implications. Thus in *P. memnon* dominance proves to be complete between most sympatric forms but generally absent between allopatric ones. Also, the resemblance of the mimics to their models is greater in the gene-complex of a race in which the allele controlling them occurs as a polymorphism than in hybrids with a race in which it is absent. The significance of these facts in respect of the evolution of dominance and of mimicry will now be clear from what has already been said on these subjects.

Furthermore, the studies of *P. memnon* by Clarke *et al.* provide valuable information on super-gene formation in that species. Outside Palawan they have analysed genetically seven female mimetic forms of *P. memnon*. All but two copy tailless models and are tailless. The exception is provided by the two *achates* forms which mimic the tailed *A. coon* and *A. aristolochiae* and are themselves tailed due to the action of the allele T which acts only in the female. Since this is within the main super-gene, the shape, (the tailed or tailless condition) segregates within the appropriate colour-pattern.

Clarke *et al.* have carried out a penetrating investigation of crossing-over within this super-gene. At Hong Kong the two common chromosome types are: (1) $TW^{al} B^Y$; this produces *achates* having tails in mimicry of *A. aristolochiae* and a body marked with yellow; (2) tWb for *agenor*; tailless with a black body.

A few tailed *agenor* with black bodies (known as *titania*) have been found in Taiwan. They represent the cross-over TWb. An example of its reciprocal, $TW^{al}b$, has been found at Hong Kong. Clarke *et al.* successfully bred from it and showed that it is inherited as a single unit.

Titania is occasionally found also in Java where, in addition, a further cross-over produces $tW^d B^Y$, that is to say, a tailless *achates*, has been recorded. It has, however, seldom been seen for it does not resemble any distasteful species. In Borneo, however, this same cross-over exists and the form to which it gives rise (known as *anura*) is common, since what seems to be a suitable model for it inhabits that island. This is the great Bird Wing Swallowtail, *Troides amphrysus*. It should be noticed, however, that though this belongs to the Troidini, a group generally regarded as distasteful, there is no evidence to show that *T. amphrysus* is so. The Hon. Miriam Rothschild informs me that she and Professor Reichstein have failed to obtain from it even a trace of aristolochic acid, found in others of the group, nor does it contain magnoflorin, the toxic alkaloid from *Aristolochia* upon which it feeds. But in view of its warning coloration and affinity, it is likely nevertheless that it possesses some undetected repellent substance. It seems that we have here an illustration of the fact that cross-over types remain rare in habitats where they break down a useful combination of characters; but that one of them has become common, and further improved by polygenic adjustment, in an island region where it can be used in mimicry.

As already mentioned, the *memnon* population in Palawan is exceptional in being invariably tailed in both sexes. This is due to another single gene (*Pt*) for tails, which, in contrast with that previously discussed, is not sex-controlled nor fully dominant (at least in the hybrids) in its action while, unlike the other gene (*T*) for tails, it has not been brought into the super-gene: for it is not polymorphic and does not have to take part in controlling a mimicry segregating for co-adapted shape and colour-pattern. Clarke *et al.* have here provided us with a remarkably satisfying example of evolutionary genetics.

They are at present also engaged in analysing the polymorphic mimicry of *Papilio polytes*. It seems that the suggestion of control by a pair of alleles affected in their action by a specific modifier, as proposed for *P. memnon* by de Meijere (1910) and adopted for *P. polytes* by Fryer (1913), is erroneous. *P. polytes* is normally tailed in both sexes, the males being monomorphic and non-mimetic. There are a number of female forms, three of which have been studied in detail. One, *cyrus*, is male-like and is the bottom recessive. The other two, the *polytes* form and *romulus*, are mimics of *Atropheneura aristolochiae* and *A. hector* respectively. It is now known that three multiple alleles are acting here, *romulus* being the top dominant. That is to say, *cyrus* $= cc$, *polytes* $= C^p$- and *romulus* $= C^P$-. As so often in mimicry, the red pigments of the model and the mimic are chemically different (Ford, 1944*a*); it is only necessary that the colours should deceive the eye of a predator and they are not the outcome of similar genetic situations.

It has frequently been remarked that mimetic butterflies copy the slow displaying movements of their warningly coloured models on the wing but reserve the more powerful flight normal to their group for escape at the last moment if their appearance has not protected them. Rothschild (1970) has raised a further important issue in this respect. she points out that the mimic must fly slowly enough for a predator to recognize its resemblance to its model, but that it should fly a little more quickly and cunningly, so that it is the model which is caught when both are on the wing together. Without formulating that interesting conclusion, I have myself noticed this difference in the tropics, especially in comparing *P. polytes* and *A. aristolochiae* in India.

One further general comment may be made here. Small brightly coloured patches, often red, are placed on the body near the bleeding points of many poisonous Troidinae which act as models. Yet the marks ('epaulettes') corresponding to them in the mimics are on the wing surface at the base of the wing. Rothschild (1970) has produced a convincing interpretation of this difference. She stresses that mimics must avoid examination at close quarters, revealing the absence in them of a tough integument and toxic blood. However, a predator catching sight of the conspicuous 'epaulettes' may well be reminded at a distance of the unpleasant coloured bleeding points on the model.

There is thus conclusive evidence for the evolution of super-genes in mimicry: there is evidence also for the establishment in it of another general aspect of polymorphism; that is to say, heterozygous advantage. As in all other circumstances, the various switch-genes are certain to have effects additional to those on colour or pattern, influencing viability. As already explained (p. 101), those that are advantageous will become dominant and those that are disadvantageous will become recessive. By this means, as well as by the accumulation of recessive lethals close to the switch-gene, the heterozygotes will evolve into the most favoured genotype, so providing a mechanism for the maintenance of polymorphism additional to the one supplied by the mimicry itself.

The truth of that concept is established by the existence of mimetic polymorphism in the absence of the appropriate models, as in *Hypolimnas misippus inaria* in India, or in their effective absence as in *Papilio dardanus* Polytrophus. The great rarity of the species copied by the latter race (p. 270) in the mountainous regions it inhabits, and the breakdown in the colour-pattern of the mimics there but not elsewhere, clearly demonstrates that the phases are not being maintained on account of their mimicry in that area but that they are so maintained in other parts of Africa.

The fact that the accurate adaptations of the mimics are often lost in

Polytrophus does not of course affect the fact that the switch-genes responsible for them are polymorphic; such imperfections are a consequence of the evolution of mimetic resemblance by selection operating upon the gene-complex within the limits of their unifactorial control. The genes themselves are at a frequency far above mutation-rate in Polytrophus just as are those for the non-mimetic phases of *P. dardanus*, such as *leighi* and *natalica*. There are of course numerous other butterflies, *Papilio memnon* and *P. polytes* for example, in which mimetic and non-mimetic forms exist together. The survival of the latter indicates heterozygous advantage, as does the existence of polymorphism in many species that are not mimetic at all.

It has already been explained in Chapter Six that the evolution of superior viability in the heterozygote may be responsible for the persistence of a polymorphism in the absence of the conditions which originally established it; as we see, for instance, in Polytrophus. C. A. Clarke has drawn my attention to what may well be another and interesting example of this kind in *Papilio dardanus*. He points out that some of the non-mimetic forms are very different from male-like females and closely resemble certain mimics. Thus *natalica* and *leighi* are very similar to modifications of *hippocoonides* and *cenea*, respectively obtained by placing the genes for those forms (h and H^C) in an unusual gene-complex. Yet their genetics are quite different from this, for they are each controlled by a major-gene substitution at the H locus (p. 271). Genetically they may most reasonably be interpreted as cross-overs, or mutations, within the super-gene responsible for the mimics to which they correspond.* This would explain the evolution of these forms, which would at first be established in the population on account of their mimicry. This would give an opportunity for them to develop heterozygous advantage, owing to which they would persist after a mutation in the super-gene had broken down their accurate mimetic colour-pattern.

Monomorphic mimicry

Many mimics, both among the Lepidoptera and other organisms, are monomorphic. One finds for example two sympatric butterflies closely resembling one another; one a member of a strongly distasteful group,

* E. R. Creed points out to me that a mutation is the more probable of these alternatives since both *leighi* and *natalica* are of a colour not found in any other sympatric form.

the other not protected in this way but quite unlike the species nearly related to it. As already mentioned (pp. 236, 274), the analogies of mono-morphic mimicry sometimes bridge gaps which, taxonomically, are very wide. In the Sesiidae we find moths which mimic Hymenoptera, and since this general resemblance affects all the species of the family it seems to have arisen remotely, in a common ancestor.

Sheppard (1961*b*) rightly points out that we may expect the initial stages of monomorphic mimicry to be controlled by a single major gene, occasionally by two such genes. For small changes in the original pattern and its associated effects, which must be adjusted by selection, are almost certain to be harmful. Only a considerable step, producing something near enough in appearance to a protected form to give an advantage, is likely to become established. This could then be perfected by selection acting on the gene-complex.

Sometimes the selective adjustment operating upon the monomorphic type is evident because the mimic follows rather accurately the geo-graphical variation of its model. Thus the moth *Paranthrene tabaniformis* (Sesiidae) resembles the appropriate races of the Hymenopteron *Polistes gallicus*, which indeed it copies in colour-pattern and in behaviour: both its movements and flight (Templado, 1961). However, the genetics of monomorphic mimicry can be investigated only by making species- or racial-crosses and the necessary research along these lines has seldom been carried out.

The butterfly *Limenitis arthemis* throws light on the evolution of mimicry. This insect has two major forms, long considered specifically distinct. The northern, *L. arthemis arthemis*, found principally in Canada, is non-mimetic. It is a typical 'White Admiral': blackish, with reddish-brown sub-marginal spots on the upper side of the hind-wing and wide-spread markings of a similar colour below. On both surfaces, the fore- and hind-wings are crossed by a conspicuous white band.

From southern New Jersey and New York State southwards, the *arthemis* form is replaced by *L. arthemis astyanax* which copies the highly distasteful *Battus philenor* (Trodini) in which Rothschild *et al.* (1970*b*) have now identified aristolochic acid. In this mimic, the reddish-brown spots are absent above and are reduced below roughly to correspond with those of the model. The prominent white band of the northern race has disappeared, removed by a single major gene incompletely dominant in effect (Remington, 1958). Bluish-green scales on the upper side of the hind-wings, absent from *a. arthemis*, much improve the mimicry. Platt and Brower (1968) find that they are controlled multifactorially, as is the dorsal red spotting of *arthemis*. They have undertaken a widespread field study of this species, and have shown that the two races intergrade within a relatively narrow zone of 2 degrees of latitude across the north-eastern United States and Ontario. Within this belt, all the three features principally associated with the mimicry (the white bands, dorsal blue-green iridescence, and dorsal red spotting) are extremely variable. Platt and Brower find that the intergrading forms result from a shift in favour of mimicry in the south to a disruptive pattern in the north, as the model becomes first rare and then absent.

Here we have indeed two alternative types of adaptation; the southern and mimetic one being due to a combination of characters controlled respectively by a major gene and on a multifactorial basis. This is possible since the mimicry is monomorphic. As Platt and Brower rightly remark, it is unlikely therefore that these three characters will ever come under the control of a super-gene.

Clarke and Sheppard (1955) investigated the North American *Papilio polyxenes* which is invariably a blackish but sexually dimorphic butterfly. Its females mimic *Battus philenor* which its males, though blackish also, do not closely resemble on the upper surface though they do so fairly well below. Crosses with *Papilio machaon* and *P. zelicaon*, species that are always yellow, have shown that the black of *P. polyxenes* is due to the action of a single autosomal gene dominant in effect. Its sexual dimorphism is also controlled on a unifactorial basis, the presence of that condition being dominant to its absence, as demonstrated by crosses with *P. brevicauda*: the latter insect is black, its sexes similar and of the type found in the male of *P. polyxenes*. This sexual difference in the distribution of a blackish ground colour cannot be detected in the yellow *P. machaon* and *P. zelicaon* yet they carry the gene responsible for it. Moreover, there is an apt comparison here with the situation in *Limenitis a. astyanax*. The gene producing the sexual dimorphism, and therefore the mimicry, in *P. polyxenes* is not linked with that for the suffusion of black pigment, which chiefly ensures the similarity of the females to *B. philenor*. Here again, though to be effective it is necessary for these genes to interact, they have not been built into a super-gene since the mimicry for which they are responsible is monomorphic.

In general terms, it is apparent that the evolution of polymorphic and monomorphic mimicry must follow very similar lines. It would seem possible therefore for a polymorphism, whether mimetic or non-mimetic, to be imposed upon the latter type should the mutation of an appropriate major-gene chance to take place. Its effects could then be improved gradually by selection.

After many years of discussion, mimicry has been analysed by Clarke and Sheppard in such a way as to disclose the mechanism of its evolution both when monomorphic and polymorphic. The results have demonstrated incontestably the truth of the explanation put forward by Fisher in 1927 which involves the selective adjustment of the mimetic forms, operating within the scope of one or more major-genes which may act as switches in controlling them when multiple. The polymorphism so produced exhibits the tendency for the creation of super-genes and for the establishment of heterozygous advantage. Evidently the time has now come for further studies on the ecology of mimicry in the field, carried out in the light of these discoveries.

CHAPTER FOURTEEN

Transient Polymorphism and Industrial Melanism

Transient polymorphism

If a rare gene gains and preserves some over-all advantage it will spread until its former normal allele is reduced to the status of a mutant. While that process is taking place, but not before or after, it will generate a transient polymorphism. Such an occurrence is not possible unless a genetic reconstruction has occurred in the population or it has expanded into a new and different habitat, or else the environment has changed. For the recurrent nature of mutation ensures that any species will have had repeated opportunities for incorporating into its genetic outfit those genes which have proved useful to it in the conditions it has so far experienced.

When polymorphism was originally defined (Ford, 1940a) the possibility of restricting it to the balanced type was seriously considered. However, it seemed a practical necessity to frame the definition so as to include the transient condition also, for there must sometimes be real difficulties in distinguishing between the two situations until rather long-continued observations have been made. Moreover transient polymorphism may not always be carried to completion. Before the former rare mutant becomes transformed into the normal allele, its progress through the population may be arrested because it has had the opportunity of acquiring an advantage in the heterozygous state (pp. 100–4).

The spread of melanic forms

There is much evidence to show that predators hunting by sight destroy their prey selectively, killing a greater proportion of those individuals that do not match their background well. In so far as their colour-pattern is inherited, that process must improve the adaptations of cryptic species. Many instances of the kind have been discussed already in this book. Here it will be worthwhile first to mention another group of such observations which relate to the subject of this chapter.

292

Dice (1947) studied dark and light strains of the deermouse *Peromyscus maniculatus*. Individuals belonging to each type were put into cages with a floor of dark or light soil respectively and subjected to predation by two species of owls. Throughout the experiments those mice which showed up less conspicuously against their background proved to have a significant advantage in escaping capture.

Selection of this kind has doubtless been responsible for the situation recorded in many animals in which darker populations are associated with darker habitats. For instance, Dice and Blossom (1937) analysed colonies of the mouse *Perognathus intermedius* which inhabits isolated outcrops of rock rising out of the sandy desert of Arizona. These range in shade from a rather pale disintegrated granite to blackish lava, and in the eighteen localities examined there was a close correlation between the colouring of the mice and that of their background.

A comparable situation has been detected among the moths inhabiting industrial regions blackened by smoke. Since the middle of last century, some of the species have become decidedly darker there than elsewhere, the difference being polygenic; *Gonodontis bidentata* in the centre of Birmingham provides a good example of them. Yet the reaction of many others to these same conditions is very different and of a far more spectacular kind. Each of these has become black or blackish in the polluted areas owing to the spread through its population of a previously rare major-gene responsible for excess melanin production. That is to say, they have all been subject to transient polymorphism. Over a hundred species have been affected in Britain alone as well as large numbers in other countries also; events which constitute the most striking evolutionary change ever actually witnessed in any organism, plant, or animal.

A few of the principal British species involved in such 'industrial melanism' are illustrated on Plate 13. They are not all closely related but belong to several distinct families of the Lepidoptera. One thing, however, they have in common. They rest fully exposed upon tree trunks or rocks and derive their protection from a cryptic resemblance to their background. Warningly coloured forms are not involved nor are those which hide themselves in crevices. In general, therefore, the affected species match wood or bark and, especially, lichens, to which they bear a resemblance that would astonish those who have only seen the imagines spread, as in Plate 13; of this the photograph of the typical *Biston betularia* on Plate 14 will give some indication.

Anyone who has studied the awful change that has befallen the vegetation, indeed the whole countryside, in manufacturing districts will realize that such adaptations as these must be useless there. For lichens are particularly sensitive to pollution becoming rare at an early stage of industrialization; while, later on, the tree trunks and branches are blackened, the leaves are covered with soot and many plants and animals disappear, unable to survive in such conditions.

Beginning in centres of this kind about the middle of the nineteenth century, industrial melanism has affected first one and then another species and new forms of it are still appearing today. These increase relatively slowly at first, then rapidly, and later more slowly again. Accordingly it is found that their frequency in the population falls into a sigmoid curve when plotted against time (Kettlewell, 1958a). The process is, however, surprisingly fast. Thus the first black specimen of *Biston betularia*, the earliest species to be affected, seems to have been caught in 1848 at Manchester and, though it has but one generation in the year, by 1895 about 98 per cent of the population in that area were black, the form *carbonaria*: a change which must be due to a 50 per cent advantage over the typical form (Haldane, 1924).

New occurrences take place in fresh areas from time to time owing sometimes to mutation but occasionally to dispersal from places where they are already established; for Kettlewell (1958b) has shown that males of *B. betularia* regularly fly about two miles a night, though some of the other species affected are much more lethargic. Starting from manufacturing districts the black insects have spread far across rural England, mainly in an easterly direction (pp. 306–7), colonizing unpolluted as well as the slightly polluted countryside because selection has favoured them there (pp. 306–8).

Once the results of breeding from industrial melanics had accumulated sufficiently to make generalizations possible, certain facts about their genetics began to emerge. The more subtle and curious of these had indeed to await detailed study (p. 296), but at least it became clear that the black forms were almost always unifactorial and completely dominant to the pale cryptic ones which they replaced. Exceptions exist but they are few indeed: three dominants acting quantitatively darken *Lymantria monacha* to an increasing extent; also in *Polia nebulosa* the heterozygote, known as *robsoni*, is intermediate between the pale typical form and the homozygote *thomsoni* which is a melanic with white fringes (see also pp. 314–16).

The spread of black moths in industrial areas had begun to attract some attention by the end of last century and several naïve suggestions of a 'Lamarckian' type were then put forward to account for it. The first to attack the problem experimentally seems to have been Hasebroëk (1925). However, he entirely misunderstood it and contended that air-pollution so altered the physiology of the organism as to produce an excess of black pigment. Accordingly he exposed pupae of Lepidoptera to certain gases: hydrogen sulphide, ammonia, and pyredin, in various doses. He used eight species in his studies, four of which were butter-flies, a group in which the phenomenon investigated has never occurred. His illustrations show that the abnormal forms which appeared in his treated stocks were not melanics at all and he did not study their genetics.

Heslop Harrison, working about the same time upon somewhat similar lines, concluded that the increase of black moths in manufacturing dis-tricts is due to mutation-pressure, not to selection by predators which he regarded as negligible, though its importance has now been fully estab-lished (pp. 297–302). Salts of lead and manganese are present in the soot which collects on the leaves in some areas, and he suggested that these substances cause the mutation of genes for melanin-production but of no others (Harrison and Garrett, 1926; Harrison, 1928). He used *Selenia bilunaria* and *Tephrosia bistortata* (Selidosemidae) as material on which to test this hypothesis. The larvae were fed upon leaves which had in-corporated the salts in question and melanics duly appeared in his treated stocks but, as these were recessive, Harrison was investigating a highly unusual phenomenon, for recessive industrial melanism has occurred only in two species *Lasiocampa quercus* (pp. 314–16) and *Lycia hirtaria* (p. 316).

It is of course difficult to ensure that such genes, being recessive, have not been introduced from the wild population, in which their existence has in fact been demonstrated, and Fisher (1933) showed that Harrison's controls were not adaquate to exclude that possibility: had he worked with dominants this particular difficulty would have been avoided. Harrison, moreover, seems never to have considered the mutation-rate involved, which would amount to 8 per cent of the loci concerned were his claim substantiated (Fisher, *l.c.*). Thus we should have to assume that the mutagenic effect of these salts is many thousands of times greater than that of the most powerful doses of penetrating radiation which can be used without producing sterility.

McKenny Hughes (1932), as well as Thomsen and Lemche (1933), have repeated Harrison's experiments, breeding, respectively, 3,265 and 1,920 moths in the process. They were entirely unable to confirm his results: no melanics at all appeared in their treated stocks.

It was an obvious possibility that the black and the pale specimens are at an advantage in manufacturing districts and in unpolluted country respectively owing solely to their resemblance to the differing backgrounds in such places. This view, based on simple crypsis, failed like Harrison's to take into account other possible effects of the genes concerned. Yet some indication of these was beginning to emerge from the various breeding results which had gradually accumulated when I reviewed the genetics of the Lepidoptera in 1937, for it had begun to appear that the successful black forms might be hardier than the pale ones, which nonetheless they had never replaced save in manufacturing areas. I therefore suggested that cryptic resemblance and differential viability were both operating to produce the phenomenon of industrial melanism (Ford, 1937). The evidence for the second of these two components, that of differential viability, was suggestive but not conclusive at that time, so it was imperative to test it by direct experiment.

Cleora repandata (Selidosemidae), a well-known industrial melanic, was used for that purpose and a non-significant excess of black specimens was obtained in the segregating families that were reared. It seemed necessary therefore to enhance the effect of any difference in viability, if in reality such existed, and this was done by resorting to an unusual technique. Further broods were purposely subjected to semi-starvation, a state in which superior hardiness might be expected to have a decisive effect. That possibility was realized, for the black individuals survived better under these conditions and to a significant extent. The physiological advantage of successful heterozygous melanics over the normal pale form was thus established for the first time (Ford, 1940*b*). It has subsequently been detected in other species (pp. 302–3).

The theory put forward to account for the spread of black moths in industrial areas (Ford, 1937) received considerable support from these breeding results and it was slightly elaborated later (Ford, 1945). Briefly it may be summarized as follows. Those genes which confer greater viability than their alleles will have spread through the population unless checked by some counterbalancing disadvantage such as the obliteration by excess melanin of the cryptic colour-pattern upon which the safety of the insect depends. Such blackening will, however, cease to be a handicap, it may even become an advantage, in the blackened countryside of industrial areas where indeed it will provide a new cryptic colouring. Consequently moths will be able to avail themselves in such places of those genes conferring superior hardiness which, previously, they were debarred from using. Moreover, since the polluted environment is less favourable, such genes will be at a greater advantage in it.

Though this theory may be said to underlie the modern concept of industrial melanism now that the essential features of that situation have been established, it is clear that in framing it I attributed much to little importance to the selective effect of predation in manufacturing areas, though I gave this full weight in normal conditions. In saying 'black colouring will not be such a handicap, perhaps even an asset, in industrial districts', I had supposed predation to be much less severe in these places than in reality it proves to be, regarding them as denuded to a considerable extent of normal wild life, as indeed they certainly are; for in 1937 I much underestimated the great selective forces operating in nature, the magnitude of which has only been realized as the result of further work on ecological genetics during the last quarter of a century. Moreover, I certainly attached too much relative importance to the greater viability of the successful melanics. For there is now reason to think that this may generally have developed or increased during the period of adjustment which the local abundance of these forms has made possible (p. 302). One point should here be mentioned which was stressed as clearly in 1937 as it should be now. That is to say, the physiology associated with excess melanin-production is not in itself correlated with improved viability, for many rare black varieties are much less hardy than are their normal forms. These mutants are recessives, having always been at a disadvantage and remaining so today.

From 1951 onwards the study of industrial, and indeed also non-industrial, melanism has made great advances. It now provides one of the most complete pictures of evolution in progress that has so far been obtained: an achievement due to the originality and enthusiasm of H. B. D. Kettlewell. His own researches have demonstrated to an outstanding degree the value of combining laboratory genetics with accurate and controlled observations in the field, while he has organized on an extensive scale the supply of information from moth collectors scattered throughout Britain.

Selective elimination by birds

It will have been noticed that this theory to account for Industrial Melanism assumed heavy elimination by predators, those insects which match their background least well being the most vulnerable. Here a difficulty was encountered. Entomologists and ornithologists had by no means worked in collaboration, but on one matter they were fully agreed: that birds hunting by sight do not destroy resting moths selectively, or indeed to any appreciable extent at all. Kettlewell realized that a carefully planned investigation of this subject had never been carried out, though butterflies and moths had been more extensively collected, and birds more widely observed, than any other animals: a circumstance from which we may perhaps assess the contribution made by these activities to the study of wild life. He therefore proceeded to repair this singular omission by methods which should be a model for such field-work in the future (Kettlewell, 1955c, 1956).

For this purpose Kettlewell used the moth *Biston betularia* (Selidose-midae), which is greyish white sprinkled with black dots, and its two industrial melanics. (1) *Carbonaria*. This is completely black except for a small white dot at the base of the fore-wings and another where each antenna joins the head. It is a simple and complete dominant. At the present day (pp. 308–9) it is almost invariable, its two genotypes are alike in appearance and it can always be distinguished phenotypically from the ordinary form with complete certainty. (2) *Insularia*. This looks like an intermediate between *betularia* and *carbonaria*, being black with a speckling of white scales. It also is dominant to the pale form but rather variable. It will be discussed on pp. 311–13.

It was necessary for Kettlewell to carry out one aspect of his studies in a rural area, for which he chose Dean End Wood, Dorset, working in 1955. The moth rests fully exposed upon tree-trunks which are light-coloured in this locality, lichen is abundant, and leaf washings even in late summer showed only a very slight degree of pollution. The frequencies in the wild population, assessed on 314 specimens, were typical 297 (94·6 per cent), *carbonaria* 0, *insularia* 17 (5·4 per cent).

Kettlewell had built up a stock of three thousand pupae of *B. betularia*, providing large numbers of the typical and *carbonaria* forms, also a few *insularia*. He established an electric generator to work a mercury-vapour trap in the central ride of the woods and lined the periphery of the whole site with cages of virgin females. All the moths released were marked with a dot of cellulose paint, placed underneath so as to be invisible to predators, in order to distinguish them from the wild specimens and to indicate, by the colour of the paint, the date on which they were liberated.

The work was conducted in two ways: first by direct observation. Equal numbers of female *betularia* and *carbonaria* (an average of about 50 per day) were released on to tree-trunks and they were replaced when all the specimens of one phenotype had disappeared, a direct watch, using binoculars, being kept from hides to detect what predation, if any, occurred. It was found that insectivorous birds hunting by sight preyed upon the moths. The species observed to do so were the Spotted Flycatcher (*Muscicapa striata*), Nuthatch (*Sitta europaea*), Yellowhammer (*Emberiza citrinella*), Robin (*Erithacus rubecula*), and Song Thrush (*Turdus ericetorum*). To the human eye the *carbonaria* were extremely conspicuous and the pale specimens very difficult indeed to see (Plate 14). In all, 190 specimens of *B. betularia* were captured by birds and, though the two forms were released in equality, 164 of them were *carbonaria* and 26 were typicals.

The second type of study consisted in liberating large numbers of the marked moths and recapturing them by means of mercury-vapour traps and assembling to caged virgin females. These two methods of trapping were used in case there should be any inequality in the way in which the phenotypes are attracted by one or other of them. Only male moths could be used since females are rarely caught at lights and the assembling technique of course has relevance to males only.

A total of 799 males was released: 393 typicals, 406 *carbonaria*. 13·7 per cent (64) of the typicals but only 4·7 per cent (19) of the *carbonaria* were recaught, indicating relative elimination of the *carbonaria* form. There was no difference between the frequencies of recapture by ultraviolet light and by assembling to the female scent ($\chi^2_{(1)} = 0.32$).

It was obviously a matter of great interest to carry out corresponding experiments in the reverse situation, that of an industrial area, and Kettlewell conducted this work in 1953 and 1955 in the Christopher Cadbury Bird Reserve, Rubery, Birmingham. Here the tree-trunks are blackened by soot, lichens are now totally absent, and leaf washings in late summer produced large amounts of pollution. The frequencies of the three forms of *B. betularia* in this area, obtained from 621 wild specimens, were typical 63 (10·1 per cent), *carbonaria* 528 (85·0 per cent), *insularia* 30 (4·8 per cent).

In these conditions *carbonaria* is well protected as it rests on the tree-trunks, though not so perfectly as are the typical *betularia* in Dorset; while the latter phase, the pale one, is here very conspicuous on the dark sooty bark (Plate 15). Thus the whole situation was reversed compared with that found in the rural setting of Dean End Wood. The two techniques used there were employed at Rubery also. First, equal numbers of typical *betularia* and of *carbonaria* were released on to the trees, replaced as before when all the specimens of one phenotype had disappeared, while a watch was kept upon them from hides. Here also they were preyed upon by birds: Hedge Sparrows (*Prunella modularis*), Robins (*Erithacus rubecula*), and Redstarts (*Phoenicurus phoenicurus*) were observed searching the tree-trunks and capturing the moths, and other species certainly did so too. Concentrating upon the Redstarts, these birds were seen to eat 43 typical *betularia* but only 15 of the equally numerous *carbonaria*.

Secondly, large numbers of specimens, marked as in the Dorset locality, were recaptured at mercury-vapour light traps, two in this instance, and by assembling to caged virgin females surrounding the area of experimental woodland. The work was carried out in two seasons, with the results shown in the following table, in which they are compared with those obtained in Dorset. The reasons for the higher rate of recapture in 1955 are not known, perhaps the arrangement of the traps was more efficient. It will especially be noticed, however, that, based on percentage differences, the relative recovery of the two forms was remarkably similar to that obtained the previous year; also to the Dorset figures, where the frequencies were in the opposite direction.

TABLE 13

		Typical	*carbonaria*	Total
Dorset	Released	496	473	969
1955	Recaptured	62	30	92
	% of releases recaught	12·5	6·3	
Birmingham	Released	137	447	584
1953	Recaptured	18	123	141
	% of releases recaught	13·1	27·5	
Birmingham	Released	64	154	218
1955	Recaptured	16	82	98
	% of releases recaught	25·0	52·3	

Recovery of the pale and melanic forms of *Biston betularia* liberated in an unpolluted area in Dorset and one polluted by smoke near Birmingham. Approximately twice as many of the appropriately coloured type were recaptured, whichever this might be.

As Kettlewell points out, the results in either locality considered alone might be interpreted in any of four ways.

1. Melanics might be attracted to light and to virgin females to a different extent from the typicals, though evidently the two methods are equally effective.

2. The life-span of the two forms might be different.
3. They might wander or migrate to different extents.
4. Typicals and *carbonaria* may be subjected to differential predation.

Kettlewell provides detailed evidence in each locality to eliminate all but the latter alternative. Here it will suffice to point out that since the frequency of recovery is almost exactly reversed in Birmingham and Dorset, alternatives one to three are automatically eliminated.

These studies prove that birds hunting by sight actively search tree-trunks and capture in large numbers moths resting upon them. Also that those insects which best match their background, whether pale cryptically-coloured forms in rural districts or melanics in industrial ones, have much the better chance of escaping such predation, so that their colour-pattern operates as a powerful agent in natural selection. Its effect is much enhanced by the fact, demonstrated experimentally by Kettlewell (1955*b*), that moths tend to take up positions in which their colouring matches their background. This does not mean that pale specimens will fly to and settle upon pale trees and dark specimens upon dark ones but that, having alighted, they will move about slightly so as to accord better with their immediate surroundings. It seems that this is done by recognizing the contrast between the colour of the circumocular tuft of scales and that of the bark or lichen on which they rest.

Thrushes scan the tree-trunks from the ground and dart up to take resting moths which they have marked down. Hedge Sparrows and Robins make their observations from twigs and bracken, while Yellowhammers hover in front of the tree searching the surface from the air, and other species have other techniques. In all of them, however, the act of capturing a moth is exceedingly quick. The insect is taken in a flash, while the birds are very easily disturbed, so that the process, common as it is, can only be detected by careful observation. This doubtless is the reason why it had passed unnoticed until 1953 but it does not exonerate generations of moth-collectors and bird-watchers from failing to observe it. Indeed so unexpected did Kettlewell's results appear that at first many naturalists did not believe them. When, however, Tinbergen accompanied him and not only photographed birds taking the moths from tree-trunks (Plate 16(1)) but secured a cinematograph film showing the process in detail, it became impossible to deny it.

An attempt made by Harrison (1919–20) to study the predation of resting moths failed, from which he reached the unjustified conclusion that birds are of negligible importance in this connection. The reasons for his mistakes are evident. He made observations of the moth *Polia chi*

(Agrotidae) and its dark form *olivacea* on three types of wall in the neighbourhood of Newcastle, Northumberland. He says that he examined 'up to 300 examples daily and never was there any diminution of numbers in which more *olivacea* vanished than type *chi*. As a matter of fact, we used to consider it a marvellous thing if even a single one had disappeared.' This is not surprising since *olivacea* may not be an industrial melanic at all. The normal pattern is present on an olive grey, instead of a whitish, ground colour and the two forms are almost equally inconspicuous upon the same backgrounds.

Additional effects due to genes for melanism

The selective importance of birds in eliminating those moths which, sitting exposed, do not match their surroundings well has been fully established, as has the part taken by these predators in producing industrial melanism. It is necessary therefore briefly to consider other known effects of the genes responsible for the black phase which has spread so effectively. One of these has already been mentioned: that is to say, the superior hardiness of the successful melanics compared with normal pale forms. This curious fact, first fully proved in *Cleora repandata* (p. 296), but indicated in the breeding results obtained with many of the species (Ford, 1940b), has now been established also in *Biston betularia*. There is, however, some evidence that though *carbonaria* is more viable than the typical form today the reverse was true at the beginning of this century. Kettlewell (1957a) gives the data on four large back-cross families bred between 1900 and 1906 from specimens obtained in London. They are homogeneous and the totals amount to: typicals 255, *carbonaria* 217. Those of his own back-cross broods of industrial origin and reared on unwashed food between 1953 and 1956 produced: typicals 65, *carbonaria* 108. The difference between these two sets of results, tested by means of a 2 × 2 table, is heavily significant ($\chi^2_{(1)}$ = 13·05, P < 0·001). It suggests that an evolutionary change adjusting the gene-complex to interact favourably with *carbonaria* had not yet taken place in the London area where at the beginning of this century that form was still uncommon.

In Britain the larvae of *Biston betularia* are normally to be found from June to October. Kettlewell (1961a) has shown that in certain broods of industrial origin those which feed rapidly and pupate in early September produce a relatively large proportion of the typical form, while an excess of *carbonaria* emerge from those still feeding in early October. He suggests that there has been selection in favour of rapid growth in those producing the less hardy pale form, for this does not survive well when the leaves are polluted by the soot which gradually accumulates on them throughout the summer. Alternatively, or in addition, he also points out that slow feeding may allow *carbonaria* to get rid of toxic materials so that this habit may be favoured in that form. Moreover, he remarks that the closely related Canadian species *B. cognataria* feeds up very quickly: a circumstance that may plausibly be connected with the earlier onset of Canadian than of British winters. Such correlated adaptations may well favour linkage and the formation of a controlling super-gene in these species.

Thus it seems clear that the gene for melanism affects larval habits in *B. betularia*, and it is possible that another aspect of this operates in *Colocasia coryli* (a moth belonging to a very different family, the Cara-drinidae). Melanics of that species began to appear in Hertfordshire, England, about 1926 and there is some evidence that the larvae destined to produce them tend to feed near the tops of trees (*Fagus sylvatica*) while those that give rise to the normal form are more often to be found upon the lower branches.

There is also a decided indication that typical and *carbonaria* females attract typical males differently, in a way that depends upon climatic factors. The pale females receive relatively more visits on cool nights and the black on warm ones (Kettlewell, 1957*b*). In the circumstances, this is more likely to be due to different female behaviour in these two forms than to a distinction between the stimulating scents liberated by them.

It is evident therefore that the genes controlling the typical and melanic phases affect several characters as well as colour. The physiological distinction of unequal viability between them is fully proved, the others less completely so.

Mutation and the occurrence of melanism

The spread of melanics, due to the cryptic and physiological advantages of the black form in industrial areas, was first recorded in *Biston betularia* in Manchester beginning, as already mentioned, in 1848. In most of the species, approximately a hundred, now affected in Britain, the change started in industrial areas of northern England. In a few instances, however, it began elsewhere. Thus the deep chocolate form of *Hemerophila abruptaria* established itself first in London, appearing in 1897; it is now spreading in parts of the Midlands also, probably arising from a separate mutation there. Normally this insect is beautifully adapted for resting upon light-coloured tree-trunks and fences (Ford, 1955*c*, Plate 20). A dark form, *perfumaria*, had replaced the ordinary pale *Alcis rhomboidaria* in the London area by the early part of this century but, though the species is common and widespread in the British Isles, the completely black phase *nigra* was long restricted to the Norwich district and has only recently begun to appear elsewhere.

This fact reminds us that in general moths have at first to await the mutation of the gene responsible for the successful melanics, since these are dominants. Sedentary species are dependent upon that event in each separate locality, but in others of a more wandering habit melanics may reach distant areas by the dispersal of the imagines; they have certainly done so to some extent in *B. betularia*. A few further instances in which the inception of industrial melanism is thus limited by the occurrence of the appropriate mutation are worth considering briefly.

The coal-black form of *Gonodontis bidentata* has almost completely replaced the normal greyish-brown one in the Manchester area but has not yet appeared in Birmingham, where the insect has had to resort to polygenic darkening (p. 293), atlhough the dominant melanic is found only twenty miles away in the colliery area of Cannock Chase in Staffordshire. This species is not one with any considerable powers of flight.

There are at least two black forms of *Phigalia pilosaria*, both industrial melanics. One is, as usual, an autosomal dominant, the other is maternally-inherited. Presumably it is due to a gene in the non-pairing region of the Y chromosome and, since the females of this species are wingless, it must arise by mutation in each new locality.

The distribution of *Ectropis consonaria* has been somewhat reduced by the growth of industrialism in the southern half of England, to which this species is restricted in Britain, while the black form *nigra* has only maintained itself in two widely separated woods. One of these, near Maidstone, Kent, may be ecologically exceptional. Rare localities of this type would provide a source of melanics without waiting for mutation should a manufacturing area engulf them.

The second stronghold of black *E. consonaria* is a wooded cleft in the Cotswolds, rather funnel-shaped and sloping down to open into the Severn plain. Air currents from the south-west bring soot from Bristol, thirty miles or so away, and deposit it on the trees in this curious valley, though the surrounding country is almost clear of pollution.

The *nigra* form of *E. consonaria* has become a highly successful industrial melanic in Germany; but not, in the normally accepted sense, in England where it has never arisen in circumstances appropriate for doing so. Yet should the mutation take place in, for example, the area affected by the smoke-pall of London, the moth might recolonize districts in which it has recently become extinct. This has in fact occurred in two other species, *Procus literosa* and *Apamea charcterea*, which after becoming black returned to the neighbourhood of manufacturing towns where their normal pale forms had long ago disappeared (Kettlewell, 1958*b*).

These facts might suggest a particularly high mutation-rate for some melanics. Yet we have, in fact, no reason to think it excessive. If the population of a species in areas affected by soot amounts to no more than 10^6 imagines per generation, and in many of the relevant instances it would be much larger, the opportunities for one of these mutants to appear in conditions suitable for it to spread would be quite considerable in fifty years: the majority of the moths concerned have one generation annually, though a few have two or else a second partial one.

Industrial melanism outside Britian

Industrial melanism has made notable progress in other countries besides Britain, where it first appeared. It was soon reported in the great manufacturing region of north-west Germany and it now affects a large number of species there. More recently it seems to have become widespread in the principal centres of industry in eastern Europe: in Yugoslavia, Czechoslovakia, and Poland. The phenomenon has attained impressive proportions also in North America. Kettlewell (1957*c*) reports that more than forty species are already involved in and around the large towns along the St Lawrence in Canada. In the U.S.A., Owen (1961) describes the change to black forms, which in many instances have reached 90 per cent or more of the population, round New York City, Detroit, Philadelphia, and elsewhere: over 100 species are already affected in the Pittsburg district (Kettlewell, 1961*a*).

The spread of industrial melanism in these countries follows the same lines as in England, in so far as we have any information on its progress. But everywhere its history is the same: one of lost scientific opportunities. In these remarkable events, evolution has been taking place before our eyes, and among the very organisms most favoured by collectors. Yet their activities have told us deplorably little about the process. Moreover, the specimens they have preserved are of little value since they do not represent random samples, which would throw light upon the population-structure of these moths in the past. On the contrary, the melanics were amassed merely as curios, while contemporary records giving frequencies of the phases, which could easily have been obtained and would have been so informative, are almost lacking. Some records showing the spread of industrial melanism were indeed brought to-gether by Doncaster (1906) for an inquiry into the subject organized by the Evolution Committee of the Royal Society in 1900, but it would be difficult to describe the result as scientific. The co-operation of a large number of entomologists was enlisted but they were so badly briefed, or their results so ill recorded, that the frequencies of the melanic forms are merely given in percentages without the numbers upon which these were calculated, save in two instances. As a result, the information contained in the report is largely useless.

Industrial melanism has hardly been studied experimentally outside Britain and there the interpretation and analysis of it is due almost entirely to Kettlewell and, to a smaller extent, to a few other students of ecological genetics. We must therefore turn again to their labours to illustrate this remarkable instance of evolution in progress.

Melanism in rural areas

Some melanics having established themselves at a high frequency in industrial areas have spread thence far into the apparently normal countryside, where in the species concerned they may comprise 90 per cent or more of the population. This tendency has chiefly been studied in *Biston betularia* and Kettlewell (1958a) is largely respon-sible for analysing the population-structure of that moth in Britain. He obtained records of over 20,000 specimens from 82 localities, sup-plied by collaborators and by himself. His results are illustrated in Map 7 which shows that the black forms have usually spread eastwards from centres of industry; a fact well illustrated by the distribution of the *B. betularia* forms outwards from the Liverpool and Manchester areas, stretching on one hand along the north coast of Wales and on the other across Yorkshire and Lincolnshire.

The Department of Scientific and Industrial Research has shown that smoke travels for considerable distances across Britain blown by the prevalent south-west winds. It pollutes some of the relatively normal countryside near manufacturing areas on the way, not to the extent that desolates the woods and indeed the whole landscape in what is well-named the 'black country', but in a fashion that can easily be traced (Kettlewell, 1957*b*). Thus it kills foliose lichens, which are now become a rarity in eastern England in contrast to the way in which they cluster in masses round the tree-trunks and boughs in the west. Where the

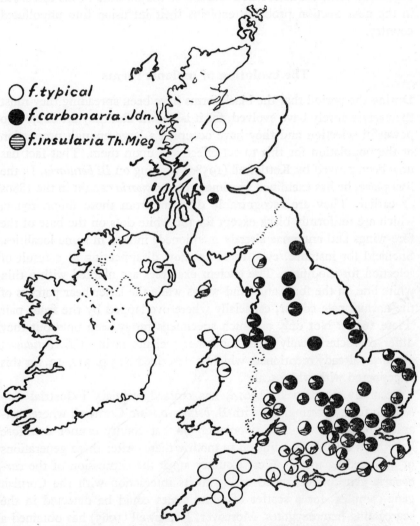

○ *f. typical*

● *f. carbonaria. Jdn.*

◒ *f. insularia Th. Mieg*

MAP 7. *Map showing the sites in Britian where the frequencies of the pale and melanic forms of* Biston betularia *have been ascertained by H. B. D. Kettlewell in his correspondents. These results are based upon an examination of over 20,000 specimens. The melanic forms are found in industrial areas and to the east, but less to the west, of them owing to the drift of pollution by the prevailing westerly winds* (insularia *cannot be detected in the presence of the* carbonaria *gene*).

pervading smoke exists, it can be detected by the dark rain-runnels down the trees. Also by means of leaf washings in the latter part of the summer; the quantity of soot which accumulates is of course greater towards the end of a dry spell and when *Aphis* secretion has made the vegetation sticky. The recent evolution of the melanic forms discussed in the next Section probably explains their intrusion into unpolluted country.

The evolution of melanic forms

During the period that the black forms have been spreading they must themselves surely have evolved, for it is known that they are subject to powerful selection and they have occupied a large enough proportion of the population for this to act effectively upon them. That fact has now been proved by Kettlewell (1958*a*) working on *B. betularia*. In the first place, he has examined specimens of *carbonaria* caught in the 1880s or earlier. They are recognizably different from those found today, which are uniformly black except for the white dots on the base of the fore-wings and antennae already mentioned; indeed in some localities, Sheffield for instance, even these are now disappearing as a result of selection for modifiers. The ancient examples are marked with a thin white line on the fore- and hind-wings while the latter bear patches of the normal pale colour, especially where overlapped by the front pair (Plate 16(2)). Not only are such specimens rarely seen now but they differ characteristically from the other melanic form of *B. betularia*, *insularia*, already mentioned, which is described on pp. 311–13, for this is peppered with white scales.

Secondly, Kettlewell (1961*a*) has crossed ordinary industrial *carbonaria* from Birmingham with *B. betularia* from Cornwall where melanics are unknown. The population in that county cannot therefore have been selectively adjusted to modify them. After three generations of outbreeding to this West Country stock the expression of the *carbonaria* gene had changed owing to its interaction with the Cornish gene-complex, for a scatter of white scales could be detected in the segregating heterozygotes. Moreover, Kettlewell (1965) has obtained a complete breakdown of the *carbonaria* pattern after three generations of out-crossing with the North America, but closely related, *B. cognataria* from regions where its own melanics do not exist (p. 318); the forms are illustrated in his article.

It seems likely that in Britain selection has favoured the spread of another allele of the *carbonaria* gene with a rather more extreme effect than the original one while some further modification of the gene-complex has also taken place. The latter process is of course responsible for the main adjustment of *carbonaria* and for its dominance, as shown by the results of outcrossing to the American species.

The black phase of *B. betularia* has therefore not only spread and become the predominant one in many parts of Britain but has *evolved* in the process, being genetically and physiologically adjusted by selection acting on the gene-complex. This may well be a factor in enabling it to colonize country that is rural though affected by smoke. Thus we cannot be sure that it could have spread beyond industrial areas when it first appeared. Doubtless these statements are in general true of other successful melanics.

The conversion of transient to balanced polymorphism

There is considerable evidence that the process of transient poly-morphism is not carried to completion in industrial melanism. Typical *B. betularia* had already been reduced to about 2 per cent of the popula-tion in the Manchester area by 1895 and they are no rarer today. At this value therefore they have stabilized and the polymorphism has now passed into a balanced state.

Haldane (1956) has analysed the rate of spread of the *carbonaria* gene (C) and the relative fitness of the three genotypes on the basis of the information supplied by Kettlewell. We may assume random mating, and give the relative fitness of CC, Cc, and cc (the latter producing the typical pale form) as $1:1-k:1-K$, and the gene-frequencies as $pC + qc$. Before the Industrial Revolution, C was a rare disadvantageous mutant. If μ were the mutation rate of c to C, Haldane shows that p was about $\dfrac{2\mu(1 - K)}{k - K}$ for which he reasonably suggests a value of say 10^{-5}; perhaps three or four times that of μ. Not less than 1 per cent of the recessives are found in industrial areas, where indeed their frequency in well-established melanic species does not usually exceed 10 per cent; consequently q may be taken as 0·2 approximately. Haldane shows that if $1-K = (1-k)^2$, so that the fitness of the heterozygote is the geometric mean of that of the homozygotes, the time needed for p to increase from 10^{-5} to 0·8 is thirty-seven years if $K = \frac{1}{2}$ as Kettlewell's data suggest. This represents a rate of change which would well accord with his findings.

It is, as already explained, clear that modifying factors have been selected which make heterozygous *carbonaria* as dark as the homozygotes, the appearance of which they do not affect. Haldane remarks 'such genes will only confer an advantage if present in *Cc* animals' and finds that there have been too few of these to explain the selection of such modifiers. He points out, however, that the situation becomes explicable if the heterozygotes are at a physiological advantage. On the assumptions already made, this could lead to the end of transient polymorphism (in the Manchester area) about 1890, producing a balanced polymorphism from then onwards with p about 0·86 or so, giving an initial population structure of 74 *CC*, 24 *Cc*, and 2 *cc* per cent per generation. Haldane finds that Kettlewell's results are consistent with the fitness of *cc* being about half that of the heterozygotes. This would mean that the fitness of *CC* must be about 92 per cent that of *Cc*. In other words, we again have evidence of heterozygous advantage once a balanced polymorphism, that of *carbonaria*, has become established. Indeed in the light of modern breeding results there appears to be no doubt of the advantage of *Cc* over the other two genotypes today. As already indicated, there is some suggestion that such heterosis only evolves after *carbonaria* has become common (see p. 302).

Clarke and Sheppard (1966) have obtained evidence on the selective values of the three genotypes in a survey extending from the high pollution of Liverpool down the less polluted northern part of the Wirral Peninsula and north Wales which, after an industrialized area at the eastern end, becomes rural: the prevailing west wind blows towards the region of industry. They show that this change is reflected by the lichen on the trees, which decreases from heavy infestation in north central Wales to almost complete absence near Liverpool.

That sequence is also followed in the frequency of the *carbonaria* form of *B. betularia*. This increases from about 3·0 per cent at Bangor through 12 per cent at Old Colwyn and 53 per cent at St Asaph to 73·5 per cent at Flint. That is to say, the change in the selective value of *carbonaria* in passing from an area of low to high pollution is a large one.

Clarke and Sheppard (*l.c.*) working at Caldy on the Wirral, have made two estimates of the relative selective values of the three genotypes there prior to 1959. One is based upon captures at light-traps and the assumption that the selective advantage of *carbonaria* remains as it was before the smokeless zone was introduced. It is 0·92, *CC*; 1·0, *Cc*; 0·76, *cc*. The other arises from captures at light-traps together with an interesting experiment in which moths killed and preserved at $-20\,°C$ were put out in life-like positions on tree trunks in a wood at Caldy and exposed for 24 hours to bird predation. The values obtained by these means are given as 0·85, CC; 1·0, Cc; 0·61, cc. These results are to be compared with Haldane's estimates from the more polluted area of industrial Manchester, where the typical form appeared to be subject to a somewhat greater handicap (at about 0·5).

From Clarke and Sheppard's Table 4, it appears that the percentage disadvantage of typical *betularia* (cc) before 1963 lay within the range 41 to 55. After 1963 it was 21 to 23. A smokeless zone was introduced during 1962 and 1963 and it seems that the melanism of *B. betularia* has responded to this by a decrease in the frequency of *carbonaria*. Up to 1963 this seems to have been increasing, since the selective disadvantage before 1959 is given as 24 to 39. We have here important evidence of the accurate selective adjustments of the melanic forms of this insect and of their response to the measures taken to reduce pollution (p. 327).

It would be of great value if it were possible to distinguish the *carbonaria* heterozygotes and homozygotes, by paper chromatography or other means, so that the actual population-structure of *B. betularia* in regions where *carbonaria* has spread or is spreading could be ascertained. Indeed it is very likely that the quantity of the black pigment present in Cc individuals is substantially less than in CC. All that is required of the dominance-modification in this respect is that the two genotypes should look alike: one can considerably dilute a bottle of ink without altering its shade.

The *insularia* forms

A curious problem is posed by the existence of the *insularia* forms of *B. betularia* (Plate 13, Figure 3), 298, which occur in Continental Europe as well as in Britain. They are blacker in some areas than others. Indeed, certain of them may be difficult to distinguish from typical *betularia*, others from *carbonaria*, though a condition intermediate between these two extremes is the most frequent. All are dominant to the normal pale phase and none can be distinguished in the presence of *carbonaria*. In so far as relevant breeding results have been obtained, *insularia* has proved to be due to other alleles at the *carbonaria* locus. This applies to rather pale forms from Oxford and from the Wirral, Cheshire (Clarke and Sheppard, 1964); also to a darker one from Ross-on-Wye

(Lees, 1968). Unfortunately we do not yet know the inter-relationship of the two, nor the genetics of the darkest form of all, that approaching *carbonaria* in appearance.

At the time when Kettlewell was studying the melanism of *B. betularia* both in unpolluted and polluted country (pp. 298–301), he and his colleagues devised a numerical scoring system to indicate the degree of conspicuousness of the forms. The normal pale insects were very obvious on the blackened trees in the Birmingham district and wonderfully well concealed on the light lichen-covered ones in Dorset, *carbonaria* being the reverse. *Insularia* gave an intermediate score in both situations.

The facts in regard to the occurrence of *insularia*, given by Kettlewell (1958a), seem to be these. It often appears before *carbonaria* but does not exceed a phenotype frequency of 40 per cent of the population. It is particularly well adapted to resting on boughs covered with *Pleurococcus* and crustose but not foliose lichens. A high proportion of *carbonaria*, 65 per cent or more, makes it difficult to assess the frequency of *insularia*; but though it has then lost its cryptic advantage, it maintains itself in these circumstances and may even increase. It is, however, significantly heterogeneous in such populations, taking sometimes high and sometimes low values. The greatest frequencies of *insularia* are associated with 10 to 30 per cent of *carbonaria* and the lowest with 40 to 65 per cent of the latter form. When *carbonaria* begins to spread, *insularia*, if present, stabilizes at about 15 per cent of the population no matter whether it were common or rare previously.

This suggests that *insularia* is favoured in the early stages of pollution when *Pleurococcus* and crustose lichens are present on the trees but foliose lichens can survive no longer. Such conditions give it an opportunity not only to spread for a period owing to its semi-melanic colouring but enables it to exploit a decided heterozygous advantage. This favours *insularia* even in individuals in which it is phenotypically obliterated by *carbonaria*. That situation is counterbalanced by a heavy disadvantage of the *II* homozygotes and leads to a balanced polymorphism: one which cannot establish itself in a truly uncontaminated countryside but can do so when the smoke-pollution first begins to have an effect. It is subsequently maintained, purely on physiological grounds, when *carbonaria* has, owing to its cryptic superiority, reached a high frequency in the population.

It is likely also that other environmental agencies in addition to crypsis must be controlling the *insularia*, as the *carbonaria*, frequencies in any area. This is indeed suggested by the fact that both these melanics may exist in very different proportions in two or more localities isolated from one another where the degree of pollution is approximately similar: eastern East Anglia and some parts of the Severn Valley, for example. Moreover, it is obvious enough that *insularia* cannot be treated as a single entity in this matter. The contribution of crypsis to its maintenance must be different in the paler compared with the darker forms.

Industrial melanism in larvae

One apparently little known fact must be mentioned at this point; namely that industrial melanism has affected the larvae as well as the imagines of moths. Apart from recent work by Kettlewell on *Lasiocampa quercus* (pp. 314–15), this subject has hardly been advanced since Harrison reviewed it in 1932. For though black larvae of certain species have become common in some manufacturing areas in the English Midlands, information on their genetics is extremely limited and we are ignorant of any associated effects of the genes controlling them.

Black larvae of the moth *Meganephria oxyacanthae* are abundant in some districts. They and the black imagines of the same species are both simple dominants and the two conditions have spread in areas polluted by soot in the north of England and the northern Midlands. I have illustrated them in colour, together with the corresponding pale cryptic forms (Ford, 1955c, Plate 31).

Melanic larvae of the moth *Lymantria monacha* are also unifactorial and dominant to the ordinary type, which is whitish with a rather complex pattern. Both have spread in and around industrial areas in Continental Europe but not apparently in England, where indeed the species does not seem to occur in such an environment. Black larvae are known in *Selenia bilunaria* and in *Abraxas grossulariata*, but these are rare recessives in both species, as are the black imagines.

In all these instances, the genes controlling larval and imaginal melanism are distinct. Thus the black larvae can give rise to normal moths, while the black moths can be the product of normal larvae. This, as I have pointed out in the past (Ford, 1953b), is to be expected. The larvae and imagines of the Lepidoptera lead such utterly different lives that ecologically they correspond to distinct groups of organisms. It would be an impossible situation if the same set of genes were to control the corresponding characters of both, for these must clearly be capable of varying independently of one another. Indeed the instance of *Lasio-*

313

campa quercus about to be mentioned is not an exception to this statement, for the genes producing melanic imagines and larvae in that species are distinct, though linkage has evolved between them in one locality. The gene producing blackish fore- and hind-wings (*fumosa*) and black larvae in *Arctia caja* (Kettlewell, 1959) is merely a disadvantageous mutant.

Recessive melanism

Attention has already been drawn to the fact that almost all industrial melanics are dominants. Recessive melanics have indeed been recorded in many species, but they are merely rare mutants less viable than the normal pale forms. Yet it would not be surprising if these occasionally chanced to interact favourably with the new type of environment generated by a smoke-laden atmosphere: just as the vestigial mutant of *Drosophila melanogaster*, which is a disadvantageous recessive, proves to be superior to the normal form in conditions of severe drought. Three such exceptional instances have indeed been detected in melanic moths and two of them have been studied in detail by Kettlewell (1959). They occur in *Lasiocampa quercus*, which is a moorland insect in northern England and Scotland but of wider distribution in the south. The males are about 6 cm across the expanded wings and are reddish-brown marked with conspicuous yellow bands. They fly actively by day. The females are even larger, about 8 cm in expanse, and of a pale yellow colour. They rest among heather and fly only at night.

The form *olivacea*, in which the imagines are greenish black, has long been known as a rare recessive. It has, however, lately begun to spread on several of the Yorkshire moors that are heavily polluted with soot, also in manufacturing districts in Lancashire and Cheshire. Thus Kettlewell reports that its frequency on Rombold's Moor, near Ilkley, was 4·7 per cent (23 out of 493 moths) in 1957. Consequently about 35 per cent of the normal specimens must have been heterozygotes at that time, assuming equal viability of the genotypes.

The larvae of *L. quercus* are normally brownish, velvety black between the segments and covered with brown hair; but a recessive black form, which in addition spins black cocoons in place of the normal pale-brown ones, is now spreading in the Yorkshire localities where *olivacea* is becoming established. On Rombold's Moor it amounted to 4 per cent of the population (82 out of a sample of 2,000), in 1956. Though the larval and imaginal melanics are controlled by different genes, they are associated; for about half the black larvae produce *olivacea* imagines. It looks as if linkage is being evolved between the two loci concerned. If so, at the frequency established, about 2 per cent of normal larvae should give rise to *olivacea*, and this proves to be correct.

In other areas where *olivacea* is polymorphic, including north-east Scotland (p. 316), the mutation producing larval melanism has apparently not yet taken place, or has not done so sufficiently often to provide an opportunity for the resulting gene to become established. In these circumstances, melanic imagines are of course always the product of normal larvae.

The imagines are destroyed by birds in great numbers. Thus Black-headed Gulls, *Larus ridibundus*, which, in recent years only, have appeared in large flocks inland, may be seen flying slowly over the moors. In 1956 when *L. quercus* was very abundant, they were constantly dropping into the heather to eat the females. They also capture many of the males upon the wing, fast flying though they are, as do Swifts, *Apus apus*, and Wheatears, *Oenanthe oenanthe*. The larvae are sometimes heavily parasitized but, being thickly covered with irritant hairs, they are free from predation by birds except by Cuckoos which take them in some numbers.

Rombold's Moor is only some fourteen miles from Leeds and Bradford and is heavily polluted by soot. There is good reason to think that the black forms of *L. quercus* are at a considerable cryptic advantage there. To the human eye *olivacea* is much less obvious than the typical insect both as flying males and resting females. Kettlewell (1959) was unable to undertake extensive mark, release experiments of the two imaginal forms owing to a shortage of the melanics. He reports, however, that on one occasion he released 44 normal and 2 *olivacea* females and that one of the latter was the only moth surviving after two hours. This also indicates the heavy elimination taking place when the numbers reach a high level, as they did in 1956. By means of marking, release, and recapture, Kettlewell estimated the total male population in 1957 as 10,000 on the moor where he was working. This represents a great reduction from the swarming mass of larvae the previous summer (in Yorkshire and northwards this species has a two-year life-cycle). In general, *L. quercus* is subject to extreme numerical fluctuations which must greatly accelerate its evolutionary adjustments (Chapter Two).

A form phenotypically similar to *olivacea*, but sometimes distinguishable in the heterozygote, is found with a frequency of up to 70 per cent on moors in the north-east corner of Scotland. It is due to a different gene from that occurring in Yorkshire: matings between the melanic insects, respectively from the two localities, produce an F₁ of typical light specimens only. Probably the Scottish melanism has been long established, for it is adjusted to an area where black peaty soil is widely exposed among stunted heather, giving the moors a very dark appearance, and here the species is subject to heavy predation by Gulls. However, the mutation producing melanic larvae has not yet taken place in this area or has not done so sufficiently often to provide an opportunity for the resulting gene to become established.

The northern *olivacea*, from Caithness, are of a deeper shade than those found in Yorkshire where both they and the typical imagines have been gradually darkening during the last twenty years; evidently owing to the selection of modifiers which operate on both phases of the polymorphism. Thus *olivacea* has not only become relatively common recently but has also *evolved*, in a way to be compared with *carbonaria* in *Biston betularia*. The darkening of the normal forms of *L. quercus* on Rombold's Moor is an example of the same process but has no parallel in *B. betularia*, though it represents a change similar to that undergone by the non-melanic *Gonodontis bidentata* in Birmingham (p. 304).

The third instance of the spread of recessive melanism is provided by the moth *Lycia hirtaria* (Selidosemidae) in its intensely black form *nigra*. This has reached a frequency of 83 per cent in the tree-lined streets and squares of London, where it has been studied by Cadbury. It does not seem to have established itself elsewhere.

Cadbury (personal communication) has been able to cross the *nigra* form of *L. hirtaria* with the normal pale specimens of a moth in a different though related genus, *Nyssia zonaria*; a species in which melanism is unknown. Yet though completely recessive in *L. hirtaria*, the melanism was clearly detectable in the F₁ progeny; a situation at once intelligible when one recognizes that dominance and recessiveness are the product of selection operating on the gene-complex, not of the genes responsible for the characters which display these qualities.

The origin of dominance in melanic moths

The fact that nearly all industrial melanics are complete dominants when they begin to spread poses a considerable problem. Kettlewell (1957c) has suggested that their colouring must have had some advantage in the past, one for which it was selected during a long period of time. Though they are eliminated in woods of lichen-covered deciduous trees, he thought it possible that the black forms had a cryptic value in the pine forests which covered much of Britain during the Pre-Boreal and Boreal Periods, extending from rather before 8000 to about 5000 B.C. He accordingly investigated the Black Wood of Rannoch in Perthshire (Kettlewell, 1957*b* and *c*, 1958*b*), one of the few surviving indigenous pine-forests in this country. Here he found seven species of moths with melanic phases in a condition of balanced polymorphism. In his analysis he concentrated upon one of them, *Cleora repandata*, in which 10 per cent of the imagines are melanic at Rannoch though the region is remote from pollution. The form occurring there is known as *nigricata*. Phenotypically it resembles the dominant *nigra* which has proved highly successful in the manufacturing districts of northern England and the Midlands.

Cleora repandata rests upon pine-trunks and Kettlewell reports that 50 per cent of the specimens may take flight during the day; they travel considerable distances and in the end nearly always alight upon another pine-tree. In that locality they are disturbed either by the sun or by ants, *Formica rufa* and others, which run over the bark in numbers.

The pale form of *C. repandata* is the less conspicuous on the pine-trunks, which carry a certain amount of lichen, and indeed it is exceedingly difficult to detect them. On the other hand, the melanics are much the better concealed on the wing when the insects are preyed upon by birds: Kettlewell and his colleagues witnessed this on three occasions. The black specimens are almost impossible to follow in flight and are invisible at a distance of twenty yards, while in similar conditions the normal form could easily be watched a hundred yards away. Thus the pale and dark phases each have advantages and disadvantages. This may be one of the factors which have maintained them as balanced polymorphisms in unpolluted country long before the onset of industrial conditions, to which, however, the melanics might, in such circumstances, be 'pre-adapted'. Disruptive selection would ensure the evolution of dominance in this situation since there would be two optimum phenotypes but three genotypes (pp. 114–15). Moreover, having regard to the family as a whole (Selidosemidae) to which *C. repandata* belongs, one cannot doubt that the pale phase has been the 'normal' one and that in the recent past the melanics arose at some stage by mutation. Yet if a rare mutant gains an advantage and spreads up to some frequency, it and not the original form will become dominant; for it is the mutant which will for long be present in single dose, so that it must be selected as a heterozygote, not as a homozygote, to produce the optimum effect.

Kettlewell (1965) has indeed obtained evidence that certain 'modifiers' responsible for the production of dominance in these insects during the Boreal Period have persisted in the gene-complex up to the present time, so giving rise to the fully dominant condition which is so striking and curious a feature of industrial melanics on their first appearance. Presumably such modifiers have been preserved because built into the gene-complex on account of co-adapted effects for which they have been selected: they are palaeogenic. Kettlewell's evidence on this point is provided by the following results.

As already indicated (p. 308), he had largely broken down the dominance of *Biston betularia carbonaria* by three generations of black-crossing that form, obtained in industrial areas in Britain, with pale specimens of *Biston cognataria* from a Canadian locality where its melanic phase is unknown. He had expected that it would be possible gradually to build up dominance again by the reverse process, that of crosssing randomly chosen intermediate heterozygotes so produced with typical British *B. betularia*. To his surprise, the results of that experiment did not retrace the steps of the original one: for full dominance was immediately achieved, in the first generation. Evidently, therefore, the necessary modifiers were present in the normal *Biston betularia* which he used. It was in consequence necessary to inquire if they were widespread. The typical pale specimens employed in this cross were all from an industrial area; the Birmingham district. He therefore repeated the work by mating intermediates, which had segregated gradually in his racial cross, with normal pale insects from the south-west of England where *carbonaria* has never been reported. The result proved to be the same: that is to say, full dominance was achieved in a single generation; so indicating that the modifiers necessary to establish it exist in the normal British population. It is true that we have no direct evidence that their efficacy in producing dominance today would have been equally great when interacting with the gene responsible for *carbonaria* in the past (p. 309). However, dominance for a given effect of one allele seems generally operative for that of another at the same locus.

Kettlewell (1961*b* and *c*) has also studied another aspect of non-industrial melanism: one which characterizes certain populations of moths at high latitudes. This phenomenon increases progressively northwards from about 50° to 60° on both sides of the Atlantic. He decided to analyse it in the Shetland Isles. These are a long narrow group running north to south for fifty-four miles. The sixtieth parallel passes through them ten miles from their southern extremity, which is one hundred and seven miles from the Scottish mainland. They comprise many small islands and three larger ones: Unst the most northerly, Yell in the middle, and the 'mainland' of Shetland in the south.

Sixty-two indigenous species of 'macro' moths occur here, 27 of which have given rise to local races, 18 of them melanic. In a few of these the whole population is affected, as in *Diarsia festiva*, while in others, such as *Hepialus humuli*, there is continuous variation from the pale to the dark form. In some species, however, clear-cut pale and melanic phases constitute a balanced polymorphism and it was one of these, *Amathes glareosa* (Agrotidae) and its dark form *edda*, that Kettlewell decided to investigate. This is a locally common moth throughout Britain and parts of Western Europe. The fore-wings of the imagines are normally of a soft slate-grey colour, often tinged with pink in England, especially in the south-west, and in Continental Europe. They bear three conspicuous narrow black marks running back from the costa about one-third of the way to the inner margin, which are certainly disruptive in effect. The thorax and abdomen are of the same grey shade; the hind-wings are white but are hidden when the insect is resting.

The melanic form *edda* has never been found except in Shetland, Fair Isle, and Orkney. It varies from dark brown to black, the principal markings on the fore-wings being outlined in a paler shade. The hind-wings, exposed only in flight, are whitish but the abdomen is dark below as well as above; an unusual feature in melanic moths.

This is generally a moorland insect. The larvae feed on low-growing plants, especially the Bluebell, *Endymion nonscriptum*, and the imagines obtain honey from the flowers of heather and ling, for they are on the wing in August and September, and there is but one generation in the year.

Edda is under unifactorial control and Kettlewell has recently shown that it is a dominant. The heterozygotes comprise the slightly paler individuals and are more variable than the homozygotes. Consequently most of the specimens are black in the north of the Isles where *edda* is much the commonest form, but are dark brown in the south where it is rare. There is an indication, moreover, that *edda* is more fully dominant in Unst than further South (Kettlewell *et al.*, 1969).

The frequency of the melanic and non-melanic phases of *A. glareosa* has been established by Kettlewell, using mercury vapour traps on a large scale. His most northerly samples were obtained at Baltasound on Unst; they comprised 97 per cent of *edda* (10,356:292 typicals). His most southerly ones were from the Dunrossness area, and here *edda* amounted only to 2 per cent of the population (100:4,913). These two places are fifty miles apart in a direct line, and Kettlewell obtained collections from seventeen localities between them. He demonstrated that, in passing southwards from Baltasound along his irregular trap sequence, the frequency of *edda* falls by 40 per cent in forty-five miles, then by 35 per cent in eight miles, and finally by 2 per cent in the remaining twelve miles before reaching Dunrossness. The relatively sudden change of 35 per cent in eight miles includes the Tingwall Valley, one to two miles wide, which consists of fertile agricultural land quite unsuited to this insect. Thus it seems that there is a cline, subject to relatively sudden change at one or more barriers, between situations which predominantly favour the melanic in the north and the pale form in the south (Kettlewell and Berry, 1969).

Kettlewell (1961c) has shown that bird-predation is an important factor in maintaining a high frequency of *edda* in north Shetland, for this gives a selective advantage to the cryptic colouring of that form. He and his colleagues constantly saw Common Gulls, *Larus canus*, feeding by day on the hillsides and found that *A. glareosa* comprised up to 60 per cent by volume of their stomach-content. Taking into account its abundance and large size, this gull must certainly destroy immense numbers of the moth, as must other birds, though to a lesser extent. Yet the insect is not caught when flying, a conclusion suggested indeed by the conspicuous hind-wings which, however, are invisible when the moth is at rest. For Kettlewell impaled equal numbers of melanic and pale specimens on small fish hooks and allowed them to fly attached to a fine nylon strand: neither form was attacked by birds.

Edda is much better concealed than the pale specimens on dark peat, its usual resting place in the north, and when seen with a background of this soil feeding on heather flowers at night. This is true even if the moth be hanging upside down, owing to the dark underside of the abdomen which is thus adaptive.

Kettlewell (1961c) undertook extensive marking and recapture experiments, releasing the two forms together. For this purpose he had to arrange for daily supplies of pale and dark specimens to be exchanged from opposite ends of the islands and transferred along their entire length by omnibus and boat. This was necessary owing to the rarity of one or other phase at Baltasound and Dunrossness respectively.

The results showed first that *edda* flies less than the typical form, or possibly the Unst race as a whole does so, compared with that from Dunrossness for, owing to the high frequency of one or the other type in each locality (p. 320), it was not possible to distinguish between these two alternatives. This is probably an adaptation to escape being blown out to sea by the very high winds of north Shetland.

Secondly, the recapture of marked specimens indicated a $7 \pm 6\cdot5$ per cent differential survival in favour of *edda* in Unst, though it will be seen that this was not significant on the numbers available. No indication of the kind was obtained at the Dunrossness releasing site, where predation by birds was not observed though carefully looked for by the methods which so easily detected it in the north. The frequencies do not appear to be directly connected with latitude, for *edda* is decidedly commoner in a small sample from Orkney (27 per cent, being 15:40), eighty miles or so to the south-west, than at Dunrossness (2 per cent).

Haldane (1948) devised a mathematical model of a cline in frequency of an autosomal 'dominant gene' compared with its allele. He takes the situation in which the range of the species is divided by a barrier into two parts, such that the advantage of the recessive homozygote is $1 + K$ times that of the dominant form on one side of the barrier and $1 - k$ on the other. Kettlewell and Berry (1961) have compared this with the results obtained by studying *A. glareosa* in Shetland.

Such a cline may be due to the fact that different phases are favoured at opposite ends of a continuous distribution. Selection will then tend to adjust their proportions until they are in equilibrium, while random migration will oppose that result. In order to compare the observed cline, from melanic to pale *A. glareosa* in Shetland with Haldane's model, Kettlewell and Berry took his formula

$$\sqrt{k} = 1\cdot27 \, m/d^\star$$

where d = the interquartile range (that between the points where the frequencies of the recessive character are 25 and 75 per cent), m = the average distance over which individuals migrate at random, and $1\cdot27$ = Haldane's constant where both quartiles are on the same side of the barrier.

Considering the cline from the maximum *edda* population towards typical *glareosa*, being that from Baltasound to the northern barrier of

---

* Kettlewell and Berry omit the root sign on their p. 410, but this is a mere printing (or typing) error for they employ it in their calculations.

the Tingwall Valley, Kettlewell and Berry give $d = 48$ miles. In their paper they state (p. 410) that they have as yet no accurate estimate of m, but they consider the situation when three possible distances are attributed to it and calculate k as follows.

$$m = 0\cdot25 \text{ miles} \qquad k = 0\cdot00004$$
$$m = 1 \text{ mile} \qquad k = 0\cdot0007$$
$$m = 5 \text{ miles} \qquad k = 0\cdot0175$$

On the first assumption, and choosing a value for K which puts the frequency for the recessive gene at 90 per cent at the northern barrier of the Tingwall Valley, they compare the observed data northwards to Unst with Haldane's model. The result was a good fit, with $\chi^2_{(13)} = 24\cdot16$ (for which P lies between $0\cdot05$ and $0\cdot02$). In these conditions, that is to say, the total advantage of either phenotype is minute. Kettlewell and Berry, however, stress that the existence of heavy visual selection is not thereby excluded, but suggest that this is counterbalanced to an almost equal extent at some point in the life-history. Larger values of m of course require a greater total selective intensity in order to maintain the cline.

Kettlewell's capture-recapture experiments (1961c) could provide independent estimates of k and K. As already indicated, he calculated that *edda* was at an approximate advantage of $7 \pm 6\cdot5$ per cent at Baltasound but that neither form was detectably favoured at Dunrossness. Thus no significant value for the selective advantages was obtained.

Kettlewell and Berry point out that powerful but balanced selective forces may be operating upon the organism in ways other than that resulting from imaginal colour to maintain a cline such as this, along which one phase largely replaces the other. They mention five of these possibilities. (1) Differences in behaviour between *edda* from Unst and south Shetland. Of this there is some indication. (2) The tendency, described previously (p. 188), for birds to prey to a disproportionate extent upon the commonest form of the polymorphism, so giving an advantage to the rarer one whichever it may be. (3) It is conceivable that, as in *Panaxia dominula* (p. 140), mating may not be at random between the two phases of *A. glareosa*. (4) The existence of heterozygous advantage in the control of this polymorphism, the probability of which is very strong. (5) The effects of a gradual change in conditions from north to south in Shetland.

It will be noticed that this comparison, of the frequency data obtained by field-observations with those expected on the basis of a mathematical model, indicates only the average advantage of the phases along a cline. This may be very slight even if powerful but approximately balanced selective forces are operating upon them. That situation would ensure rapid evolutionary adjustments to environmental or genetic changes. It would therefore differ profoundly in effect from one in which the selective forces themselves were small.

Indeed I am myself far from convinced that the fitting of field observations to theoretically derived models is a useful technique. It is subject to serious defects of several kinds. In the first place, the required parameters are almost invariably affected by such errors that they could accord with different types of models (pp. 194-5). They are seldom likely to be better quantified than in the *edda* cline in Shetland. For the figures are derived from exceptionally skilful and extensive field-work, involving the capture of approximately 20,000 moths and mark-release experiments upon 7,232 specimens. Yet even here the errors involved in the required constants are relatively large.

Furthermore, serious over-simplifications of fact may be necessary in order to describe such a model at all and they occur in this one. To take one of these, it has had to be assumed that the density of the species is effectively equal throughout the area studied (Kettlewell and Berry, *l.c.*, p. 409). This is a most improbable assumption at all times when the distance involved is considerable compared with the range of individual migration. Indeed in this instance it much distorts the truth, since Kettlewell tells me he has evidence that the density of *Amathes glareosa* is probably ten times greater on Unst than elsewhere in Shetland.

Moreover, it seems that such models often require serious over-simplifications of theory also (see p. 140). For in this instance, we are asked to consider a cline in a 'dominant gene' and to reject a model based upon the supposition of no dominance (Kettlewell and Berry, *l.c.*). But it may well be inquired what this means. Dominance is a property of genetic characters, not of genes. Nor are we in this cline concerned with a dominant *character* (that of melanic over pale colouring) but with a gradation in the frequency of the alelles controlling it, as Kettlewell and Berry rightly make clear. It has constantly been stressed in this book (e.g. pp. 16-17, 101, 302-3) that genes have multiple effects, of which some may be dominant, others recessive, and others, perhaps, without dominance. The same will be true of the situation 'involving no dominance'; for this merely means that at least one effect of a given gene, that selected for special study, is without dominance. Yet we are told that this latter situation is of no concern relative to the *edda-glareosa* cline, although this is one in *gene-frequency*.

Dr D. Y. Downham (1970)* has been so good as to inform me that he finds an error in the basic equation of Haldane's paper. A still more important criticism of it is that, starting from the same assumptions, Downham derives a quite different equation from them. Thus no valid conclusions can be drawn from Haldane's model of a cline.

Though then Haldane's model does not contribute to our knowledge of the *edda* cline in Shetland, rather the contrary, there is no doubt that Kettlewell's field-studies have done so. In general, he has shown that the melanism of *A. glareosa* produces a cryptic effect in Unst, where it matches the prevalent blackish soil. This may be an asset on that island owing to the intense predation by Common Gulls to which the species seems subject there. So too may be reduced flight, which diminishes the danger to any moth of winds at gale force so prevalent in that far northern habitat. These considerations outweigh the normal advantages of increased dispersal by more active flight and of disruptive colouring on a diversified background which seems to obtain at Dunrossness. It is reasonable to suppose that conditions resembling those of north Shetland may have existed far to the south in Britain after the last Ice Age.

Thus working in two very different types of habitat, those represented by the Black Wood of Rannoch and by Shetland, both likely to have been widespread in the past, Kettlewell has obtained confirmation for his view that melanism in moths was well established at a former period. We may thus have an explanation for the initial dominance of nearly all industrial melanics. Yet one aspect of that phenomenon still remains mysterious when we consider the extremely powerful selection-pressures that are operating: why, in the long period after the disappearance of the widespread indigenous pine-forests of Britain, the melanics adapted to live in them did not retain their dominant physiological advantages but evolved recessive colouring. One may make the tentative suggestion that some of the steps in (palaeogenic) melanin-production facilitate the development of the type of viability which these forms display. The conclusion that high viability is therefore a necessary outcome of melanin-formation is of course not a logically (or developmentally) valid generalization, while in this instance we know it to be incorrect; there are, as already mentioned numerous recessive melanics in these

* Department of Computational and Statistical Science, The University, Liverpool.

and related species that are markedly inviable. It will be important to determine whether or not initial dominance, partial or complete, is a feature of melanic forms should they arise in a manufacturing area in the tropics or sub-tropics. Evidently certain further adjustments to the conditions imposed by smoke-pollution are needed in those species which have now been affected by them, and of this there is evidence in *Biston betularia*. Considering the problems reviewed in this chapter in all their aspects, it seems that the methods of ecological genetics have, in Kettlewell's hands, provided a comprehensive analysis of one of the striking evolutionary events of our Age.

Melanism in the beetle *Adalia bipunctata*

The ladybird *Adalia bipunctata* (Coleoptera, Coccinellidae) is subject to widespread polymorphism in colour and pattern which is partly, but not entirely, controlled by a series of multiple alleles. Four of these may be considered here. Arranged in their dominance-order (from the 'top dominant' downwards) they are: a black ground-colour with two red spots (one on each elytron), black with four spots, black with six spots, and the 'typical' form: that is to say, with red elytra each bearing one black spot.

Timofeeff-Ressovsky (1940) counted the numbers of the red phase, and of the three others combined, in a locality near Berlin in April and October for five, not consecutive, years. The reds proved to be relatively common after the winter and rare after the summer. Thus his totals for the whole period give the proportion of blacks (based on 2,848 individuals) as 37·4 per cent in April and 58·7 per cent (based on 5,488) in October. The changes were in the same direction, though varying in amount, each season. That is to say, the red type survives hibernation the better, a fact established also by an examination of the beetles which died during that time, while the black had a great advantage in the summer. This conclusion received strong support from the earlier, though less precise, work of Marriner (1926) which, apparently, was unknown to Timofeeff-Ressovsky. Useful confirmation of these field-observations has been obtained by breeding *Adalia bipunctata*, for Marriner found that the black specimens flourished at a temperature high enough to kill the red.

The frequencies of the different phases vary greatly from one region to another; and, judged at a given period in the season, they often retain their characteristic values at a given place. These facts were established by Hawkes (1920, 1927) and additional European data on the subject have now been provided by Lusis (1961), from which he draws two deductions. (i) The black forms are the commoner in places with a maritime or humid climate. This statement is not consistent with the facts given by Hawkes in her 1927 paper which he quotes, while he does not refer to her earlier one. Furthermore, recent work carried out on a large scale by Creed (1966) does not support it. (ii) Lusis concludes that black specimens of *A. bipunctata* are more frequent in large cities, especially those with highly developed industry. The frequency-distribution of this polymorphism has now been studied in eighty-three localities in England and Wales by Creed (*l.c.*). In southern and eastern England, including London, the black forms rarely exceed 2 per cent, except that they may amount to as much as 10 per cent in some populations to the south-west of London. In Birmingham the frequency was as high as 80 per cent in the early 1920s (Hawkes, 1927) though the highest level found by Creed was 69 per cent. In the heavily indus-trialized region round Manchester the black phases may reach over 90 per cent. Similar values are found in Glasgow and Edinburgh; thirty miles to the south-west of Glasgow, at Ayr, the frequency is down to 28 per cent and at Haddington, eighteen miles to the east of Edin-burgh it is 8 per cent (Creed, 1969).

The polymosphism of *Adalia bipunctata* is of great interest since it provides the only known instance of industrial melanism in any animal with warning coloration. This, indeed, advertises powerfully protective qualities in this beetle; for Frazer and Rothschild (1960) have shown that birds and other potential predators find it quite exceptionally distasteful. As already explained, the industrial melanism of the Lepidoptera depends upon a combination of cryptic and physiological advantage; in *A. bipunctata* only the latter is operative. One is reminded of the some-what similar comparison provided by the genes for colour and banding in *Cepaea nemoralis*. For in the ordinary countryside these exert a dual control, the physiological component of which overrides crypsis in the regions where 'area effects' are operative (pp. 196–7).

There is evidence indeed that the genes controlling the ladybird polymorphism influence the adjustment of the organism to its environ-ment in a number of ways. This is clear from the differential survival of the phases in summer and winter which, though apparent in Berlin, can hardly be detected in England. Creed therefore suggests that re-lative heat-absorption may be more important in places with a con-siderable temperature range, both seasonal and diurnal, than in the relatively constant conditions of an Atlantic climate. Furthermore, Lusis (*l.c.*) finds that differential mating is apparent in this polymorphism; such that there is a significant excess of black × black pairings, at least in Russia. Creed points out that several factors seem to be at work here, and it may well be that toxic substances have a differential effect upon the genotypes. He found a correlation between the frequency of melanics in his Scottish samples and the local concentrations of smoke, though none with the sulphur dioxide levels. This is also borne out in Birmingham, where decreases of around 20 per cent in the frequency of melanics have occurred between 1960 and 1969 at the same time as the introduction of smoke-control areas (Creed, 1971), a situation to be compared with that in *Biston betularia*, pp. 310–11; though these rules relating to pollution have resulted in much lower smoke levels, sulphur dioxide has been much less affected by them. A comparable situation is found in both *Drosophila melanogaster* and *D. pseudoobscura*. For in these species Kalmus (1942) showed that the pale mutants (e.g. *yellow*) are more, and the dark mutants (e.g. *black*) less, susceptible to toxic substances such as sprays of petroleum oil or tar emulsion. Creed's analysis, which indicates that different components of pollution have distinct effects, is likely to be of wide application.

CHAPTER FIFTEEN

Isolation and Adaptation

Mortality in small localities

It was pointed out in Chapter Four that populations can be adapted in detail to the special environments of restricted habitats provided that they are isolated, but only to the average conditions of relatively extensive and diversified ones. Those situations were illustrated by the butterfly *Maniola jurtina* on large and small islands in the Isles of Scilly (pp. 58–74). Furthermore, clines may be formed if a large region is subdivided ecologically in such a way that the various parts are substantially greater than the range normally exploited by the individuals of a species (p. 45).

Evidently, then, the formation of isolated groups inhabiting relatively small territories favours the adaptation of a species to local conditions and the evolution of distinct races. On the other hand, it involves certain disadvantages which will often be reflected in increasing mortality. There are several reasons for this.

In the first place, a large diversified terrain will provide adequate shelter in most circumstances while a little one may be seriously exposed to storms from certain quarters. A hazard which threatens certain species, particularly insects, is loss from the periphery of a site, the length of which becomes relatively greater with decreasing area. Moreover, should an insect be in serious risk of straying from its habitat if within, say, thirty feet from the boundary, the whole of a small locality may be within the danger zone which, however, will merely represent a narrow belt round a large one. Moreover, difficulties arising from shortage of food are likely for a variety of reasons to be less acute in a considerable than in a limited stretch of country even with populations of similar proportions.

The study of *Maniola jurtina* on Tean, Isles of Scilly (pp. 64–7) provides an instance, almost the only one, in which the mortality of a wild population has been related to the size of its territory. It was explained in Chapter Four that up to 1951 three colonies of that butterfly inhabited this island, and we are here concerned with estimates made during the period 16 August to 2 September 1946 by the technique of marking, release, and recapture (pp. 134–7). It will be recalled that in order to calculate population-size by this method, the average expectation of life of the individuals must be assessed: a procedure which may in itself provide information of the utmost value.

The full data upon which our analysis was based may be obtained from Dowdeswell, Fisher, and Ford, 1949. The total number of imagines in area 1, which is approximately 5·3 acres, exceeded 3,000, the largest number alive on any one day being 1,400. This includes both sexes which, however, having slightly different times of emergence and possibly different expectations of life, must be treated as distinct populations. In the males, assuming a daily elimination rate of 11 per cent the expectation of life is eight days, and the total days exposure of marks recaptured was 186·4 expected and 200 observed. Reworking with a daily elimination rate of only 7 per cent, giving an expectation of life of 13·3 days, the value is 198·0 expected, to the 200 observed. In the females, an 11 per cent daily elimination gave the total days exposure of the marks recaptured as 196·8 expected compared with an actual value of 213.

In parenthesis, the phrase 'total days exposure of marks recaptured' perhaps needs some explanation. As is well known, the method of estimating population-size referred to here involves the capture, marking, release, and subsequent recapture of specimens. The marks must be made to indicate dates which, at least in the Lepidoptera, can conveniently be done by dots of cellulose paint, varying their colours and positions. We are concerned here with the time in days each mark has survived up to the final recapture of the insect, summating for all marks on all specimens. This information is obtained from the triangular tables in which the data are entered. An example of these, giving captures, releases, and recaptures of *Panaxia dominula*, is shown in Figure 9. Using this as an illustration, it will be found that in this *dominula* instance the observed 'total days exposure of marks recaptured' is 880. The number of insects recaught will equal the number of marks or it may be less, as some specimens may have been recaptured, and marked, several times. The method for calculating the corresponding expectations; that is to say, the length of such exposures assumed on a particular death-rate, is described by Fisher and Ford (1947) and by Ford (1953a).

The population in the large area 3 (20·8 acres) was very much greater than in area 1; exceeding 15,000 imagines, combining the sexes, while the largest number alive on any one day was about 7,000. Using a daily elimination rate of 11 per cent the total days exposure of marks recaptured on the males was estimated as 72·6, while 76 were observed. Unfortunately in this large area it proved impossible to collect a sufficient number of females recaptures to provide even a rough indication of survival, for only 15 marked specimens were obtained.*

The third habitat, area 5, is much smaller, being about 1·8 acres. The total population of imagines there was only about 500 (both sexes) with a maximum number of living insects which falls between 100 and 250. Here the death-rate was considerably heavier. It can conveniently be assessed by the device of using the same elimination-rate as before, 11 per cent, when the observed days exposure of marks released falls decidedly below expectation: 202·0 days calculated for the males, with 181 actually obtained. The figures for the females (based again on 11 per cent elimination) are, respectively, 278·5 compared with an observed value of 207.

TABLE 14

Area	Total imagines (both sexes)	survival of marks (observed ÷ expected) × 100	
		♂	♀
1, 5·3 acres	> 3,000	107·3	108·2
3, 20·8 acres	>15,000	104·7	?
5, 1·8 acres	500	89·6	74·3

Area of habitat, population-size, and survival-rate (at 11 per cent daily elimination) in three colonies of the butterfly *Maniola jurtina* on Tean, Isles of Scilly. Survival is assessed in terms of the number of days exposure of marks released (observation ÷ expectation) × 100.

These results are summarized in Table 14. The elimination of specimens, with which we are concerned, can be compared by dividing total days exposure of marks observed by that expected on a fixed basis (a death-rate of 11 per cent), and multiplying by 100, as in the last two columns. It would be unwise to place too much reliance on them when based on a single season only, but they certainly show that the colony inhabiting the smallest area has the highest death-rate. The fact that this

* These, in fact, suggest a heavy elimination; days survival of marks released = 65·0 expected to 42 observed at 11 per cent daily survival. But a result based upon such small numbers is meaningless.

is confirmed in the two sexes treated separately gives considerable confidence in the result; moreover, the males accord very well in the two
larger colonies.

It is also worth noting that in the course of other work, the elimination-rate and population-size of an isolated colony of *Maniola jurtina*,
that at Top Rock Valley, St Martin's, Isles of Scilly, was studied in
1957 (Dowdeswell, Ford, and McWhirter, 1960, p. 349). The area is
approximately 3 acres.* It seemed that the total population of imagines
(both sexes) could not greatly exceed 1,000 there, with a maximum of
100 specimens flying at one time, the estimated daily death-rate being
16 per cent. This falls between that calculated for the large areas 1 and
3 on Tean (11 per cent) and for the small area 5 there (25 per cent):
the St Martin's site is intermediate between these also, both in area
and in the density of its *M. jurtina* population. Thus, though the Top
Rock Valley results are based upon less extensive data than those obtained on Tean, they are certainly in accord with them.

It should be emphasized that this method of employing the mark,
release, and recapture technique is of the utmost potential importance
in ecological genetics: making use, that is to say, of the estimate of
elimination-rate which must in fact be obtained in calculating population-size. It would evidently allow us to determine whether two polymorphic forms of a species have similar survival-values in nature, and
to compare average lengths of life at different seasons or in different
localities. We may thus be able to assess the effects of selection in the
wild. Little attempt has yet been made to exploit these possibilities.

The evolution of local races and species

We have seen some of the advantages of small isolated populations and
some of their drawbacks. A further hazard to which they are exposed
results from immigration, which may seriously hinder their micro-
evolution. That fact, obvious in itself, can be illustrated by many examples, of which two will suffice here.

Camin and Ehrlich (1958) find that the Water Snake, *Natrix sipedon*,
is almost invariably banded in Ontario but that unbanded forms are
actually in excess on islands in Lake Erie where, moreover, there is
differential elimination during growth. One form does not change into
the other, yet a much higher proportion of juveniles than of adults is

* Not the whole area at Top Rock Valley shown on Map 3, p. 339, of Dowdeswell *et al.* (1960) is occupied by *M. jurtina*.

banded (combining data from the three neighbouring Bass Islands, the difference between the frequency of banding in juveniles and adults is measured by $P < 0.001$).

These islands, which vary in extent from about an acre to fifteen square miles, are wooded and without inland water. The snakes are therefore confined to the coastlines where they feed upon *Necturus* and storm-killed fish, and rest upon flat limestone rocks. In this situation the plain form is very difficult to see while, to the human eye at least, the banded is extremely conspicuous: its predators seem to be gulls, herons, and raptorial birds, which evidently destroy the two phases differentially. It is impossible to explain this difference in pattern from the mainland by means of random drift. An excess of unbanded snakes is found on all the islands, while the populations are far too large for drift phenomena to operate. The species is indeed extremely abundant there: three collectors captured 234 on South Bass in four hours.

The tendency to produce a race of unbanded snakes in these islands, where that form has an evident advantage, is opposed by the flow of migrants from the mainland whence only the banded form can come. Consequently a proportion of the latter type, disadvantageous in the insular habitat, persists there: it amounts to about 60 per cent on Kelly's Island and about 20 per cent on Middle and Pelee Islands.

A second example of the way in which migration prevents the evolution of a local race, this time on an extensive scale, is one to which I have already drawn attention elsewhere (Ford, 1955c, p. 80). The Agrotid Moth *Eurois occulta* is found up to the north coast of France and Belgium but is established in Britain only in the northern half of Scotland and in Norfolk, where the populations differ from one another and from the Continental form. The latter occurs in southern England but is sporadic and rare. Apparently this species, which is powerful on the wing, has not established itself and differentiated there because migrants from across the Channel prevent its isolation and local evolution.

These two instances are illustrative of migration in general. In view, however, of the laboratory experiments of Thoday (p. 93) and the field-work of Bradshaw (pp. 357–9), it is clear that local races can evolve in the face of considerable gene-flow if the selection be powerful.

In circumstances of this type there is, however, at least the possibility of introgressive hybridization (Anderson, 1949). In general, intermediate hybrids tend to be eliminated since they must be less well adapted than either of the parental forms; however, those resembling one or other of the latter to an exceptional degree may persist, especially if the habitat be slightly abnormal. Back-crossings may then result in a closer approach to individuals of the local race, in which a few new genes have been included; the types of variation to which these give rise may occasionally prove useful in adjusting the organism to a changed environment. Instances in which such introgression has been established with fair certainty have mostly been detected in localities recently modified by Man, as in the example of *Iris fulva* and *Iris hexagona* and their hybrids in the Mississippi Delta (Riley, 1938). The phenomenon has certainly been recognized more often in plants than in animals, perhaps because exceptionally favoured plant hybrids can sometimes exist for long periods as clones capable of vegetative reproduction, giving rise to various crosses with the local race from which something useful may eventually accrue.

We have in Chapter Five been faced with a situation which presents an interesting contrast with ordinary hybridization. It was pointed out that a single population of a species can form clear-cut local races with no isolation at all, as in *Maniola jurtina* on the Devon–Cornwall border and on Great Ganilly in the Isles of Scilly. Doubtless such an event has but rarely been recorded because the type of observations capable of distinguishing it have so seldom been made. It seems explicable only in view of the exceptionally high selection-pressures now known to operate in nature: so high indeed that they force us to envisage micro-evolution upon new lines. It is probably far easier for a single population to build up and use alternative adaptations to slightly different local conditions when powerful selection is constantly opposing intermediates than it may be to do so against the gene-complex introduced by a trickle of migrants. The situation is no doubt facilitated in *M. jurtina* because that species does not wander far within an area suited to it, though the individuals are of course constantly crossing the line where one spot-stabilization changes into another. That much pressure is operating to maintain the local forms near the boundary between them is indicated by the 'reverse cline' effect which has occurred there (pp. 83–4, 85–6).

A further relevant instance is provided by the two grasses *Agrostis tenuis* and *Anthoxanthum odoratum* which can gain tolerance to heavy metals and can therefore grow on spoil-tips where the ores have been mined in Wales (McNeilly and Antonovics, 1968; Antonovics, 1968), see also pp. 357–9. In such places there is selection for earlier flowering and increased self-fertilization. The difference, at any rate in flowering-time, between the normal and tolerant plants is greatest on small mines or near the edge of large ones. We have here good evidence for the evolution of isolating mechanisms in order to reduce both gene-flow and the production of ill-adapted types. It will be noticed that this leads to the establishment of a reverse cline on the large mines. Incidentally, it may be observed that selection to alter a well-balanced adaptation such as flowering-time may run a plant into difficulties imposed by other aspects of its ecology. For example, Breedlove and Ehrlich (1968) find that in certain Californian localities heavy damage by the larvae of the Lycaenid butterfly *Glaucopsyche lygdamus* has forced the lupin *Lupinus amplus* into flowering so early that it is adversely affected by frost.

A very different situation to that just exemplified in *Maniola jurtina* exists where two forms of a species evolve in isolation and subsequently extend their ranges and meet. Each will have built up a gene-complex adjusted to produce characters suited to their respective environments. A mixture between two such balanced genetic systems is bound to be an ill-adapted one. In these circumstances the hybrids, even if freely produced, will be at a relative disadvantage and constantly removed by selection, so preserving the races intact although in contact with one another. Recombination will then ensure a high degree of variation where the two forms interbreed. This condition may arise in another way also. It is likely that a species inhabiting very different, though contiguous, environments may not evolve sharp alternative adjustments as in *Maniola*. For it may acquire such considerable and diverse adaptations that these break down to produce an intermediate form, constantly subject to elimination where the two types adjoin even though they have never been isolated. Thus a similar result can arise in distinct ways, both of which may be illustrated by examples.

Cockayne (1912–13) has suggested, no doubt correctly, that the two British races of the moth *Bupalus pinarius*, Selidosemidae, evolved in isolation during the last Ice Age. One of these belongs to the small Scandinavian form in which the pale areas of the male are white and the very distinct female is brownish. This is found throughout Scotland and northern England. The other, inhabiting central and southern England, is of the west European type, similar to that occurring in France and Germany. It is larger, the pale areas of the male are ochreous and the female has an orange tint. The two races meet about the latitude of Cheshire and Shropshire where they give rise to a belt of hybridization with considerable variability. This, however, is relatively narrow and in spite of it the populations remain distinct. It is probable therefore that *B. pinarius* is an inter-glacial relic in the north and a Holocene invader in the south, and that the two forms met at their present interface after the retreat of the ice. Their genetics are unfortunately unknown though the differences between them seem clearly polygenic.

Ingles and Biglione (1952) studied two subspecies of the Gopher *Thomomys bottae*, *pascalis* and *mewa*, where these meet for about ten miles north and east of Clovis, Fresno County, California. *Pascalis* is adapted to live in the San Joaquin valley and *mewa* in the foothills of the Sierra, from which, however, it extends for about three miles out into the plain. The two forms occur together in a zone which is rarely more than half a mile wide. Hybrids are found within but not beyond this narrow region, which follows very closely the sharp demarcation between the wild grasslands and the cultivated and irrigated country. The narrowness of the belt of hybridization indicates that the intermediate forms are eliminated. Moreover, Ingles and Biglione transplanted 14 specimens of *pascalis* and 15 *mewa* between localities well within their respective territories and none was recovered at these places six months later: an observation which at least gives some further ground for supposing that neither form could adapt itself to the environment of the other.

A careful analysis of the conditions where two races meet, so leading to an extremely steep cline, was carried out long ago by Sumner (1926 to 1930) who studied three subspecies of the mouse *Peromyscus polionotus* in Florida. An almost white form, *P.p. leucocephalus*, occurs on Santa Rosa Island, a narrow bank of extremely white sand running for fifty miles parallel with the Florida coast. It bears a little sparse vegetation and at one point approaches to within a quarter of a mile of the mainland where, along the shore, the conditions are quite similar. Here *P. polionotus* is also represented by a markedly pale form, though a less extreme one than *leucocephalus*. This is *albifrons* which, curiously enough, extends its range on to blackish soil, for it spreads forty miles inland to a region where it is replaced by the normal dark *P.p. polionotus*.

The whitish *P.p. leucocephalus* is of uniform appearance over the whole of Santa Rosa Island. The typical dark form is also very constant throughout its extensive range, though it becomes slightly lighter where it approaches the coast most nearly. *Albifrons*, on the other hand, shows a cline in increased pigmentation.

Sumner found that the differences in colour must be determined on a fairly simple genetic basis. Laboratory stocks of the three forms failed to converge after several generations in the same environment. Crosses

335

between them lead to segregation in intensity of pigment which must be controlled by very few genes since the original colours were recovered in quite small F_2 families.

There can be no doubt that *leucocephalus* is adapted to living on white sand, on which the typical *P.p. polionotus* is very conspicuous; as are both of the light forms upon dark soil. It seems clear that the adaptation of this species to the conditions of Santa Rosa Island has been perfected by its isolation there. On the other hand, the accurate adjustment of *albifrons* to a similar coastal habitat has been prevented by its continuity with the dark inland race. It is noteworthy that the cline in its pigmentation is related to the gradient in soil-colour, while other variables, affecting size and ear-length, are not.

The fact that *albifrons* spreads for a considerable distance inland must mean that having first differentiated on the coastal sand it subsequently extended its range to ground which it does not match; a consideration which suggests that its evolution has included physiological adjustments which are themselves of advantage to the mouse until balanced by the unfavourable background colour of the environment. Information collected forty miles from the coast, where *P.p. albifrons* and *P.p. polionotus* meet, shows that the zone of inter-gradation between the two forms is only three miles wide: a situation in accord with the view that selection in that narrow belt is eliminating intermediates between two balanced gene-complexes.

The last three examples provide instances in which genetic isolating mechanisms allow the maintenance, and no doubt the evolution, of distinct races in continuity with one another. Sir Julian Huxley has drawn my attention to a further interesting aspect of that situation; one in which ecologically adaptive devices and those which prevent interbreeding are accentuated where closely related species overlap. This is due to selection against the necessarily less well-adapted hybrids in favour of those forms in which the adaptive characters are intensified in different directions, so reducing competition. A good example is provided among birds by two species of Nuthatch: the western *Sitta neumeyer*, from Dalmatia to Iran and an eastern one, *S. tephronota*, found from northern Iraq to the extreme west of Pakistan. The two overlap widely south of the Caspian and, though normally they are extremely similar, they diverge markedly there in wing-length and facial stripe, but especially in beak-size. This averages 26 mm with extremes at 23 and 28 mm in *S. neumeyer*. In *S. tephronota* the extremes reach about the same values but the mean is slightly less: about 25·5 mm. In the region of over-lap beak-length falls between 22 to 25 mm with a mean of 23·5 mm in *S. neumeyer* and between 27 and 31 mm, with a mean of 29 mm in *S. tephronota* (Vaurie, 1951). There is a parallel here with the 'reverse cline' in the butterfly *Maniola jurtina* (pp. 83, 85–6) and in the snail *Partula taeniata* (pp. 202–3) in which selection eliminates individuals that are most intermediate between two types of adjustment.

A further instance of the same phenomenon may be mentioned to illustrate its operation on the subspecific level. This is provided by the Three-spined Stickleback, *Gasterosteus aculaeatus*, studied by Hagen (1967) in the Little Campbell River, British Columbia, which is $16\frac{1}{2}$ miles long. The species exists in a fresh-water form, *leiurus*, and a marine one, *trachurus*, which nevertheless enters the lower reaches of streams in order to reproduce in fresh water. Between its breeding ground, $\frac{3}{4}$ mile in length near the estuary, and that of *leiurus* which is $4\frac{1}{2}$ miles higher up, the fish is not absent but is numbers are much reduced. Here mating occurs between the two forms though they possess easily distinguishable characteristics. Laboratory work has shown that the differences between them are largely polygenic and that crosses in all combinations are successful. In studying them, no behavioural or genetic isolating mechanisms could be detected.

However, the *leiurus* and *trachurus* forms possess many adaptations to the markedly different environments they frequent, and in nature the distance of $4\frac{1}{2}$ miles between their main habitats provides a barrier sufficient to maintain their clear-cut characteristics. Moreover, a reverse cline, in which the distinctive features of each form become accentuated, is apparent on approaching from either direction the narrow zone of their natural hybridization, which extends for one mile above the reach where *trachurus* occurs. That situation provides good evidence for strong selection against the fish of an intermediate type.

In addition, a genetic isolating mechanism has been detected in this species. McPhail (1969) shows that though the male nuptial colour is normally bright red, there are populations in which it is deep black. The two types are usually allopatric, but may be sympatric. Females from populations of the red type rarely accept a black male when they have a choice. Others from allopatric populations with black males also show a preference for the red, but not in a community immediately upstream from a black one, where mating is at random. This colour discrimination is reinforced by a degree of partial infertility of the hybrid between the two types. The agency favouring the black males in certain communities is provided by a carnivorous fish *Novumbra hubbsi* which is differentially attracted to the red form. Moreover, young of the black type have an innate behaviour which makes them the less vulnerable to this preda-ion. Thus an intricate balance of selective forces here operates to favour an isolating mechanism.

A special aspect of isolation is provided by species at the edge of their range; and an example of this considered from another point of view has already been mentioned in the moth *Euplagia quadripunctaria* (pp. 31–3). In the first place, we have to recognize that any organism is likely to meet extreme difficulties at the frontiers of its distribution, unless indeed these be imposed by a simple geographical barrier, such as a coastline. An obvious way to solve these problems is by accurate adaptation to one particular type of environment; a device which often enables a plant or animal to survive in a region where its normal range of habitats is becoming too rigourous for it to endure.

Indeed we frequently find species at the edge of their range restricted to conditions which do not normally confine them. Thus the moth *Malacosoma castrensis* reaches its extreme north-western limit in England, where it is a Holocene colonist. Here it is found only in salt-marshes: those of the Thames estuary and along the coast of Essex and Suffolk, although it eats common low-growing plants of many kinds. But this species is widespread on heaths, in woods, and in other places in Continental Europe, where it has no special connection with the shore. Dr H. B. D. Kettlewell has pointed out to me a remarkable adaptation of the British race to its quite special type of habitat. That is to say, it has developed an exceptional form of dispersal by means of salt water, to which the eggs have become resistant. These are laid in batches on floating debris. At spring tides this is washed round the salterns and may eventually be left above the average tide-mark where, on hatching, the young larvae may establish themselves. Further instances of a corresponding kind are provided by two butterflies occurring only in a few localities on the south coast of England: *Thymelicus acteon*, which never flies more than a mile or two inland, and *Melitaea cinxia* which is not found further than a quarter of a mile from the sea. Yet neither species is particularly maritime within its normal European range. There are numerous similar instances in many other groups of animals and plants. Dr E. F. Warburg has kindly mentioned to me an excellent botanical example. The Box, *Buxus sempervirens*, reaches its extreme north-western limit in the south of England. Both here and in other places at the edge of its range in northern Europe it can only grow on chalk downs; whereas in southern Europe, where it is widespread, it is indifferent to the pH of the soil.

It is obvious enough that animal species which can exist only in specialized environments should take precautions to prevent themselves from scattering to unsuitable surrounding country. A number of devices may be employed to this end, including selection for habit. Thus the Swallow-tail butterfly, *Papilio machaon*, tolerates extremely diverse habitats in continental Europe: from sea-level to 5,000–6,000 feet, while it is to be found in many types of environment. Moreover, it wanders widely and is often to be seen ranging over the ordinary countryside. In Britain it is limited to Cambridgeshire (where since the Second World War it has become extinct in its one locality at Wicken) and to a number of the Norfolk Broads. Indeed it is wholly restricted to the Fens, feeding only upon *Peucedanum palustre* instead of numerous Umbellifers, including cultivated carrot, as it usually does. Moreover, powerful as it is upon the wing, it never leaves such localities and is hardly ever seen outside their actual confines: a complete difference in habit from the Continental form, one which must be of great importance to a race which has specialized itself ecologically in this way.

A further instance in which a species proves to be sharply restricted to a number of colonies is provided by the butterfly *Euphydryas (Melitaea) editha* (Nymphalinae), which has long been studied by P. R. Ehrlich and his associates. In the San Francisco Bay region, the species occurs only in scattered populations always associated with outcrops of serpentine rocks. To these the food plant, *Plantago erecta*, is by no means confined: it is indeed abundant and widespread. Yet Johnson *et al.* (1968) have shown that the curious localization of the insect is not the result of a limitation of the larvae to *P. erecta* especially adjusted to serpentine soils. Moreover, since no physical barriers prevent interchange between the populations, Ehrlich (1961) concludes that minor adaptations to each habitat must play an important part in preventing the scatter of individuals.

Ehrlich and Mason (1966) have demonstrated short-term evolutionary changes in the colour-pattern of *E. editha*, and they point out that these seem due to powerful selection fluctuating in its direction. In one of the colonies the means for three out of the four characters studied continued to behave in a uniform manner during the period 1959 to 1966, while a fourth diverged sharply from the rest in 1955–6. The results were consistent with the view that selection was operating on several polygenic complexes. As already indicated (p. 13), it may be held as a fundamental proposition that organisms tend to become more variable as their numbers increase: a conclusion first tested in a colony of *Melitaea aurinia* (pp. 14–18). To that species *E. editha* is closely related; it is likely to prove very suitable for the study of fluctuation in numbers, as it has done for a general investigation of population dynamics.

Carson (1955) has drawn attention to another important aspect of marginal populations. These may be exploratory, in the sense that only more generalized forms may be able to invade new territories and establish themselves there. Consequently types highly adapted to varied conditions in their main area of distribution will not be represented peripherally. Thus Carson finds that *Drosophila robusta*, and other species of the genus, are less polymorphic at the edge of their range than centrally.

In many situations those restrictions which limit a species to special conditions at the edge of its range tend to facilitate its evolution. They do so because they promote isolation, as in several of the instances just discussed. This, however, is often achieved in plants by autopolyploidy, which prevents the peripheral and specially adapted populations from crossing with, and being swamped by, the main body of the species. That is one of the reasons why selection favours polyploids in marginal communities, as it does, for instance, in *Primula*; another, pointed out by Darlington (1956a, pp. 62–72), is that chromosome change facilitates colonization. As he says,

> Different requirements of soil, and of climate, and of relationships with other organisms push diverging stocks or races, if they are genetically isolated from one another, into different regions. These in turn become the centres of new diversity from which they spread over characteristic distributions in space.

Indeed special polyploid races include a greater range of types than those of lower ploidy from which they were derived, because they are genetically isolated from them. It is to be noticed also that species at the edge of their range may actually extend their distribution to a selection of habitats wider than normal if a closely related and competing species be absent: the Arctic Hare in Ireland provides an instance of this kind.

As pointed out by Darlington (1956a, pp. 62–72), polyploidy may allow an invading plant to make fertile crosses with a native one from which great, and occasionally advantageous, variation may arise. The most famous example of the kind is provided by the American *Spartina alterniflora* (2n = 70) which was introduced into southern England about the middle of last century. It then crossed with the native *S. stricta* (2n = 56) to produce a male-sterile hybrid (2n = 63), known as *S. townsendii*, which, in the main, reproduces vegetatively and still survives in some places. Chromosome doubling converted this into a new and fertile, though as yet unnamed, species (2n = 126) completely sterile with both its parental types to which also it has proved ecologically superior. As is well known, it has been, and still is, spreading so as to become a pest through choking shallow estuaries such as Poole Harbour; also stretches of Southampton Water, where it was first recorded in 1878.

Darlington (*l.c.*) cites a further and revealing instance of this kind from the studies of Skalinska (1950). *Valeriana officinalis* (n = 7) exists as a diploid across the great plain of Europe and into Asia. Becoming a polyploid at the western edge of its range, only tetraploids have reached England where they are adapted to dry calcareous soils. Here they have produced a number of octoploids which, however, have the same chromosome-number as another British species, *V. sambucifolia*. They have therefore been able to interbreed with it. The resulting hybrid types are extremely variable and have even been able to invade wet and acid soils. As their fertility is reduced owing to polyploidy, selection has, as usual in these circumstances, favoured plants which can propagate easily by vegetative reproduction, which is indeed a notable feature of this race.

The remarkable forms found in animals at the edge of their range can often be crossed successfully with one another and with the main body of the species because, generally, they are all diploids. Unfortunately, however, they have very rarely been investigated genetically.

We have little information on the earlier stages of speciation or on the evolution of geological races. However, the studies of protein variation in *Drosophila* by Hubby and Throckmorton (1968) are relevant to this matter. Using electrophoresis, they examined 18 loci in 27 species, so chosen as to fall into nine groups. Each of these contained two that are siblings and one that is closely related to that pair but morphologically distinct from it. It was found that the sets of siblings differ from one another at an average of 50 per cent of their loci, while each member of the pair differs from the relevant morphologically distinct one at about 82 per cent of loci. That result may be considered in relation to the work of Selander *et al.* (1969), who also used electrophoresis to examine cryptic diversity in the mice *Mus musculus musculus* and *M. m. domesticus* in the same localities in Jutland. They showed that these two subspecies differ at 32 per cent of the 41 loci at which they were compared. Selander *et al.* are surely correct, therefore, when they suggest that identical alleles acquire differing selective values and, in consequence, differing equilibria in closely related species and in subspecies. Also that it looks as if speciation involves major reorganizations of the gene-complex.

Dobzhansy and Pavlovsky (1967) reported the development of hybrid male sterility in a laboratory stock 'New Llanos' of the superspecies *Drosophila paulistorum*. This now produces sterile male, though fertile female, hybrids with the Orinocan race or incipient species, with which a few years earlier the F_1 was fertile in both sexes. The situation shows similarity with hybrid sterility induced by infection in other strains of *D. paulistorum* (Williamson and Ehrman, 1967). It seems that the New Llanos individuals can induce 'infectious' sterility in various races of the species, suggesting that they may differ from their ancestors in carrying some symbiont which while allowing fertility within the Llanos gene-complex causes sterility with other types. As Dobzhansky and Pavlovsky remark, we have here a situation presenting certain features in common with a phenomenon suggested in the butterfly *Maniola jurtina* by McWhirter and Scali (1966) in which the bacterial flora of the larval gut is strikingly distinct in populations in the Isles of Scilly, compared with those in England and the Channel Islands. They found that the larvae are strongly selective as to intestinal bacteria. The grass which forms their food has, as expected, a rich micro-flora, but representatives of only three Orders of bacteria have been found in the larvae and these are markedly distinct in the Scyllonian and English races. Thus Pseudomonadales are found only in the larvae in England and the Channel Islands and the Actinomycetales only in those from Scilly, though both Orders are abundant on the grass in all three areas. Moreover, there is a sharp distinction between the bacteria of the larval gut from two localities, well isolated from one another on the island of St Mary's.

Work on the inception of speciation needs to be further developed by studies on the comparative genetics chiefly of local races. For instance, a number of Lepidoptera, as of other groups, inhabit Ireland and southern England but are unknown elsewhere in the British Isles. Examples of the kind are provided by *Malacosoma neustria* and *Ectropis consonaria*. An investigation of their comparative genetics is much needed, for the Irish and English populations must be the descendants of inter-glacial and Holocene colonists respectively.

The effects of isolation in producing local races have often been studied in detail. Here it is possible only to add a few examples which illustrate somewhat distinct types of evolution. (1) That in which one species can arise from another, though in the centre of its range, by chromosome changes other than polyploidy. (2) The situation in which a limited number of major-genes produce a local form restricted to a particular area of distribution and subsequently adjusted by selection. (3) What is essentially the same condition, but arising repeatedly where the right type of environment occurs. (4) Local races differentiated from the rest of the population by a single major-gene.

(1) Lewis and Raven (1958) find that *Clarkia franciscana* (Onagraceae), a distinct species, is known only in one isolated population on a single serpentine hillside at San Francisco, just south of the Golden Gate. Here it exists as an enclave within the distribution of *C. rubicunda* which it resembles morphologically. It is also closely related to *C. amoena*, which replaces *C. rubicunda* farther north. All three species are diploids ($n = 7$). *C. amoena* is chromosomally very variable but the others are not. The chromosome distinctions of *C. franciscana* are greater than those between other *Clarkia* species and of an entirely different kind. Such extreme cytological disparity, yet indicated by trifling morphological characters, when taken in conjunction with the geographical features of the colony, suggest that *C. franciscana* has arisen from *C. rubicunda* at its present site: a conclusion supported by its adaptation to the serpentine. It is likely that it has done so owing to the occurrence of a genotype which has produced extreme chromosome breakage. We have here, in microcosm, a type of situation which, though doubtless an exceptional one, must repeatedly have promoted the sudden origin of plants through genetic isolation; a condition to be adjusted later by selection.

Wahrman *et al.* (1969) have discovered a situation in which heteroploidy has produced local races reaching the point of speciation. These have arisen in *Spalax ehrenbergi*, a Mole Rat inhabiting Palestine. Four principal forms have been detected, characterized by distinct chromosome numbers in which $2n = 52, 54, 58$ and 60, respectively. They seem to be the result of whole-arm changes and pericentric inversions. Though occupying distinct geographical regions, these are contiguous, yet very few hybrids have been discovered.

The races are related to climate. The $2n = 52$ and 54 types occur in

343

the north; that is to say, the Mount Hermon region. The form in which 2n = 58 occupies the area west of the Sea of Galilee and thence west-wards and south-westwards to the Mediterranean, while the 2n=60 group is found from Samaria to Negiv. These districts roughly coincide with areas of increasing aridity, so that these chromosome races cer-tainly seem adaptive, though they do not affect the appearance of the animals. As Wahrman *et al*. point out, *Spalax* has but poor powers of dispersal and this has no doubt facilitated the evolution of its chromo-some races, which are evidently producing genetic isolation.

(2) Kettlewell (1942) has studied the geographical races of *Panaxia dominula* (pp. 128–46) in Italy, which he describes in an article dealing with this moth throughout its range. Its status in that country is rather complex but, as he points out, it can be analysed in the following terms. In the central and southern region, the species belongs to the race *per-sona*, which differs very strikingly from normal *dominula* found in wes-tern Europe in general, including the south of England. It does so principally in two ways. First, the red of the hind-wings is replaced by yellow. This distinction is controlled by a single factor-pair, the hetero-zygotes being intermediate; unlike the yellow mutant form *lutea*, which appears as a rarity from time to time in normal *dominula*, for this is completely recessive. Secondly, the body is always steely blue, instead of red (or yellow) with a central dark stripe. There is a reduction in the size of all the pale fore-wing spots and an extension, sometimes very great, of the amount of black on the hind-wings. That is to say, the whole insect is much darkened. This appears to be due to the operation of three alleles. What Kettlewell has described as the 'primary melanistic factor' converts *dominula* to the somewhat darkened form *P.d. persona italica*.* This is thought to be a unifactorial distinction. Goldschmidt (1924) has shown that two pairs of alleles are responsible for modifying it to the dark *persona persona* and the extremely dark *P.d. persona nigradonna*. These darkening factors affect also the shape of the wings, which are more pointed than in *dominula*.

In Italy, isolated, as in the main it is, by the Alps, there has then developed a race of *Panaxia dominula*, *persona*, in which the pale areas are always reduced, but to a variable extent due to the segregation of two pairs of alleles acting as modifiers. In addition, the red pigment is replaced by yellow and several ancillary changes have occurred. This isolation is, however, not complete; as Kettlewell points out, races have developed in which one or other of the two main characteristics

* It will be appreciated that quadrinomials are used in this example.

of *persona* exist alone. Thus normal *dominula* occurs in Piedmont, maintained no doubt by invaders between the Alps and sea. There, however, the moth has acquired the gene for yellow pigment from the population to the south. In north-eastern Italy, on the other hand, the primary melanistic form of *persona* is found, but with red pigment in addition; it is dissociated, that it to say, from the gene for yellow coloration characteristic of the species in peninsular Italy. *Persona*, pushing north into central Lombardy, interbreeds with these north-western and north-eastern types to produce a mixture with all combinations of darkening, together with red and yellow body and hind-wings. Partial geographical isolation, with selection for a few major-genes, has thus achieved considerable changes in *P. dominula* over an extensive range. This has necessitated also adjustments in the gene-complex which impede full fertility between the *dominula* and *persona* races; for their F_1 hybrids are subject to considerable mortality. Here we have an instance in which the subdivison of a population is known to have promoted one of the first decisive steps towards speciation. This may be manifested, when races are crossed, by reduced viability, by actual hybrid sterility, or by the onset of intersexuality; as shown by the well-known work of Goldschmidt on *Lymantria dispar*. A good example of this last phenomenon was obtained by Kettlewell (personal communication) in his study of two populations of the large and powerful moth *Lasiocampa trifolii*. Both are from the south of England but they are isolated from one another ecologically and by distance (about 300 miles). In that from Cornwall, the imagines are dark reddish-brown; in the other, from Dungeness, Kent, they are of a pale putty colour (Ford, 1967, Plate 10). Though this distinction appears to be unifactorial, the dark shade being dominant, the insects have already taken a major step towards genetic isolation in these two areas. For when crosses were obtained between them the female progeny, being heterogametic in the Lepidoptera, proved to be intersexual.

In such circumstances, we may indeed expect the frequent evolution of genetic isolating mechanisms (pp. 331–8); as in the polyploidy just discussed, or in the homostyle form of plants (pp. 208–11). In insects, this may be achieved by a change in the scent responsible for bringing the sexes together. An example of that situation seems to be provided by the *Panaxia* races just discussed, for Standfuss (1896) reports that their assembling scents have also diverged. A further instance of the same type has been mentioned to me personally by Professor Esko Suomalainen. He tells me that female *Lymantria monacha* attract males strongly in Germany and in Finland, but that Finnish males do not 'assemble' to German females. Incidentally, it is a scent difference which prevents pairing between *Drosophila pseudoobscura* and *D. similis* in natural conditions.

(3) Many dreary discussions have taken place on the taxonomic status of *Pieris napi* and *bryoniae* to decide whether these two butterflies represent distinct species, or forms of the same: as if we were here confronted with clear-cut alternatives to which a decisive answer should at least always be possible. To the biologist it is, of course, a platitude to say that, considering the effects of evolution, intermediate stages of speciation are often to be found. Such a question therefore tends to be meaningless when applied to populations adapted to live in distinct local conditions and it appears to be so in this instance.

It seems to have been established that the chromosome numbers of these two Pierids differ: n = 25 in *napi* and 26 in *bryoniae* (Bowden, 1966a). *P. napi* has an immense distribution, throughout the Palaearctic and Nearctic regions; moreover, it is found from sea level up to a high altitude. *Bryoniae* on the other hand, is restricted to various discontinuous Alpine and Boreal habitats: there it replaces *napi* in parts of the Alps and Carpathians, in Scandinavia, and in some localities in central Asia and Canada. These two 'species' (Bowden, 1966c) are separated by a number of features which tend to maintain their reproductive isolation: habitat, differences in time of flight, low viability of the hybrids, and mating behaviour (Petersen, 1963). The males closely resemble one another though they can just be separated but the females are very distinct. *Bryoniae* differs from *napi* principally owing to the action of several major genes (Lorkovic, 1962). (1) *B*, which is autosomal and nearly dominant in effect, spreads melanin along wing veins. (2) *Y* gives rise to the brownish-yellow ground colour of the *bryoniae* females. This is semi-dominant.

In addition, many populations of *bryoniae* are polymorphic for a dominant whitish colour on the undersides of the hind-wings and tips of the fore-wings, due to a gene *W*. It is allelic with those controlling the two recessive yellow forms of *P. napi* included within the name *sulphurea* (Bowden, 1966b). It has been said that the polymorphism of the insects with white undersides (*subtalba*) is partly maintained by lethality of the homozygotes. Bowden (1967) has produced some evidence for this but has also obtained homozygous whites.

The genes *Y* and *W*, and probably *B*, occur as rare mutants in the normal *P. napi* population (Bowden, 1963). We have here some indication of the means by which *bryoniae* has established itself with a widely discontinuous distribution. For the genes responsible for it do seem universally available in these Pierids when needed.

The *bryoniae* complex of genes and characters evidently fit the butterfly for life at high altitudes and latitudes. Lorkovic points out that it does so in a number of ways. *Bryoniae* has an ecological preference for lower temperatures and is single-brooded at considerable altitudes, though it may have two or three broods annually in southern-facing valleys and there it often hybridizes with *napi*. The latter always has two to four generations in the year; its larvae feed upon a wide range of plants while those of *bryoniae* are restricted to *Biscutella laevigata* at the greater heights, with *Arabis* and *Thlaspi* lower down.

These Pierids compare in an interesting way with the Lycaenid butterflies *Aricia agestis* which has two broods and *A. artaxerxes* which has one. They were not distinguished until 1935 and there has been much discussion of their taxonomic status. The northern limit of *A. agestis* is reached in southern England and northern Denmark while *A. artaxerxes* inhabits northern Britain, Norway, Sweden, and Finland (Frydenberg and Høegh-Guldberg, 1966). Each includes several more or less distinct races the characteristics of which must not be confused with those of the species as a whole.

Thus *artaxerxes artaxerxes* from eastern Scotland is easily separated from *artaxerxes salmacis*, which it meets in north-eastern England, by the presence of a white spot in the centre of the fore-wings and the absence of a black dot within the white spots on the underside. These features are controlled by a single major gene recessive in effect (Høegh-Guldberg, 1968).

A number of characters help to separate *A. artaxerxes* and *A. agestis*, some of which are unifactorial. One of them is provided by the heavily marked larvae of the latter, another by its large orange lunules on the upper side of the imaginal wings. Yet these and the other distinguishing traits overlap (Høegh-Guldberg, 1968; Høegh-Guldberg and Jarvis, 1969). Indeed the double and single brooded habit is held to be the only reliable guide for separating these butterflies. It should be noticed that this difference is by no means universally a specific one in the Lepidoptera (it is not so in the butterfly *Maniola jurtina* in southern Italy compared with its single-brooded condition further north). Yet in *Aricia* it seems to have been effectual in building up some degree of infertility between the two groups.

The distinction between the single-brooded condition of *A. artaxerxes* and the double-brooded one of *A. agestis* seems to depend upon combined environmental and genetic influences, the latter apparently multifactorial. Their rather complex interaction has not been fully analysed as this insect is difficult to breed in varying environments which depart from the optimum. If *A. artaxerxes* is reared in southern England it remains single-brooded. The multiple-brooded character of *A. agestis* apparently depends on a combination of temperature and hours of daylight operating on the gene-complex of that form. It will produce a third brood if exposed to extended artificial lighting in the laboratory (Jarvis, 1963).

In experimental crosses, there was no differential mortality between the sexes when *artaxerxes* was the female parent, but all the male offspring have died before emerging from the pupae when the female parent was *agestis*. Evidence on the cytology of *Aricia* seems to be lacking, but Remington (1954) questions if the females are heterogametic in the Lycaenidae. A number of instances, indeed, are known in hybridizing races and species in which the direction of the cross affects viability or pairing. Thus it is easy to mate male *Smerinthus ocellatus* and female *Lathoe populi* (Lepidoptera; Sphingidae) and produce an F_1 brood, but the reverse cross has very rarely succeeded.

It is certainly possible, as Høegh-Guldberg says, that a long warm period may have favoured the evolution of a bivoltine from a univoltine species, and that both *agestis* and *artaxerxes* have been able to spread northwards in the wake of the retreating last glaciation; the univoltine *artaxerxes* being the better adjusted to higher latitudes, where it has undergone further divergent evolution.

(4) The moth *Cycnia mendica* (Arctiidae) is sexually dimorphic throughout the greater part of its range, the males being sooty-brown and the females white; both have small black dots scattered over the wings. Throughout the whole of Ireland and in a few places in Continental Europe the males, on the other hand, are also whitish or cream-coloured (the females being indistinguishable from the normal *mendica*). This is the form *rustica*. It is produced by a single major-gene without dominance. The heterozygotes are very variable but the mode is of a sandy colour (Onslow, 1921; Adkin, 1927; Ford, 1967, Plate 10). Here, therefore, we have a local Irish race produced by a single factor-difference, but probably adjusted by the selection of polygenes.

The two sexes are in the normal form subject to a difference in behaviour related to their colouring. The females frequently sit exposed and may fly during the day, possibly protected by their resemblance to the White Ermine moth, *Spilosoma lubricipeda*, which Rothschild (1963) has shown to be very poisonous. On the other hand, the blackish males never expose themselves or fly in daylight. They hide among vegetation and only become active at night. It would be interesting to know if the white males in Ireland behave as the females which they resemble. If so, we should have a good instance of the genetic adjustment of habit. Though the species is not uncommon in Southern Ireland, this simple study on its behaviour seems never to have been made there.

Ecological genetics and adaptation

The potential efficiency and rapidity of adaptation is well exemplified by the response of organisms to chemicals designed to destroy them. Resistance to D.D.T. seems consistently to arise in House Flies and Mosquitoes within about two years (Brown, 1958). Sheppard (1969, p. 263) points out that it usually proceeds in two stages. First, a resistant form, generally due to a single major gene semi-dominant in effect, starts to spread rapidly. Secondly, a marked increase in the degree of resistance produced by the mutant takes place, due to the selection of modifiers which enhance its action.

Several aspects of this matter are dealt with by the World Health Organization (1963). Bearing in mind that a species may have differently adjusted gene-complexes in its distinct geographical races, its resistance to insecticides will not necessarily take a similar form in various parts of the world. Thus the Biting House Fly, *Stomoxys*, has developed D.D.T. resistance in Scandinavia but seems incapable of doing so in Nebraska. Moreover, resistance when it does arise may involve several distinct insecticides, though not, of course, all types of them. For example, laboratory strains of body-lice which had become resistant to D.D.T. were so as well to pyrethrins but not to malathron. Furthermore, when House Fly populations have become so adjusted as to survive mild but not heavy doses of D.D.T., their reaction to the continued use of dieldrin may be different. For this can produce a plague of flies, as the drug may increase their rate of oviposition while killing other insects which compete with them or are predatory on them.

One may expect even corresponding characters in the larvae and imagines of the same insect species to be controlled genetically by different genes (pp. 313–14). That situation is illustrated by the resistance to D.D.T. of the Mosquito, *Aedes aegypti* in Trinidad. Wood (1967) finds that in the larvae this is due to a major gene in linkage group

349

II with, probably, modifiers in group III but not in group I; yet in the imagines it results from the action of a major gene in group III while genes in group II have not, so far, enhanced its effects.

By far the most widely used poison employed in the destruction of rodents is Warfarin* so that the development of resistance to it is a matter of financial consequence as well as of evolutionary interest. In the Common Rat, *Rattus norvegicus* this immunity has arisen in several areas in Britain. It seems to have spread from a centre near Welshpool, Montgomeryshire, at a rate which over the period 1962–65 has averaged 4·6 kilometres radially per year, suggesting increasing efficiency in the process. In this species, it is due to a single major gene in linkage group I, semi-dominant in effect (Greaves and Ayers, 1967, 1969). Yet in the Mouse, *Mus musculus* resistance is multifactorial, perhaps with one gene particularly important in its production. Consequently it develops more easily than in the Rat since, as P. M. Sheppard has remarked to me, even in the normal mouse population there will always be a certain number of slightly resistant individuals which will act as a nucleus upon which more complete resistance can be built up.

Warfarin is an anti-coagulant and its effects are the same as those of Vitamin-K deficiency, for it interferes with the metabolism of that substance. We may therefore expect that genetic resistance to it is widespread in mammals, and this has already been detected in Man (O'Reilly and Aggeler, 1965) due principally to one gene semi-dominant in effect, as in the Rat. As we have seen, the genetic mechanism involved may differ from one species to another but the existence of 'modifiers' provides a passage between multifactorial and major-gene control.

Warfarin resistance in the Mouse does not of course lead to polymorphism since it is multifactorial; however, it does so in the Rat due, probably, to heterozygous advantage in a condition controlled by a single major-gene. For, in places where this pesticide is used, the homozygous resistants die of vitamin-K deficiency, the (resistant) heterozygotes survive, while many of the non-resistant animals die of poisoning. At present, the results are consistent with fitness ratios for the three genotypes of 0·0, 1·0, and 0·6 to 0·7 respectively (information kindly supplied by J. H. Greaves).

Myxomatosis was introduced into the wild rabbit population of Australia in 1950. This was a planned project to destroy the species there where it was doing great damage to agriculture. The disease

* 3-(α-acetonyl)-4-hydroxycoumarin.

reached England in 1953 following the accidental release of some infected animals in France (Vaughan and Vaughan, 1968). Partial genetic resistance to myxomatosis has developed as a result of selecting laboratory strains of rabbits. The response occurred in all stocks that were studied. Heritability of the resistance was calculated both by means of inter-sire correlation, using survival-time as an index, and also by means of an all-or-none probit analysis. The two methods agreed in assessing it at about 0·35 to 0·40 (Sobey, 1969). In addition, Sobey et al. (1970) found that the response of rabbits to the disease is reduced with the age at which infection takes place.

Natural selection has increased the proportion of resistant rabbits in Australia but not, apparently, in Europe. Furthermore, in both regions there has been a spread of less virulent strains of the virus. At the time of its original introduction into wild rabbits in Australia, the death rate was about 90 per cent. Virus strains causing such high mortality can no longer be isolated from wild rabbits, in which the attenuation of the virus became evident within a year.

The difference in response of rabbits to myxomatosis in Australia and Europe, including Britain, is almost certainly due in part to its differing insect vectors and their contrasting habits in these two parts of the world. Thus the rabbit flea (*Spilopsyllus cuniculi*), which is the main infecting agent in the northern hemisphere, is absent from Australia where the virus is principally spread by mosquitoes. In South America, host and virus have become reasonably co-adapted so that the disease is no longer a killing agent there. I am greatly indebted to The Hon. Miriam Rothschild for the information she has given me on myxomatosis; she has made a special study of its insect vector (Rothschild et al., 1969).

In view of the rapid adjustment of animals to pesticides it is, as Bradshaw et al. (1965) point out, surprising that there seems little indication in botanical material of the evolution of tolerance to herbicides, though these have been in use for many years; yet plants can readily undergo genetic adaptations to changes in their environment. There seems to be an unsolved problem here.

The genetics of the various characters which fit plants and animals to special environments, sometimes in obvious ways, have been studied on a number of occasions, though seldom indeed when one considers the immense wealth of known adaptations in organisms. It would be quite inappropriate to attempt to summarize here what has been achieved in this type of inquiry; but a few instances, indicating fruitful lines of research or illustrating the concepts discussed in this book, must be mentioned at this point.

The genetic control of protective colour-patterns has seldom been investigated except in polymorphic forms. I have elsewhere illustrated in colour a particularly striking example of such adaptations (Ford, 1967, Plate 30), one which would well repay study. It is provided by the moth *Gnophos obscurata* (Selidosemidae). This is generally a coastal species. It rests exposed upon rocks or earth and is reddish-brown on the Old Red Sandstone of Devon, whitish to pure white, with small black lines, on the chalk in Sussex, and black in an exceptional inland locality on the peat at Oxshot, Surrey. *Hadena lepida* (Lepidoptera, Agrotidae), illustrated on the same plate as *G. obscurata*, has very similar though slightly less striking colour-adaptations and is the easier species to breed.

As already explained, the methods of founding artificial colonies (pp. 141–6) as well as of adding new genes to already existing ones (pp. 143–4) have produced valuable results in animal material. The use of transplants represents a corresponding botanical technique. Groups of plants growing in different habitats are interchanged or they are cultivated under uniform conditions in the laboratory. In this way, genetic and environmental modifications can be distinguished and assessed.

The prostrate habit is an obvious adaptation to exposed and windswept situations. It may be purely phenotypic or under genetic control. It is often seen in the Broom, *Cytisus* (=*Sarothamnus*) *scoparius*, and when such specimens are transplanted to a sheltered place they usually develop the ordinary erect growth. However, there is a race, *prostratus*, found on the cliffs of west Cornwall, Pembrokeshire, south-west Ireland, the Channel Isles, and parts of Europe, which retains the prostrate habit in all conditions of culture (Clapham, Tutin, and Warburg, 1959). That is to say, when selection has favoured this habit of growth, it has spread through the population a genotype suitable for producing it when this has been available. Mr J. J. B. Gill and Dr S. Walker, in a personal communication for which I am most grateful, inform me that a prostrate form of the Broom also exists at Dungeness, Kent, and that this is a separate adaptation to that exposed locality, being genetically distinct from the form found on the cliffs of the Lizard and Pembrokeshire and from the normal plant.

A comparable instance, but giving further information on selective adjustment, is provided by the work of Aston and Bradshaw (1966) on the grass, *Agrotis stolonifera*. They found that the plants growing on a cliff-face in west Anglesey had short stolons, being extremely prostrate, while those in a sheltered locality over the cliff-top had the normal habit of the species. The difference was genetic, being multifactorial. There was considerable gene-exchange between the populations due to wind-blown pollen. In spite of this, selection was powerful enough to evolve respectively the two distinct types of growth within less than ten metres: one thinks here of the 'boundary phenomenon' in *Maniola jurtina* discussed in Chapter Five.

The force of stabilizing selection acting on the transect of sudden change in *Agrotis stolonifera* is demonstrated by the fact that plants grown from seed are much more variable than the adults. Selection must annul the effects of gene-flow due to pollen blown from the exposed to the sheltered populations by the prevailing westerly winds. If it were not acting powerfully, the seedling and adult samples would resemble each other.

Dobzhansky (1951, pp. 146–7) gives an excellent account of trans-plantation methods in studying *Potentilla glandulosa* in California. Races of this plant are found at a low level in the Coast Range. Here there is a rainless summer and an equable climate, in which growth can con-tinue throughout the year. The species occurs also on the foothills of the Sierra Nevada, where there is a hot summer and snow in winter, while there are also dwarf Alpine forms at the highest levels (over 8,000 feet or so) in the mountains.

Plants of the foothill and coastal races do not thrive as well when inter-changed as they do in their own environments, though they take on the respectively active and dormant winter habits appropriate to their changed conditions. None of the other races can produce ripe seeds in the habitat of the Alpine one, which can itself survive at low levels. It remains dwarf near the coast, though here it is very susceptible to disease and to the effects of drought, though it grows tall in the foothills of the Sierra Nevada, where it seems more vigorous than in its own environment.

It is evident therefore that these races are genetically distinct. Yet the genes controlling them, which are very numerous (60 to 100 are esti-mated) can only produce their appropriate characteristics in the en-vironments to which they are adjusted. Among the great outbursts of variation in the F_2 generation between inter-racial crosses, are types adapted in various ways to the diverse localities in which this plant is found.

A rather similar situation has also been studied on several occasions

353

by the methods of ecological genetics. That is to say, selection acting upon high genetic variability to adapt a plant to different ecological conditions, but operating to form a cline in a continuous population or one broken up into groups in close proximity. It is illustrated by the work of Gregor (1938, 1939) on the Sea Plantain, *Plantago maritima*. The individual plants seem to be closely adapted to the ecologically great diversity of their habitat, in which they can colonize the water-logged mud of a salt-marsh, crevices in dry rock, and all the situations intermediate between these extremes. This has been achieved by selecting distinct and appropriate genotypes polygenically determined. The successful ones have certain recognizable characteristics affecting, for instance, spike- and scape-length. The latter is short in water-logged places (20 cm) and long (50 cm) in dry ones. The physiological adjustments required in such a range of environments are extreme; they are particularly exacting in a salt-marsh where the concentrations of sea-water may vary from saturation to extreme dilution by rain. Yet even these halophytic specimens of *P. maritima* will grow well in ordinary garden soil provided they are free from competition by other plants, which they normally escape in the exceptional places which they can colonize.

There is a contrast here with the adaptations built up respectively in the two British Bladder Campions, *Silene maritima* and *S. vulgaris* = *cucubalus*. These have been analysed by Marsden-Jones and Turrill (1957) in an extremely detailed series of ecological and genetic investigations involving inter- and intra-specific crosses between many races.

Some of the characteristics studied in these plants are purely polygenic, but many are controlled by one or more major-genes, often subject to genetic variation due to recombination within the gene-complex. For instance, the erect habit and absence of barren procumbent stems, characteristic of *S. vulgaris*, are in F_1 dominant or intermediate to the prostrate growth of *S. maritima*. A great range of variation appears in F_2 showing that this condition, which is of high survival-value, is polygenically determined. So too is leaf shape.

The form of the calyx seems due to the segregation of a number of major factors giving complex but sharply modal ratios in F_2. The presence of anthocyanin in the corolla, calyx, and other parts requires certain basic genes. Plants lacking one of these are distinguished by the pale yellowish-green colour of all their vegetative parts and calices, and by white corollas, and they always breed true. Yet crosses between two specimens of this type may result in offspring with anthocyanin; each parent having brought in an essential gene for colour lacking in the

354

other. There is great polygenic and environmental variation in the expression of this condition.

The wide-mouthed capsules with reflexed teeth of *S. maritima*, and the narrow-mouthed type with erect teeth found in *S. vulgaris*, are controlled on a unifactorial basis, the characters showing no dominance. However, in *S. vulgaris* the leaves and calyx may have either a matt or a shiny surface, determined by a single pair of alleles with the matt-type dominant. A simple switch-control of a similar kind is responsible for the two forms of seed which are distributed as a polymorphism in both species: the 'armadillo' with flat plates and the 'tubercled'. The latter is completely dominant, giving a $3:1$ ratio in F_2.

Thus in these two Campions certain characters of high adaptive value, requiring delicate adjustment to various habitats, are polygenic, giving great genetic variation on which selection can operate. Modal types adapted to different conditions are due to the operation of several major factors, the action of which is easily modified by recombination within the gene-complex; while a clear-cut feature such as matt or shiny leaves, which occurs in *S. vulgaris* only, is controlled by a single gene with full dominance. It will be noticed that in one respect dominance and recessiveness overrides a specific distinction: this is the dimorphism of seed-shape, for it occurs both in *S. maritima* and *S. vulgaris*. It will be recalled, on the other hand, that it is also possible for dominance and recessiveness to be built up differently in isolated populations relative to the effects of the same gene (pp. 147–9).

Here we have two plants capable of much variation which are closely adapted to their diverse habitats. Since many of their localities are of quite recent origin, as with wayside colonies, it is evident that powerful selection must be operating to produce such a result. Of this, a concept fundamental to this book, we have clear confirmation. *Silene maritima* and *S. vulgaris* can cross reciprocally and the F_1 so produced is fertile. Yet hybrids between the two are rare in nature and no natural hybrid swarms are known. Indeed the species are clearly separable and are maintained in Britain as populations with distinct ecological requirements. That is to say, selection must constantly be eliminating the ill-adjusted gene-complex which arises when they cross, and it is powerful enough to do so with almost complete efficiency.

Phenotypic stability and plasticity must be regarded as an important aspect of adaptation. There is certainly a tendency for organisms to be less variable, including reduced variability within polymorphic phases, in the range of conditions to which they are normally exposed. For they will have been selected to produce an optimum response to their ordinary environments, while they cannot be buffered to meet novel situations or those which they seldom encounter. For example, Kettlewell (1944) studied the action of a pair of alleles in *Panaxia dominula*, capable of producing a striking change in pattern (illustrated by Ford, 1967, Plate 14). The character is not fully dominant and both mutant genotypes, the somewhat less extreme heterozygotes and the more extreme homozygotes, are of normal appearance except when the insects are reared at a constant temperature; an environment of which the species has no experience. We do not, of course know how widespread this gene may be for it will evidently escape attention except in laboratory experiments of an unusual kind: it could, indeed, be polymorphic.

Some of the cryptic variation of organisms (p. 104), having physiological advantages in the heterozygote, will surely have been obliterated in its phenotypic effects so as not to disturb overt adjustments to the environment. That situation may well be true for a proportion of the esterase polymorphisms now being detected by means of electrophoresis (pp. 28–9, 104–6).

While many genes appear to be constant in their expression throughout the range of conditions that the organism can tolerate, others give rise to a variable phenotype. When the expression of a given genotype can be altered by environmental influences, it is described by Bradshaw *et. al.* (1965) as showing 'plasticity'.

Some aspects of that condition can usefully be exemplified from the Lepidoptera. Consider the response of the Small Tortoiseshell butterfly, *Aglais urticae*, to temperature (illustrated by Ford, 1967, Plate 35). This species is abundant in Europe, including Britain, and is remarkably invariable. Yet its reddish-brown colour becomes suffused with black scales, and its black markings enlarged, if it be reared in an ice-chest during late larval and pupal life. The contrary occurs, the normal amount of black being reduced, if the insects be forced in an incubator. The latter treatment produces a form much resembling the geographical race *ichnusa* which is the normal one in Corsica. This is partly genetic and partly also the result of high temperatures (Verity, 1950). That is to say, in this race genes are taking over the production of a pattern that is elsewhere environmental: a process described by Waddington (1953) as 'genetic assimilation'. The prostrate form of the Broom, *Cytisus scoparius* already mentioned (p. 352) provides a botanical illustration of it.

Bradshaw *et. al.* (1965) rightly stress that genetic plasticity is of special value in plants because they are static and they give numerous striking instances of it; examples of the condition will indeed be known to all geneticists. It occurs also in animals in which, as in some plants, it may even assume the status of a switch-control. It does so in the production of distinct seasonal forms. The European butterfly, *Araschnia levana*, provides a good example of this. It has two broods in the year. That flying in May (*levana*) is, on the upper side, fulvous brown with black pencilings. The second annual generation which flies in July is completely different; for the wings are blackish above and are crossed by an oblique white bar (see Ford, 1967, Plate 35 for these forms). This seasonal difference was studied experimentally by Sûffert (1924) who found that it is controlled by temperature. The dark white-banded form is produced by heat acting in late larval life or alternatively during the first 24 hours after pupation; cool conditions during this period give rise to imagines as found in the spring. Comparison with *A. urticae* is revealing though the action is there reversed, a high temperature evoking a reduction in black pigments. It shows that what may be only a response to exceptional conditions in one species can be so selected in another as to provide a switch-control in the production of seasonal variation.

A brilliant study in ecological genetics has been carried out by Bradshaw and his colleagues, demonstrating adaptations achieved by powerful selection (Bradshaw, McNeilly, and Gregory, 1965). For they have shown that plant populations of distinct genetic types can co-exist within a few yards where exceptional soil conditions occur. Old workings where metalliferous ores, lead, copper, zinc, and others have been mined, are found in Wales, their spoil-tips bare of vegetation, which cannot grow on ground poisoned by these substances. Yet Bradshaw noticed that such places were not utterly devoid of plant life: a few clumps, especially of the grasses *Agrostis tenuis* and *Festuca ovina*, were established on them (Plate 18). These specimens were shown to have acquired tolerance to the metals to which they were exposed; an adaptation that proved to be genetic. It may in some species be due to a major gene influenced by modifiers, (*Festuca ovina*); but generally it is multifactorial or at least the result of several genes (*Agrostis tenuis*). The tolerance has in all instances a high degree of dominance. Owing to the partly multifactorial nature of this situation, a few slightly tolerant specimens can be found among the normal plants growing in the ordinary countryside, and it will often be from these that adaptation to the toxic materials is built up.

The same individual plant can be adjusted to withstand two such heavy metals if present in the soil together. Such multiple adaptations must be acquired independently; except for nickel and zinc in which the immunity to one extends automatically to the other. This is curious, for these two elements are not nearly related from the chemical point of view.

Since the tolerant and non-tolerant plants take up the various metals to an equal degree, they must be rendered innocuous within the tissues of the individuals that are able to withstand them (Gregory and Bradshaw, 1965). These are not restricted to grasses, but plants such as *Silene inflata*, *Plantago lanceolata* and many others can evolve tolerance. On the other hand, numerous species, *Dactylis glomerata* and *Lotus corniculata* for instance, growing in close proximity to the mines seem unable to do so.

McNeilly (1968) has studied the distribution of tolerant populations of *Agrostis tenuis* in the immediate neighbourhood of a small copper mine, of 300 × 100 square metres, at Drws y Coed in north Wales. The shape and alignment of the valley in which it is situated ensures that the wind there is channelled to a marked degree from the prevailing direction, the west. He examined two transects, passing from toxic to normal soil, respectively downwind and up. In the latter position, the adult plants changed from normal to fully copper-tolerant within one metre, exactly at the boundary of the mine. Downwind, however, tolerance gradually faded out over 150 metres of uncontaminated ground. We here have evidence not only of the effect of wind-blown pollen but of exceedingly powerful selection against unprotected plants on land heavily charged with copper and of weaker selection against copper-tolerant individuals on ordinary soil (and see pp. 333–4).

It would at first sight be tempting to assess the time taken to achieve tolerance to heavy metals from the age of the mines where this has become established. But as Bradshaw wisely points out, such an inquiry might be misleading since it is likely that the areas in question were already somewhat contaminated, and the plants growing there correspondingly adjusted to them, before the mining operations started.

Some light on the speed of these evolutionary processes may be obtained from a situation discovered by Snaydon (quoted by Bradshaw, McNeilly and Gregory, 1965). A. S. Watt erected an enclosure of galvanized wire netting at Lakenheath, Suffolk, in 1936; it was renewed in 1958. The plants of *Agrostis canina* and *Festuca ovina* growing at the foot of it were found by Snaydon to be significantly zinc-tolerant compared with the normal plants a few feet away. Evidently the process had occupied less than thirty years. In general terms, Bradshaw's laboratory studies on the heritability of tolerance suggest that the condition could be fully established in about five generations. It is likely that the selection-pressures operating in nature are often powerful enough to achieve this.

The capacity shown by some plants to gain genetic tolerance to relatively toxic soils has obviously proved valuable in natural conditions. One thinks of the races and species adapted to the high magnesium content of Serpentine rocks, while still more striking instances exist: a notable one is provided by *Viola calaminaria*, which is restricted to ground rich in zinc.

Moreover, the existence of genetic variation in tolerance opens up the possibility of reclaiming the waste heaps from the mining of heavy metals both in Wales, where the workings have now closed down, and in other parts of the world where they are still in use (Smith & Bradshaw, 1970). Any reclamation of these sites that has so far been attempted has taken the form of blanketing with normal soil or with sewage sludge. This is expensive, costing over £300 per acre, and is only temporarily effective. But the genetically tolerant plants grow well on the metal-contaminated soil provided that abundant fertilizer be added. It now appears that these waste areas can be reclaimed by seeding with appropriate stocks of grasses and by the use of fertilizers at about the cost of establishing grassland on ordinary soil; perhaps about £50 per acre. Large plots of plants suitable for this purpose are now being established by Bradshaw and his colleagues.

The work on the evolution of tolerance to heavy metals demonstrates with exceptional clarity the nature of an adaptation and of the selection responsible for it. That this is very powerful is a fact emerging from the extreme localization of the adjustments. On the other hand, the researches on *Maniola jurtina* described in Chapters Four and Five and on industrial melanism in the Lepidoptera (Chapter Fourteen) have so far provided more accurate information on the possible speed of evolution in wild populations. Here we again encounter one of the most striking discoveries achieved by the techniques of ecological genetics: that is to say, the great selection-pressures normally operating in nature, which have proved so much more powerful than previously realized. As a result, it is no longer possible to attribute to random genetic drift or to mutation any significant part in the control of evolution.

Conclusion

I am not attempting to provide a summary of this book like that generally appended to a scientific paper, for it would occupy too much space even were it restricted to the conventional 2 per cent of the whole. Yet it seems appropriate, here at the end of this work, to draw attention to a few of the most important conclusions to which the techniques of ecological genetics have led. This may in part be done by correcting a misconception; one not widely held indeed but tending in the minds of a few biologists to reduce the value of such methods. It has been suggested that the problems raised by the evolutionary adjustments of different species are individual to each of them, interesting in themselves but not of general significance. That point of view is erroneous. In fact it would be impossible to read this book without being struck by the universality of the principles revealed by an analysis of variation and selection in the most diverse forms, whether in higher plants, butterflies, snails, or men. So much so indeed is this true that by applying these fundamental concepts it has been possible to make genetic predictions from one group to another and subsequently to find them verified (e.g. pp. 116–19).

One of the most far-reaching results of recent work on ecological genetics is the discovery that unexpectedly great selective forces are normally operating to maintain or to adjust the adaptations of organisms in natural conditions. As already mentioned, when R. A. Fisher wrote *The Genetical Theory of Natural Selection* in 1930, he envisaged selective advantages of up to about 1 per cent in wild populations. When in 1924 Haldane calculated a high selective value for the *carbonaria* gene of *Biston betularia* during its spread in Manchester (see this book, p. 294), his result was regarded as entirely exceptional; being a response to the abnormal situation in an industrial area. We now know that values of 20 or 30 per cent and more are common and indeed usual: a general conclusion derived from the study of a wide range of forms whatever the control of the variation investigated may be; whether polygenic, multifactorial, unifactorial, or by means of super-genes. That consideration forces us completely to readjust our thoughts on evolution and to recognize that a population may adapt itself very rapidly to changing conditions.

It should, however, be noticed that the direction of selection may sometimes alter with time. In these circumstances, a particular character may indeed have a total advantage of small magnitude when considered over a number of generations even when the selective forces acting upon it are very powerful. Yet their high values make it possible for such a quality to be adjusted rapidly in different ways during the periods when it is respectively harmful and beneficial. It is true that balanced polymorphism is maintained by contending advantages and disadvantages in equilibrium. Yet here again, swift adaptations to changing conditions are possible since the opposed selective forces are generally great.

Furthermore, the fact that such powerful selection is normally operating reduces to negligible limits the effects of random drift upon changes in the frequency of widespread genetic qualities: the time when the operation of chance may indeed be critical is not then but when a mutant begins to spread or, as Darlington has pointed out, when an inversion occurs in a suitable place. Similarly, if ever it could have been thought that mutation is important in the control of evolution, it is impossible to think so now; for not only do we observe it to be so rare that it cannot compete with the forces of selection but we know that this must inevitably be so. One of the triumphs of Mendelism is the demonstration that there exists a system which promotes both great heritable variability and great heritable stability; for extremely permanent genes can be recombined in an infinite variety of ways and there are mechanisms, such as inversion and other devices for evolving close linkage, by which new combinations of value to the organism can be preserved. It is a platitude to say that the essential permanence of the genetic units would be lost were they to contaminate one another when brought together in the same cell; it would be lost also unless they had a high degree of intrinsic stability because their copying process is very perfect or, to put it in another form, unless mutation were an extremely rare phenomenon. Situations in which evolution is controlled by mutation, and that envisaged in Lamarckism is one, are the more clearly impossible now that we appreciate the intensity of selection for advantageous qualities (its intensity against disadvantageous ones has been obvious since the time of Darwin).

Mutation is of course originally responsible for the diversity of the genetic units. But living organisms are the product of evolution *con-*

trolled not by mutation but by powerful selection. Moreover, to reach and to pursue the evolutionary path with which we are acquainted, the selective process has been precisely determined by the contingent factors of its setting. That is to say, the conditions required to produce the living systems that confront us, or of which we have any knowledge, have had throughout geological history to be of the correct kind to a degree astronomically exact. We may, then, equate the incredible improbability of repeating such a sequence, even in broadest outline, with the rarity of planets in the Universe upon which it could be enacted. When we do so, it seems idle to hope that life, in the sense in which we know it, exists on another World, to continue when Dictators and Politicians have in their folly used the means now coming into their hands to destroy it on this one.

Among the aims of ecological genetics, the study of evolution in wild populations is surely outstanding. Yet that project is possible only when the process takes place rapidly. This it has been found to do (i) when marked numerical fluctuations affect isolated communities; (ii) when polygenic characters can be studied either (*a*) in populations inhabiting ecologically distinct and isolated areas or (*b*) even in the absence of isolation if subject to very powerful selection-pressures; (iii) when a species successfully invades a new country or territory; (iv) in all types of genetic polymorphism.

It proved essential to define the latter condition precisely. Once that had been done, its analysis could be carried out in species belonging to very distinct groups; a procedure which has disclosed two of its fundamental properties. First, since the co-operation of several major-genes is often necessary to promote advantageous adjustments, the evolution of close linkage between them is favoured, so holding together the appropriate characters for which they are responsible. Consequently polymorphism is often controlled by means of a super-gene. Secondly, when a major-gene spreads in a population, owing to changes in the environment or in the gene-complex, we have seen that it will tend to develop heterozygous advantage. It and its allele will therefore be stabilized at the best attainable frequencies: that is to say, polymorphisms are then generated, so that they must be very widespread and important phenomena.

That is one of the reasons, and they are many, why genetics is in the future likely to play an increasing part in medicine. For, owing to the powerful selection-pressures now known to operate, polymorphisms must be associated with considerable, though balanced, advantages and disadvantages, the existence of which they indeed advertise. Thus, however trivial the characters may be by which we detect their alternative phases, these must have reached their observed frequencies because the genetic unit controlling them is of importance to the organism. The fact that a quarter of the population in western Europe is unable to taste phenyl-thio-urea for long seemed a negligible matter to the medical geneticist. Its association with thyroid disease has altered that point of view now: and what of the isoniazid polymorphism?

The advances brought about by a combination of field-studies and laboratory genetics can be appreciated by looking to the past. Major Leonard Darwin told me of a conversation with his father, the great Charles Darwin, who expressed his belief that by choosing the right material it might be possible actually to detect evolutionary changes taking place in natural conditions at the present time. For this purpose he said that long-continued investigations and careful records would be needed, extending over a period which he estimated at perhaps fifty years in species reproducing annually. As usual, Darwin was right, but on this occasion he was too pessimistic. The techniques of ecological genetics have made it possible not only to observe evolutionary changes in three or four generations in such forms but to evaluate the force of selection in wild populations and to analyse its immediate effects.

PLATE I. The Meadow Brown, *Maniola jurtina* (natural size), showing the spots on the underside of the hind-wings used in making a quantitative study of variation in this butterfly (Chapters Four and Five). Numbered downwards in four vertical rows: *1–6*, males with 0 to 5 spots. *7–12*, females with 0 to 4 spots. *9* and *10* show females with 2 spots differently placed.

PLATE 2(1). Tean, Isles of Scilly, showing the boundary between two ecologically distinct regions, subsequent to 1953. On the left is Area 4 consisting of grass, now grown long enough to provide a habitat for *M. jurtina*; on the right, Area 3 covered with a dense mass of gorse, bramble, and bracken, constituting a barrier to the butterfly. (The hill in the distance, separated from Tean by sea, is on the island of St Helen's.)

PLATE 2(2). The field-hedge at Larrick which, in 1956, formed the boundary between the 'Southern English' and 'East Cornish' stabilizations of *Maniola jurtina*. The photograph is taken looking eastwards from 'West Larrick'. Tall grass in the hedge gives the impression of a crop of wheat. However, the field beyond is, in fact, pasture, and cattle can be seen in it.

PLATE 3(I). Primrose flowers dissected to show the sexual organs. Numbered in two horizontal rows. *1–3* represent pin, thrum, and (long) homostyle. Below are the corresponding types, in which the style has been pulled aside to show the relationship of stigma to anthers more clearly (Cadbury Camp, near Sparkford, Somerset, 23.4.62).

PLATE 3(2). Cothill Marsh, near Abingdon, Berkshire, which supports the colony of *Panaxia dominula* in which the *medionigra* gene is polymorphic.

PLATE 4. *Panaxia dominula* (natural size) illustrating the artificial Tring colony. (Numbered horizontally.) *1* and *2* are normal male and female. *3–7* are examples showing the selected stock liberated to form the Tring colony in 1949. *8* and *9* are specimens from the Tring colony in 1952 when the population had returned nearly to normal.

PLATE 5 (1). *Panaxia dominula*: the polymorphic forms, numbered horizontally. *1* and *2*: normal homozygotes, male and female. *3–6*: *medionigra* (i.e. the heterozygotes), showing variation in expression of the gene. *7*: *bimacula* (the rare homozygote). (Natural size.)

PLATE 5 (2). *Cepaea nemoralis*: polymorphic forms; yellow unbanded, yellow banded, brown unbanded. (Slightly enlarged; longitudinal diameter of left-hand shell = 17 mm.)

PLATE 6(1). *Cepaea nemoralis*: Banded and unbanded forms among long grass. (Actual size of left-hand specimen, taken horizontally from lip = 2·0 cm.)

PLATE 6(2). A 'thrush stone'. The shells of *Cepaea nemoralis* which have been broken on it are in their original positions as left by the birds.

PLATE 7. *Angeronia prunaria* (Selidosemidae): numbered horizontally. *1* and *2*, male and female of the normal *prunaria* form (recessive). *3* and *4*, male and female of the *corylaria* form (dominant). *5* and *6*, male and female *corylaria* selected for more extreme expression of the gene. *7* and *8*, male and female *corylaria* selected for its diminished expression. (Slightly reduced: actual size of *1*, from wing-tip to wing-tip = 4·1 cm.)

PLATE 8. Numbered horizontally. *1, Danaus chrysippus chrysippus. 2, Danaus chrysippus dorippus. 3–6, Hypolimnas misippus. 3,* the male. *4,* the *misippus* form of female (dominant), a mimic of *1,* the specimen is a heterozygote. *5,* a specimen showing genetic variation in the expression of the *misippus* gene. *6,* the *inaria* form of the female (recessive), a mimic of *2.* (Reduced: actual size of *1,* from wing-tip to wing-tip = 7·2 cm.)

PLATE 9. *Papilio dardanus*, the South African race.
1, the male. *2* and *3*, mimetic females, *2*, *trophonius*,
a mimic of Figure *1* on Plate 8. *3*, the form *cenea*.
4, *Amauris echeria*, the model for *3*. (Reduced: actual
size of *1*, from wing-tip to wing-tip = 9·0 cm.)

PLATE 10. *Papilio dardanus. 1* and *2*, male and female of the race Meriones (Madagascar). *3*, the form *cenea* segregating in a cross between the mimetic South African race and the wholly non-mimetic Meriones. The mimetic pattern breaks down when the *cenea* gene (the normal expression of which is shown on Plate 9, Figure *3*) is placed in a gene-complex half of which is derived from a race in which it is unknown. *4*, the abnormal effect of the *hippocoonides* gene (for the typical expression of which see Plate 11, Figure *3*) in the F_1 from the cross between the mimetic South African and the non-mimetic Madagascan races. *5*, shows the unaltered effect of the *cenea* gene (except that the gene for tails is present in addition) segregating in F_1 when a South African *cenea* female has been mated with a male of the Abyssinian race which, unlike Meriones, is polymorphic for the *cenea* form. *6*, a non-mimetic female of the Abyssinian race, Antinorii, in which (tailed) mimics also occur. (Reduced: actual size of *1*, from wing-tip to wing-tip = 9 cm.)

PLATE 11. *1* and *2*, *Papilio dardanus*. *1*, the abnormal effect of the *cenea* gene (for the typical form see Plate 9, Figure *3*) in a cross between a South African race, polymorphic for this form, and the west coast one in which *cenea* is unknown. The specimen is damaged as it had to be used for breeding. *2*, the unaltered effect of the *trophonius* gene (a South African *trophonius* is shown on Plate 9, Figure *2*) segregating in the same cross as in Figure *1*. The form occurs in both races; accordingly its effect is unaltered in F_1 from the mating between them. *3*, the *hippocoonides* form of *Papilio dardanus* form South-east Africa and, *4*, its model *Amauris niavius dominicanus*. *5*, the *hippocoon* form of *Papilio dardanus* from West Africa and, *6*, its model *Amauris niavius niavius*. (Reduced: actual size of *5*, from wing-tip to wing-tip = 11 cm.)

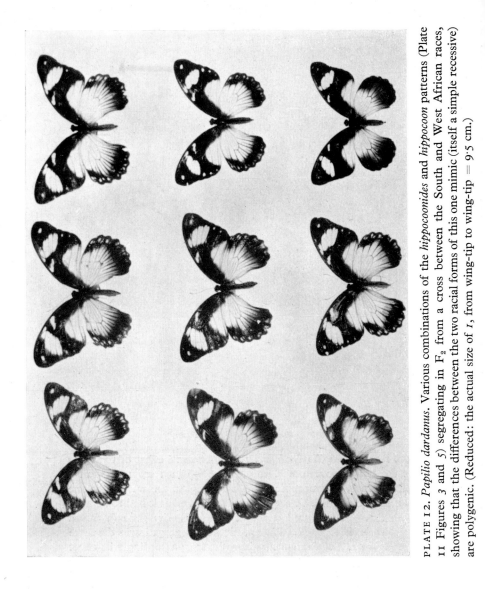

PLATE 12. *Papilio dardanus.* Various combinations of the *hippocoonides* and *hippocoon* patterns (Plate 11 Figures 3 and 5) segregating in F₂ from a cross between the South and West African races, showing that the differences between the two racial forms of this one mimic (itself a simple recessive) are polygenic. (Reduced: the actual size of 1, from wing-tip to wing-tip = 9·5 cm.)

PLATE 13. Industrial melanics (numbered vertically in two rows.) *1–3*, *Biston betularia; 1*, typical, *2*, *carbonaria*, *3*, *insularia*. *4* and *5*, *Gonodontis bidentata*, typical and melanic. *6* and *7*, *Hemerophilia abruptaria*, typical and melanic. *8* and *9*, *Colocasia coryli*, typical and melanic. *10* and *11*, *Cleora rhomboidaria*, typical and melanic. *12* and *13*, *Cleora repandata*, typical and melanic. *14* and *15*, *Semiothisa liturata*, typical and melanic. (Natural size.)

PLATE 14. *Biston betularia*, one typical and one *carbonaria* resting upon a lichen-covered tree in unpolluted country (Dorset). (Natural size.)

PLATE 15. *Biston betularia,* one typical and one *carbonaria* resting upon blackened and lichen-free bark in an industrial area (the Birmingham district). (Natural size.)

PLATE 16(1). A Robin, *Erithacus rubecula*, holding in its beak the *carbonaria* form of *Biston betularia* which it has just picked off the tree-trunk. One *carbonaria* and one typical are also in view, resting on the bark.

PLATE 16(2). *Biston betularia* v. *carbonaria*. On the left two modern specimens; on the right two specimens caught eighty to ninety years ago when the black coloration of *carbonaria* had not yet fully evolved (Actual size of top left-hand specimen, from wing-tip to wing-tip = 5·2 cm.)

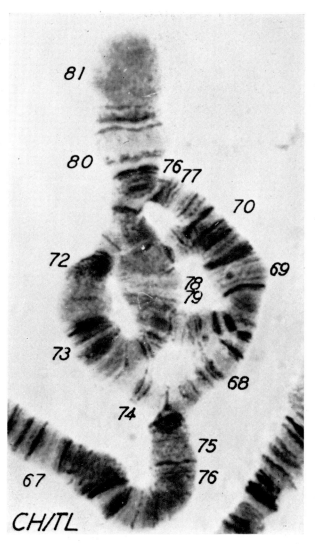

PLATE 17. Chromosome pairing in an inversion heterozygote of *Drosophila pseudoobscura*.
(Photograph kindly provided by Professor Th. Dobzhansky).

PLATE 18. Trelogan, Flintshire, North Wales. Mining ceased at this site more than 40 yea[cut]
ago. The area of waste material shown in this Plate is poisoned by zinc and remains almo[cut]
uncolonized by plants.

   In the foreground will be seen, to the right, a large tuft of a natural zinc-tolerant plant [cut]
the grass *Agrostis tenuis*. To the left are trial plantings of tolerant and non-tolerant specimen[cut]
of the same species and of *Festuca rubra*. In the distance, is a fenced area containing furth[cut]
experimental plantings. These show that the non-tolerant grass dies while the tolerant clum[cut]
grow extremely well upon this site when provided with artificial fertilizer.

   (Photograph kindly provided by Professor A. D. Bradshaw.)

References

ACTON, A. B. (1956). Crossing-over within the inverted regions in *Chironomus, Amer. Nat.*, **90**, 63–5.

ACTON, A. B. (1957). Chromosome inversions in natural populations of *Chironomus tentans, J. Genet.*, **55**, 61–94.

ACTON, A. B. (1959). A study of the differences between widely separated populations of *Chironomus* (= *Tendipes*) *tentans* (Diptera), *Proc. roy. Soc.*, B, **151**, 277–96.

ADKIN, R. (1927). [Report of Exhibit], *Proc. R. ent. Soc. Lond.*, **2**, 15–16, 66.

AIRD, I., BENTALL, H. H., and FRASER ROBERTS, J. A. (1953). A relationship between cancer of stomach and the ABO blood groups, *Brit. med. J.*, (**1**), 799–801.

ALLISON, A. C. (1954). Notes on Sickle-cell Polymorphism, *Ann. Hum. Genet.*, **19**, 39–57.

ALLISON, A. C. (1956). The Sickle-cell and Haemoglobin C genes in some African Populations, *Ann. Hum. Genet.*, **21**, 67–89.

ALLISON, A. C. and CLYDE, D. F. (1961). Malaria in African children with deficient erythrocyte glucose-6-phosphate dehydrogenase, *Brit. med. J.*, **1**, 1346–9.

ANDERSON, E. (1949). *Introgressive hybridization*, John Wiley, New York.

ANTONOVICS, J. (1968). Evolution in closely adjacent plant populations, *Heredity*, **23**, 219–38.

ANTONOVICS, J., LOVETT, J., and BRADSHAW, A. D. (1967). The evolution of adaptation to nutritional factors in populations of herbage plants, *Isotopes in Plant nutrition and physiology; International Atomic Energy Agency*, 549–67.

ARNOLD, R. W. (1968). Climatic selection in *Cepaea nemoralis* in the Pyrenees, *Phil. Trans. roy. Soc.*, B, **253**, 549–93.

ARNOLD, R. W. (1969). The Effects of selection by climate on the land snail *Cepaea nemoralis, Evolution*, **23**, 370–8.

ASTON, J. L. and BRADSHAW, A. D. (1966). Evolution in closely adjacent plant populations. II. *Agrostis stolonifera* in maritime habitats, *Heredity*, **21**, 649–64.

BAILEY, D. W. (1956). Re-examination of the diversity in *Partula taeniata*, *Evolution*, **10**, 360–6.

BARKER, J. F. (1968). Polymorphism in west African snails, *Heredity*, **23**, 81–98.

BATEMAN, A. J. (1955). Self-incompatibility systems in Angiosperms III. Cruciferae, *Heredity*, **9**, 53–68.

BATTAGLIA, B. (1958). Balanced polymorphism in *Tisbe reticulata*, a marine Copepod, *Evolution*, **12**, 358–64.

BAUR, E. (1911). *Einführung in die experimentelle vererbungslehre*, Bornträger, Berlin.

BEARDMORE, J. (1970). Ecological Factors and the Variability of Gene-Pools in *Drosophila*, *Essays in Evolution and Genetics* in *Honor of Theodosius Dobzhansky*, 299–314. Appleton-Century-Crofts, New York.

BEAUFOY, E. M. and S., DOWDESWELL, W. H., and MCWHIRTER, K. G. (1970). Evolutionary studies on *Maniola jurtina* (Lepidoptera, Satyridae): the Southern English stabilization, 1961–8, *Heredity*, **25**, 105–12.

BERG, L. S. (1926). *Nomogenesis; or evolution determined by Law*, (tr. J. N. Rostovtsow,) Constable, London.

BERNHARD, W. (1966). Über die Beziehung zwischen ABO-blutgruppen und Pockensterblichkeit in Indien dun Pakistan, *Homo*, **17**, 111–18.

BILLINGTON, W. D. (1964). Influence of immunological dissimilarity of mother and foetus and of size of placenta in mice, *Nature*, **202**, 317–18.

BIRCH, L. C. (1955). Selection in *Drosophila pseudoobscura* in relation to crowding, *Evolution*, **9**, 389–99.

BISHOP, J. A. (1969). Changes in genetic constitution of a population of *Sphaeroma rugicauda* (Crustacea: Isopoda), *Evolution*, **23**, 589–601.

BISHOP, J. A., and KORN, M. E. (1969). Natural selection and cyanogenesis in White Clover, *Trifolium repens*, *Heredity*, **24**, 423–40.

BLACKWELL, J. A., and DOWDESWELL, W. H. (1951). Local Movement in the Blue Tit, *Brit. Birds*, **44**, 397–403.

BOCQUET, C., LEVI, C., and TEISSIER, G. (1951). Recherches sur le polychromatisme de *Sphaeroma serratum*, *Arch. Zool. exp. gen.*, **87**, 245–97.

BODMER, W. F. (1958). Natural crossing between homostyle plants of *Primula vulgaris*, *Heredity*, **12**, 363–70.

BODMER, W. F. (1960). The genetics of homostyly in populations of *Primula vulgaris*, *Phil. Trans.*, B, **242**, 517–49.

BOVEY, P. (1941). Contribution à l'étude génétique et biogéographique de *Zygaena ephialtes* L., *Rev. suisse. Zool.*, **48**, 1–90.

BOWDEN, S. R. (1963). Polymorphism in *Pieris:* Forms *subtalba* and *sulphurea* (Lep., Pieridae), *Entom.*, **96**, 77–82.

BOWDEN, S. R. (1966a). *Pieris napi* in Corsica (Lep., Pieridae), *Entom.*, **99**, 57–68.

BOWDEN, S. R. (1966b). Polymorphism in Pieris: '*subtalba*' in *P. virginiensis* (Lep., Pieridae), *Entom.*, **99**, 174–82.

BOWDEN, S. R. (1966c). Sex-ratio in *Pieris* hybrids, *J. Lepidopterists' Soc.*, **20**, 189–96.

BOWDEN, S. R. (1967). Polymorphism in *Pieris:* Supposed lethality of homozygous '*subtalba*' (Lep., Pieridae): *Entom.*, **100**, 307–14.

BOYER, S. H., RUCKNAGEL, D. L., WEATHERALL, D. J., and WATSON-WILLIAMS, E. J. (1963). Further evidence for linkage between the β & δ loci governing Human hemoglobin and the population dynamics of linked genes, *Am. J. Human Genet.*, **15**, 438–48.

BRADSHAW, A. D. (1965). Evolutionary significance of phenotypic plasticity in plants, *Advanc. Genet.*, **13**, 115–55.

BRADSHAW, A. D., MCNEILLY, T. S., and GREGORY, R. P. G. (1965). Industrialisation, evolution and the development of heavy metal tolerance in plants, *5th Symp. Bri. Ecol. Soc.*, 327–43, Blackwell, Oxford.

BREEDLOVE, D. E., and EHRLICH, P. R. (1968). Plant-herbivore co-evolution: Lupins and Lycaenids, *Science*, **162**, 671–2.

BROWER, J. VAN Z. (1958a). Experimental studies of mimicry in some North American butterflies. Part 1. The Monarch, *Danaus plexippus*, and Viceroy, *Limenitis archippus archippus*, *Evolution*, **12**, 32–47.

BROWER, J. VAN Z. (1958b). ibid. Part 2. *Battus philenor* and *Papilio troilus*, *P. polyxenes* and *P. glaucus*, *Evolution*, **12**, 123–36.

BROWER, J. VAN Z. (1958c). ibid. Part 3. *Danaus gilippus berenice* and *Limenitis archippus floridensis*, *Evolution*, **12**, 273–85.

BROWER, J. VAN Z. (1960). Experimental studies of mimicry. IV. The reactions of Starlings to different proportions of models and mimics, *Amer. Nat.*, **94**, 271–82.

BROWER, J. VAN Z., and BROWER, L. P. (1965). Experimental studies on mimicry. 8. Further investigations of honeybees (*Apis mellifica*) and their dronefly mimics (*Eristalis spp.*), *Am. Nat.*, **99**, 173–88.

BROWER, L. P. (1969). Ecological chemistry, *Scientific American*, **220**, 22–9.

BROWER, L. P., and BROWER, J. VAN Z. (1962). The relative abundance of model and mimic butterflies in natural populations of the *Battus philenor* mimicry complex, *Ecology*, **43**, 154–8.

BROWER, L. P., and BROWER, J. VAN Z. (1964). Birds, butterflies and plant poisons, *Zoologica*, *N.Y.*, **49**, 137–59.

BROWER, L. P., BROWER, J. VAN Z., and COLLINS, C. T. (1963). Relative palatability and Müllerian mimicry among Neotropical butterflies of the subfamily Heliconiinae, *Zoologica*, *N.Y.*, **48**, 65–84.

BROWER, L. P., BROWER, J. VAN Z., and CORVINO, J. M. (1967a). Plant poisons in a terrestrial food chain, *Proc. Nat. Acad. Sci.*, **57**, 893–8.

BROWER, L. P., BROWER, J. VAN Z., and WESTCOTT, P. W. (1960). The reactions of Toads (*Bufo terrestris*) to Bumblebees (*Bombus americanorum*) and their Robberfly mimics (*Mallophora bomboides*), *Amer. Nat.*, **94**, 343–55.

BROWER, L. P., COOK, L. M., and CROZE, H. J. (1967b). Predator responses to artificial Batesian mimics released in a Neotropical environment, *Evolution*, **21**, 11–23.

BROWER, L. P., RYERSON, W. N., COPPINGER, L. L., and GLAZIER, S. C. (1968). Ecological chemistry and the palatability spectrum, *Science*, **161**, 1349–51.

BROWN, A. W. A. (1958). Insecticide resistance in arthropods, *Monograph Ser. No.* 38 WHO, Geneva, 240 pp.

BULLINI, L., SBORDONI, V., e RAGAZZINI, P. (1969). Mimetismo mülleriano in popolazioni italiane, *Arch. zool. Italiano*. **54**, 181–214.

BURNS, J. M. (1966). Expanding distribution and evolutionary potential of *Thymelicus lineola* (Lepidoptera), *Canadian Entom.*, **98**, 859–66.

BURNS, J. M. and JOHNSON, F. M. (1967). Esterase polymorphism in natural populations of the Sulfur Butterfly, *Colias eurytheme*, *Science*, **156**, 93–6.

CAIN, A. J. (1953). Visual selection by tone in *Cepaea nemoralis* L., *J. Conch.*, **23**, 333–6.

CAIN, A. J., and CURREY, J. D. (1963). Area effects in *Cepaea*, *Phil. Trans.*, B., **246**, 1–81.

CAIN, A. J., and CURREY, J. D. (1968a). Ecogenetics of a population of *Cepaea nemoralis* subject to strong area effects, *Phil. Trans.*, B, **253**, 447–82.

CAIN, A. J. and CURREY, J. D. (1968b). Climate and selection of banding morphs in *Cepaea* from the climate optimum to the present day, *Phil. Trans.*, B, **253**, 483–98.

CAIN, A. J., KING, J. M. B., and SHEPPARD, P. M. (1960). New data on the genetics of polymorphism in the snail *Cepaea nemoralis* L., *Genetics*, **45**, 393–411.

CAIN, A. J., and SHEPPARD, P. M. (1950). Selection in the polymorphic land snail *Cepaea nemoralis* (L.), *Heredity*, **4**, 275–94.

CAIN, A. J. and SHEPPARD, P. M. (1952). The effects of natural selection on body colour in the land snail *Cepaea nemoralis*, *Heredity*, **6**, 217–31.

CAIN, A. J. and SHEPPARD, P. M. (1954). Natural Selection in *Cepaea*, *Genetics*, **39**, 89–116.

CAMIN, J. H. and EHRLICH, P. R. (1958). Natural Selection in Water Snakes (*Natrix sipedon* L.) on Islands in Lake Erie, *Evolution*, **12**, 504–11.

CARPENTER, G. D. H. (1913): *Pseudacraea eurytus hobleyi*, Neave, its forms and its models on Bugalla Island, Lake Victoria, with other members of the same combination, *Trans. R. ent. Soc. Lond.*, 606–45.

CARPENTER, G. D. H. (1946). Mimetic Polymorphism, *Nature*, **158**, 277–9.

CARPENTER, G. D. H., and FORD, E. B. (1933). *Mimicry*, Methuen, London.

CARSON, H. L. (1955). The genetic characteristics of marginal populations of *Drosophila*, *Cold Spr. Harb. Symp. quant. Biol.*, **20**, 276–87.

CARTER, M. A. (1968). Area effects and visual selection in *Cepaea nemoralis* and *Cepaea hortensis*, *Phil. Trans.*, B, **253**, 397–446.

CASPARI, E. (1950). On the selective value of the alleles *Rt* and *rt* in *Ephestia kuhniella*, *Amer. Nat.*, **84**, 367–80.

CASTLE, W. E., and PINCUS, G. (1928). Hooded Rats and Selection, *J. exp. Zool.*, **50**, 409–39.

CHAKRAVARTTI, M. R. and VOGEL, F. (1966). ABO blood groups and smallpox . . ., *Humangenetik*, **3**, 166–80.

CHITTY, D. (1937–8). *Rep. Bur. Anim. Pop. Oxf.*

CHITTY, H. (1948). The Snowshoe Rabbit Enquiry, *J. Anim. Ecol.*, **17**, 39–44.

CHUNG, C. S., MATSUNAGA, E., and MORTON, N. E. (1961). The MN polymorphism in Japan, *Jap. J. hum. Genet.*, **6**, 1–11.

CHUNG, C. S. and MORTON, N. E. (1961). Selection at the ABO locus, *Amer. J. hum. Genet.*, **13**, 9–27.

CLAPHAM, A. R., TUTIN, T. G., and WARBURG, E. F. (1959). *Flora of the British Isles*, Cambridge.

369

CLARKE, B. (1960). Divergent effects of natural selection on two closely-related polymorphic snails, *Heredity*, **14**, 423–43.

CLARKE. B. (1962). Balanced polymorphism and the diversity of sympatric species, *Systematics Assn. publ.* **4**, 47–70.

CLARKE, B. (1970*a*). Darwinian Evolution of Proteins, *Science*, **168**, 1009–11.

CLARKE, B. (1970*b*). Festschrift in Biology, *Science*, **169**, 1192.

CLARKE, B. and KIRBY, D. R. S. (1966). Maintenance of histocompatibility polymorphisms, *Nature*, **211**, 999–1000.

CLARKE, C. A. (1961). Blood groups and disease, pp. 81–119, *Progress in Medical Genetics* (ed. A. G. Steinberg), Grune and Stratton, U.S.A.

CLARKE, C. A. (1964, 2nd. Edn.). *Genetics for the Clinician*, Blackwell, Oxford.

CLARKE, C. A., *et al.* (1955). The relationship of the ABO blood groups to duodenal ulcer and gastric ulceration, *Brit. med. J.* (2), 643–6.

CLARKE, C. A., MCCONNELL, R. B., and SHEPPARD, P. M. (1960). ABO blood groups and secretor character in rheumatic carditis, *Brit. med. J.* (1), 21–3.

CLARKE, C. A., PRICE EVANS, D. A., MCCONNELL, R. B., and SHEPPARD, P. M. (1959). Secretion of blood group antigens and peptic ulcer, *Brit. med. J.* (1), 603–7.

CLARKE, C. A., and SHEPPARD, P. M. (1955). A preliminary report on the genetics of the Machaon group of swallowtail butterflies, *Evolution*, **9**, 182–201.

CLARKE, C. A. and SHEPPARD, P. M. (1959*a*). The genetics of some mimetic forms of *Papilio dardanus*, Brown, and *Papilio glaucus*, Linn., *J. Genet.*, **56**, 236–60.

CLARKE, C. A. and SHEPPARD, P. M. (1959*b*). The genetics of *Papilio dardanus*, Brown. I. Race Cenea from South Africa, *Genetics*, **44**, 1347–58.

CLARKE, C. A. and SHEPPARD, P. M. (1960*a*). The Genetics of *Papilio dardanus*, Brown. II. Races Dardanus, Polytrophus, Meseres, and Tibullus, *Genetics*, **45**, 439–57.

CLARKE, C. A. and SHEPPARD, P. M. (1960*b*). The evolution of dominance under disruptive selection, *Heredity*, **14**, 73–87.

CLARKE, C. A. and SHEPPARD, P. M. (1960*c*). The evolution of mimicry in the butterfly *Papilio dardanus*, *Heredity*, **14**, 163–73.

CLARKE, C. A. and SHEPPARD, P. M. (1960*d*). Super-genes and mimicry, *Heredity*, **14**, 175–85.

REFERENCES

CLARKE, C. A., and SHEPPARD, P. M. (1962). The genetics of the mimetic butterfly *Papilio glaucus*, *Ecology*, **43**, 159–61.

CLARKE, C. A. and SHEPPARD, P. M. (1964). Genetic control of the melanic form *insularia* of the Moth *Biston betularia*, *Nature*, **202**, 215–16.

CLARKE, C. A. and SHEPPARD, P. M. (1966). A local survey of the distribution of industrial melanic forms in the moth *Biston betularia* and estimates of the selective values of these in an industrial environment, *Proc. roy. Soc.*, B, **165**, 424–39.

CLARKE, C. A., SHEPPARD, P. M., and THORNTON, I. W. B. (1968). The genetics of the mimetic butterfly *Papilio memnon*, *Phil. Trans. roy. Soc.*, B, **254**, 37–89.

COCKAYNE, E. A. (1912–13). Notes on *Bupalus piniarius* Linn., *Trans. S. Lond. ent. nat. Hist. Soc.*, parts 22 and 23, pp. 56–9.

COCKAYNE, E. A. (1932). A new explanation of the genetics of sex-limited inheritance in butterflies, *Entomologist*, **65**, 169–76.

COLEMAN, E. (1928). Pollination of an Australian orchid by the male ichneumonid *Lissopimpla semipunctata*, Kirby, *Trans. R. ent. Soc. Lond.*, **75**, 533–7.

COOK, L. M. (1961). Food-plant specialization in the moth *Panaxia dominula*, *Evolution*, **15**, 478–85.

COOK, L. M., BROWER, L. P., and ALCOCK, J. (1969). An attempt to verify mimetic advantages in a Neotropical environment, *Evolution*, **23**, 339–45.

COOK, L. M., BROWER, L. P., and CROZE, H. J. (1967). The accuracy of a population estimation from multiple recapture data, *J. anim. Ecol.*, **36**, 57–60.

COOK, L. M. and KETTLEWELL, H. B. D. (1960). Radioactive Labelling of Lepidopterous Larvae, *Nature*, **187**, 301–2.

COOK, L. M. and KING, J. M. B. (1966). Some data on the genetics of shell-character polymorphism in the snail *Arianta arbustorum*, *Genetics*, **53**, 415–25.

COOK, L. M. and MURRAY, J. (1966). New information on the inheritance of polymorphic characters in *Cepaea hortensis*, *J. Hered.*, **57**, 245–7.

COTT, H. B. (1940). *Adaptive Coloration in Animals*, Methuen, London.

CRAMPTON, H. E. (1916). Studies on the variation, distribution, and evolution of the Genus *Partula*: The species inhabiting Tahiti, *Publ. Carneg. Instn.*, No. 228.

CRAMPTON, H. E. (1925). Contemporaneous organic differentiation in the species of *Partula* living in Moorea, Society Islands, *Amer. Nat.*, **59**, 5–35.

CRAMPTON, H. E. (1932). Studies on the variation, distribution, and evolution of the Genus *Partula*: The species inhabiting Moorea, *Publ. Carneg. Instn.*, No. 410.

CREED, E. R. (1966). Geographic variation in the Two-spot Ladybird in England and Wales, *Heredity*, **21**, 57–72.

CREED, E. R. (1969). *Technological Injury*, pp. 119–34 (ed. J. Rose), Gordon & Breach, New York.

CREED, E. R. (1971). Industrial melanism in the Two-spot Ladybird and smoke abatement, *Evolution*, [In Press].

CREED, E. R., DOWDESWELL, W. H., FORD, E. B., and MCWHIRTER, K. G. (1959). Evolutionary Studies on *Maniola jurtina*: the English mainland 1956–57, *Heredity*, **13**, 363–91.

CREED, E. R., DOWDESWELL, W. H., FORD, E. B., and MCWHIRTER, K. G. (1962). Evolutionary Studies on *Maniola jurtina*: the English mainland, 1958–60, *Heredity*, **17**, 237–65.

CREED, E. R., DOWDESWELL, W. H., FORD, E. B., and MCWHIRTER, K. G. (1970). Evolutionary studies on *Maniola jurtina* (Lepidoptera, Satyridae): The 'Boundary Phenomenon' in Southern England 1961–1968, *Essays in Evolution and Genetics* pp. 263–87 (ed. M. K. Hecht, and W. C. Steere), Appleton-Century-Crofts, New York.

CREED, E. R., FORD, E. B., and MCWHIRTER, K. G. (1964). Evolutionary studies on *Maniola jurtina*: The Isles of Scilly, 1958–59, *Heredity*, **19**, 471–88.

CROSBY, J. L. (1940). High Proportions of homostyle plants in populations of *Primula vulgaris*, *Nature*, **145**, 672–3.

CROSBY, J. L. (1948). Population genetics in the genus *Primula*, *Dissertation for the Degree of Ph.D.*, University Library, Cambridge.

CROSBY, J. L. (1949). Selection of an unfavourable gene-complex, *Evolution*, **3**, 212–30.

CROSBY, J. L. (1959). Outcrossing on homostyle primroses, *Heredity*, **13**, 127–31.

CROSBY, J. L. (1960). The use of electronic computation in the study of random fluctuations in rapidly evolving populations, *Phil. Trans.*, B. **242**, 551–73.

CROWE, L. K. (1964). The evolution of outbreeding in Plants, *Heredity*, **19**, 435–57.

CUNHA, A. B. DA (1949). Genetic analysis of the polymorphism of color pattern in *Drosophila polymorpha*, *Evolution*, **3**, 239–51.

DADAY, H. (1954). Gene frequencies in wild populations of *Trifolium repens*, *Heredity*, **8**, 61–78.

DADAY, H. (1962). Mechanism of Natural Selection, *Ann. Rep. Division of Plant Industry*, *C.S.I.R.O. Australia*, pp. xiii, 14.

DARLINGTON, C. D. (1956a). *Chromosome Botany*, Allen and Unwin, London.

DARLINGTON, C. D. (1956b). Natural populations and the breakdown of classical genetics, *Proc. roy. Soc.*, B. **145**, 350–64.

DARLINGTON, C. D. (1958, 2nd edn.). *The Evolution of Genetic Systems*, Cambridge.

DARLINGTON, C. D. (1964). *Genetics and Man*, Allen and Unwin, London.

DARLINGTON, C. D. and DOBZHANSKY, TH. (1942). Temperature and 'Sex-ratio' in *Drosophila pseudoobscura*, *Proc. nat. Acad. Sci. Wash.*, **28**, 45–8.

DARLINGTON, C. D. and JANAKI-AMMAL, E. K. (1945). Adaptive iso-chromosomes in *Nicandra*, *Ann. Bot.*, **9**, 267–81.

DARLINGTON, C. D., and MATHER, K. (1949). *The Elements of Genetics*, Allen and Unwin, London.

DARLINGTON, C. D. and WYLIE, A. P. (1955). *Chromosome Atlas of Flowering Plants*, Allen and Unwin, London.

DARWIN, C. R. (1877). *The different forms of flowers on plants of the same species*, Murray, London.

DAWSON, C. D. R. (1941). Tetrasomic inheritance in *Lotus corniculatus* L., *J. Genet.*, **42**, 49–72.

DAY, J. C. L. and DOWDESWELL, W. H. (1968). Natural selection in *Cepaea* on Portland Bill, *Heredity*, **23**, 169–88.

DEMPSTER, J. P. (1967). The natural mortality of the young stages of *Pieris*, *J. applied Ecol.*, **4**, 485–500.

DICE, L. R. (1947). Effectiveness of selection by owls of deer mice (*Peromyscus maniculatus*) which contrast in colour with their background, *Contr. Lab. Vertebr. Biol. Univ. Mich.*, **34**, 1–20.

DICE, L. R. and BLOSSOM, P. M. (1937). Studies of mammalian ecology in Southwestern North America, *Publ. Carneg. Instn.*, 458–85.

DIVER, C., BOYCOTT, A. E., and GARSTANG, S. (1925). The inheritance of inverse symmetry in *Limnaea peregra*, *J. Genet.*, **15**, 113–20.

DIVER, C. (1929). Fossil records of Mendelian mutants, *Nature*, **124**, 183.

DIVER, C. (1940). The Problem of closely related species living together in the same area, in *The New Systematics* (ed. Julian Huxley), Oxford.

373

DOBZHANSKY, TH. (1947a). Genetics of natural populations: XIV, *Genetics*, **32**, 142–60.

DOBZHANSKY, TH. (1947b). Adaptive changes induced by natural selection in wild populations of *Drosophila*, *Evolution*, **1**, 1–16.

DOBZHANSKY, TH. (1950). Genetics of Natural Populations, XIX, *Genetics*, **35**, 288–302.

DOBZHANSKY, TH. (1951, 3rd Edn.). *Genetics and the Origin of Species*, Columbia University Press, New York.

DOBZHANSKY, TH. (1956). Genetics of natural populations, XXV, *Evolution*, **10**, 82–92.

DOBZHANSKY, TH. (1958). Genetics of Natural Populations, XXVII, *Evolution*, **12**, 385–401.

DOBZHANSKY, TH. (1961). On the dynamics of chromosomal polymorphism in *Drosophila*, *Symposia R. ent Soc. Lond.*, No. 1, 30–42.

DOBZHANSKY, TH. and LEVENE, H. (1948). Genetics of natural populations, XVII, *Genetics*, **33**, 537–47.

DOBZHANSKY, TH. and PAVLOVSKY, O. (1957). An experimental study of interaction between genetic drift and natural selection, *Evolution*, **11**, 311–19.

DOBZHANSKY, TH. and PAVLOVSKY, O. (1960). How stable is Balanced polymorphism?, *Proc. nat. Acad. Sci., Wash.*, **46**, 41–7.

DOBZHANSKY, TH. and PAVLOVSKY, O. (1967). Experiments on the incipient species of the *Drosophila paulistorum* complex, *Genetics*, **55**, 141–56.

DOBZHANSKY, TH. and SPASSKY, N. (1954). Environmental modification of heterosis in *Drosophila pseudoobscura*, *Proc. nat. Acad. Sci., Wash.*, **40**, 407–15.

DOBZHANSKY, TH. and SPASSKY, B. (1967). Effects of selection and migration on geotactic and phototactic behaviour of *Drosophila*, *Proc. roy. Soc. B*, **168**, 27–47.

DONCASTER, L. (1906). Collective Inquiry as to progressive melanism in Lepidoptera, *Ent. Rec.*, **18**, 165–254.

DONCASTER, L. and RAYNOR, G. H. (1906). On breeding experiments with Lepidoptera, *Proc. Zool. Soc. Lond.*, **1**, 125–33.

DOWDESWELL, W. H. (1961). Experimental Studies on natural selection in the butterfly *Maniola jurtina*, *Heredity*, **16**, 39–52.

DOWDESWELL, W. H. (1962). A further study of the butterfly *Maniola jurtina* in relation to natural selection by *Apanteles tetricus*, *Heredity*, **17**, 513–23.

DOWDESWELL, W. H., FISHER, R. A., and FORD, E. B. (1940). The

Quantitative study of populations in the Lepidoptera, *Ann. Eugen., Lond.*, **10**, 123–36.

DOWDESWELL, W. H., FISHER, R. A., and FORD, E. B. (1949). The Quantitative study of populations in the Lepidoptera. 2. *Maniola jurtina* L., *Heredity*, **3**, 67–84.

DOWDESWELL, W. H. and FORD, E. B. (1948). The Genetics of habit in the Genus *Colias*, *Entomologist*, **81**, 209–12.

DOWDESWELL, W. H. and FORD, E. B. (1952). The distribution of spot-numbers as an index of Geographical Variation in the butterfly *Maniola jurtina* L. (Lepidoptera), *Heredity*, **6**, 99–109.

DOWDESWELL, W. H. and FORD, E. B. (1953). The influence of isolation on variability in the butterfly *Maniola jurtina* L., *Symp. Soc. exp. Biol.*, **7**, 254–73.

DOWDESWELL, W. H. and FORD, E. B. (1955). Ecological Genetics of *Maniola jurtina* in the Isles of Scilly, *Heredity*, **9**, 265–72.

DOWDESWELL, W. H., FORD, E. B., and MCWHIRTER, K. G. (1957). Further studies on isolation in the butterfly *Maniola jurtina* L., *Heredity*, **11**, 51–65.

DOWDESWELL, W. H., FORD, E. B., and MCWHIRTER, K. G. (1960). Further studies on the evolution of *Maniola jurtina* in the Isles of Scilly, *Heredity*, **14**, 333–64.

DOWDESWELL, W. H. and MCWHIRTER, K. G. (1967). Stability of spot-distribution in *Maniola jurtina* throughout its range, *Heredity*, **22**, 187–210.

DOWNHAM, D. Y. (1970). *The Mathematical Theory of a Cline*, [in preparation].

DOWRICK, V. P. J. (1956). Heterostyly and homostyly in *Primula obconica*, *Heredity*, **10**, 219–36.

DUBININ, N. P. and TINIAKOV, G. G. (1946). Structural chromosome variability in urban and rural populations of *Drosophila funebris*, *Amer. Nat.*, **80**, 393–6.

EDMUNDS, M. (1969). Polymorphism in the mimetic butterfly *Hypolimnas misippus* L. in Ghana, *Heredity*, **24**, 281–302.

EDWARDS, V. C. WYNNE-, (1962). *Animal Dispersion in relation to Social Behaviour*, Oliver and Boyd, Edinburgh.

EHRLICH, P. R. (1961). Intrinsic barriers to dispersal in checkerspot butterfly, *Science*, **134**, 108–9.

EHRLICH, P. R. and MASON, L. G. (1966). The Population Biology of the Butterfly *Euphydryas editha*. III. Selection and the Phenetics of the Jasper Ridge Colony, *Evolution*, **20**, 165–73.

ELTON, C. (1942). *Voles, Mice, and Lemmings: Problems in Population Dynamics*, Oxford.

ETON, C. (1958). *The Ecology of Invasions by Animals and Plamts*, Methuen, London.

ELTON, C. and NICHOLSON, M. (1942). Fluctuations in Numbers of the Muskrat (*Ondatra zibethica*) in Canada, *J. Anim. Ecol.*, **11**, 96–126.

ERNST, A. (1933). Weitere untersuchungen zur phänanalyse zum fertilitäts problem und zur genetik heterosyler Primeln, I. *Primula viscosa, Arch. Klaus-Stift. VererbForsch.*, **8**, 1–215.

ERNST, A. (1936). ibid. II. *Primula hortensis. Arch. Klaus-Stift. Vererb Forsch.*, **11**, 1–280.

EUW, J. VON, FISHELSON, L., PARSONS, J. A., REICHSTEIN, T., and ROTHSCHILD, THE HON. M. (1967). Cardenolides (heart poisons) in a Grasshopper feeding on Milkweeds, *Nature*, **214**, 25–9. (April).

EVANS, D. A. P., MANLEY, K. A., and MCKUSICK, V. A. (1960). Genetic control of isoniazid metabolism in Man, *Brit. med. J.* (2), 485–91.

FISCHER, E. (1929–30). Valesina-Männchen, *Ent. Z.*, **43**, 151, 159, 184, 194.

FISHER, J. (1952). *The Fulmar*, The New Naturalist Series, Collins, London.

FISHER, R. A. (1927). On some objections to Mimicry Theory; Statistical and Genetic, *Trans. R. ent. Soc. Lond.*, **75**, 269–78.

FISHER, R. A. (1928a). The Possible Modification of the Response of the Wild Type to Recurrent Mutations, *Amer. Nat.*, **62**, 115–26.

FISHER, R. A. (1928b). Two further notes on the origin of dominance, *Amer. Nat.*, **62**, 571–4.

FISHER, R. A. (1929). The evolution of dominance; reply to Professor Sewall Wright, *Amer. Nat.*, **63**, 553–6.

FISHER, R. A. (1930a). *The Genetical Theory of Natural Selection*, Oxford.

FISHER, R. A. (1930b). The Distribution of Gene Ratios for rare Mutations, *Proc. roy. Soc. Edinb.*, **50**, 204–19.

FISHER, R. A. (1933). On the evidence against the chemical induction of melanism in Lepidoptera, *Proc. roy. Soc.*, B, **112**, 407–16.

FISHER, R. A. (1935). Dominance in Poultry, *Philos. Trans. roy. Soc.*, B, **225**, 197–226.

FISHER, R. A. (1939). Selective forces in wild populations of *Paratettix texanus, Ann. Eugen., Lond.*, **9**, 109–22.

FISHER, R. A. and DIVER, C. (1934). Crossing-over in the land snail *Cepaea nemoralis*, L., *Nature*, **133**, 834–5.

FISHER, R. A. and FORD, E. B. (1947). The spread of a gene in natural conditions in a colony of the moth *Panaxia dominula* L., *Heredity*, **1**, 143–74.

FISHER, R. A. and MATHER, K. (1943). The inheritance of style length in *Lythrum salicaria, Ann. Eugen., Lond.*, **12**, 1–23.

FORD, E. B. (1931, 1st Edn.). *Mendelism and Evolution*, Methuen, London.

FORD, E. B. (1936). The genetics of *Papilio dardanus* Brown (Lep.). *Trans. R. ent. Soc. Lond.*, **85**, 435–66.

FORD, E. B. (1937). Problems of heredity in the Lepidoptera, *Biol. Rev.*, **12**, 461–503.

FORD, E. B. (1940*a*). Polymorphism and Taxonomy, pp. 493–513, *The New Systematics* (ed. Julian Huxley), Clarendon Press, Oxford.

FORD, E. B. (1940*b*). Genetic research in the Lepidoptera, *Ann. Eugen., Lond.*, **10**, 227–52.

FORD, E. B. (1942*a*, 1st Edn.). *Genetics for Medical Students*, Methuen, London.

FORD, E. B. (1942*b*). Studies on the chemistry of pigments in the Lepidoptera with reference to their bearing on systematics. 2. Red pigments in the genus *Delias* Hübner, *Proc. R. ent. Soc. Lond.*, **17**, 87–92.

FORD, E. B. (1944*a*). ibid. 3. The red pigments of the Papilionidae, *Proc. R. ent. Soc. Lond.*, **19**, 92–106.

FORD, E. B. (1944*b*). ibid. 4. The classification of the Papilionidae, *Trans. R. ent. Soc. Lond.*, **94**, 201–23.

FORD, E. B. (1945). Polymorphism, *Biol. Rev.*, **20**, 73–88.

FORD, E. B. (1949). Genetics and Cancer, *Heredity*, **3**, 249–52.

FORD, E. B. (1950, 2nd edn.). *The Study of Heredity*, Home University Library, Oxford Press.

FORD, E. B. (1953*a*). The Experimental study of evolution, *Rep. Aust. Ass. Adv. Sci.*, **28**, 143–54.

FORD, E. B. (1953*b*). The genetics of polymorphism in the Lepidoptera, *Advanc. Genet.*, **5**, 43–87.

FORD, E. B. (1955*a*). A uniform notation for the human blood groups, *Heredity*, **9**, 135–42.

FORD, E. B. (1955*b*). Polymorphism and Taxonomy, *Heredity*, **9**, 255–64.

FORD, E. B. (1957). Polymorphism in plants, animals and man, *Nature*, **180**, 1315–19.

FORD, E. B. (1958). Darwinism and the study of evolution in natural populations, *Proc. Linn. Soc. Lond., Zool.*, **44**, 41–8.

FORD, E. B. (1960*a*). Evolution in Progress from *Evolution after Darwin* (ed. Sol Tax), **1**, 181–96, The University of Chicago Press.

FORD, E. B. (1960b, 7th Edn.). *Mendelism and Evolution*, Methuen, London.

FORD, E. B. (1961). The Theory of Genetic Polymorphism, *Symposia R. ent. Soc. Lond.*, No. 1, 11–19.

FORD, E. B. (1962, 3rd Edn., reprinted). *Butterflies*, The New Naturalist Series, Collins, London.

FORD, E. B. (1965). *Genetic Polymorphism*, All Souls, Studies Faber & Faber, London.

FORD, E. B. (1967, 2nd. Edn.). *Moths*, Collins, London.

FORD, E. B. and HUXLEY, J. S. (1927). Mendelian genes and rates of development in *Gammarus chevreuxi*, *Brit. J. Exp. Biol.*, **5**, 112–34.

FORD, E. B. and SHEPPARD, P. M. (1969). The *Medionigra* polymorphism of *Panaxia dominula*, *Heredity*, **24**, 561–9.

FORD, H. D. and FORD, E. B. (1930). Fluctuation in numbers and its influence on variation in *Melitaea aurinia*, *Trans. R. ent. Soc. Lond.*, **78**, 345–51.

FORMAN, B., FORD, E. B., and MCWHIRTER, K. G. (1959). An evolutionary study of the butterfly *Maniola jurtina* in the north of Scotland, *Heredity*, **13**, 353–61.

FRASER, J. F. D., and ROTHSCHILD, THE HON. M. (1960a). Defence mechanisms in warningly-coloured moths and other insects, *XI Int. Kongr. fr. Entom. Wien, BIII*, (Symposium 4), 249–56.

FRAZER, J. F. D. and ROTHSCHILD, M. (1960b). Defence mechanisms in warningly coloured Moths and other insects, *Proc. 11th. Int. Cong. Entom.*, **3**, 249–56.

FRÉDERIC, J. (1961). Contribution à l'étude du caryotype chez le poulet, *Arch. Biol. (Paris)*, **72**, 185–209.

FRYDENBERG, O. and HØEGH-GULDBERG, O. (1966). The genetic difference between southern English *Aricia agestis* and Scottish *A. artaxerxes*, *Hereditas*, **56**, 145–58.

FRYER, J. C. F. (1913). An investigation by pedigree breeding into the polymorphism of *Papilio polytes*, Linn., *Phil. Trans.*, B, **204**, 227–54.

FRYER, J. C. F. (1928). Polymorphism in the moth *Acalla comariana*, *J. Genet.*, **20**, 157–78.

FUJINO, K. and KANG, T. (1968). Transferrin groups in Tunas, *Genetics*, **59**, 79–91.

GAUL, A. T. (1952). Audio Mimicry: an adjunct to colour mimicry, *Psyche*, **59**, 82–3.

GAUSE, G. F. and SMARAGDOVA, N. P. (1940). The decrease in weight and mortality in dextral and sinistral individuals of the snail *Frutici-cola lantzi*, *Amer. Nat.*, **74**, 568–72.

GEROULD, J. H. (1921). Blue-green caterpillars, *J. exp. Zool.*, **34**, 385–412.

GEROULD, J. H. (1923). Inheritance of white wing-color, a sex-limited variation in yellow Pierid butterflies, *Genetics*, **8**, 495–551.

GILLESPIE, J. H. and KOJIMA, K. (1968). The degree of polymorphism in enzymes . . . in two *Drosophila ananassae* populations, *Proc. Nat. Acad. Sci.*, **61**, 582–5.

GODFREY, M. J. (1925). The fertilisation of *Ophrys speculum*, *O. lutea* and *O. fusca*, *J. Bot., Lond.*, **63**, 33–40.

GOLDSCHMIDT, R. (1924). Erblichkeitsstudien an Schmetterlinger IV, *Z. indukt. Abstamm.-u. Vererblehre*, **34**, 229–44.

GOLDSCHMIDT, R. (1945). Mimetic polymorphism, a controversial chapter of Darwinism, *Quart. Rev. Biol.*, **20**, 147–64, 205–30.

GOLDSCHMIDT, R. and FISCHER, E. (1922). *Argynnis paphia-valesina*, ein fall geschlechtskontrollierter Vererbung bei Schmetterlingen, *Genetica*, **4**, 247–78.

GORDON, C. (1935). An experiment on a released population of *D. melanogaster*, *Amer. Nat.*, **69**, 381.

GREAVES, J. H. and AYERS, P. (1967). Heritable resistance to warfarin in Rats, *Nature*, **215**, 877–8.

GREAVES, J. H. and AYERS, P. (1969). Linkages between genes for coat colour and resistance to warfarin in *Rattus norvegicus*, *Nature*, **224**, 284–5.

GREGOR, J. W. (1938). Experimental Taxonomy, II, *New Phytol.*, **37**, 15–49.

GREGOR, J. W. (1939). Experimental Taxonomy, IV, *New Phytol.*, **38**, 293–322.

GREGORY, R. P. G. and BRADSHAW, A. D. (1965). Heavy metal tolerance in populations of *Agrostis tenuis* and other grasses, *New Phytol.*, **64**, 131–43.

GULICK, J. T. (1905). Evolution, Racial and Habitudinal, controlled by segregation, *Publ. Carneg. Inst.*, **25**, 1–269.

GUSTAFSSON, A. (1946). The effect of heterozygosity on Variability and Vigour, *Hereditas*, **32**, 263–86.

GUSTAFSSON, A., NYBOM, N., and VON WETTSTEIN, U. (1950). Chlorophyll factors and heterosis in Barley, *Hereditas*, **36**, 383–92.

HAASE, E. (1893). *Untersuchungen über die Mimicry* (Vol. 2), Nägele, Stuttgart.

HAGEN, D. W. (1967). Isolating mechanisms in Threespine Sticklebacks (*Gasterosteus*), *J. Fish Res. Bd. Canada*, **24**, 1637–92.

HALDANE, J. B. S. (1924). A mathematical theory of natural and artificial selection. *Trans. Camb. phil. Soc.*, **23**, 19–40.

HALDANE, J. B. S. (1930). A note on Fisher's theory of the origin of dominance, and on a correlation between dominance and linkage, *Amer. Nat.*, **64**, 87–90.

HALDANE, J. B. S. (1948). The theory of a cline, *J. Genet.*, **48**, 277–84.

HALDANE, J. B. S. (1953). Animal Populations and their Regulation, *New Biol.*, **15**, 9–24.

HALDANE, J. B. S. (1954). An exact test for randomness of mating, *J. Genet.*, **52**, 632–5.

HALDANE, J. B. S. (1956). The theory of selection for melanism in Lepidoptera. *Proc. roy. Soc.*, B, **145**, 303–6.

HALDANE, J. B. S. (1957). The cost of natural selection, *J. Genet.*, **55**, 511–24.

HAMBURGER, V. (1936). The larval development of reciprocal hybrids of *Triton taeniatus* (and *T. palmatus*) × *Triton cristatus*, *J. Exp. Zool.*, **73**, 319–373.

HARLAND, S. C. (1947). An alteration in gene-frequency in *Ricinus* due to climatic conditions, *Heredity*, **1**, 121–5.

HARRIS, H. (1966). Enzyme polymorphism in Man, *Proc. roy. Soc.* **B**, **164**, 298–310.

HARRISON, J. W. H. (1919–20). Genetical studies in the moths of the Geometrid Genus *Oporabia* (*Oporinia*) with a special consideration of melanism in the Lepidoptera, *J. Genet.*, **9**, 195–280.

HARRISON, J. W. H. (1928). A further induction of melanism in the Lepidopterous insect, *Selenia bilunaria* Esp., and its inheritance, *Proc. roy. Soc.*, B., **102**, 338–47.

HARRISON, J. W. H. (1932). The recent development of melanism in the larvae of certain species of Lepidoptera, with an account of its inheritance in *Selenia bilunaria* Esp., *Proc. roy. Soc.*, B, **111**, 188–200.

HARRISON, J. W. H. and GARRETT, F. C. (1926). The induction of melanism in the Lepidoptera and its subsequent inheritance, *Proc. roy. Soc.*, B, **99**, 241–63.

HASEBROËK, K. (1925). Untersuchungen zum problem des neuzeitlichen Melanismus der Schmetterlinge, *Fermentforschung*, 8, 199–226.

HAWKES, O. A. M. (1920). Observations on the life-history, biology and genetics of the lady-bird beetle, *Adalia bipunctata*, *Proc. Zool. Soc. Lond.*, 475–90.

HAWKES, O. A. M. (1927). The distribution of the Ladybird *Adalia bipunctata* L. (Coleoptera), *Ent. mon. Mag.*, **63**, 262–6.

HEED, W. B. and BLAKE, P. R. (1963). A new color allele at the *e* locus of *Drosophila polymorpha* from Northern South America, *Genetics*, **48**, 217–34.

HØEGH-GULDBERG, O. (1968). Evolutionary trends in the Genus *Aricia* (Lep.), *Natura Jutlandica*, **14**, 3–76.

HØEGH-GULDBERG, O. and JARVIS, F. V. L. (1969). *Central and North European Ariciae (Lep.)*, *Natura Jutlandica*, **15**, 1–119.

HOESTLANDT, H. (1955). Étude de populations de *Sphaeroma serratum* de long du littoral de la Grande-Bretagne, *C. R. Acad. Sci.*, **240** (1), 916–19.

HOVANITZ, W. (1944*a*). Genetic data on the two races of *Colias chrysotheme* in North America and on a white form occurring in each, *Genetics*, **29**, 1–31.

HOVANITZ, W. (1944*b*). The distribution of gene frequencies in wild populations of *Colias*, *Genetics*, **29**, 31–60.

HOVANITZ, W. (1944*c*). The ecological significance of the colour phases of *Colias chrysotheme* in North America, *Ecology*, **25**, 45–60.

HOVANITZ, W. (1945). The Distribution of *Colias* in the equatorial Andes, *Caldasia (Bogota)*, **13**, 283–300.

HOVANITZ, W. (1948). Differences in the field activity of two female color phases of *Colias* butterflies at various times of the day, *Contr. Lab. Vertebr. Biol. Univ. Mich.*, **41**, 1–37.

HUBBY, J. L. and LEWONTIN, R. C. (1966). A molecular approach to the study of genic heterozygosity in natural populations, *Genetics*, **54**, 577–94.

HUBBY, J. L. and THROCKMORTON, L. H. (1968). Protein differences in *Drosophila*. IV. A study of sibling species, *Am. Nat.*, **102**, 193–205.

HUGHES, A. W. MCKENNY (1932). Induced melanism in the Lepidoptera, *Proc. roy. Soc.*, B, **110**, 378–402.

HUXLEY, J. S. (1942). *Evolution, The Modern Synthesis*, Allen and Unwin, London.

HUXLEY, J. S. (1955). Morphism and Evolution, *Heredity*, **9**, 1–52.

HUXLEY, J. S. (1956). Morphism as a clue in the study of population dynamics, *Proc. roy. Soc.*, B, **145**, 319–22.

INGLES, L. G. and BIGLIONE, N. J. (1952). The contiguity of the ranges of two subspecies of Pocket Gophers, *Evolution*, **6**, 204–7.

JACKSON, C. H. N. (1936). The use of the 'recovery index' in estimating the true density of Tsetse Flies, *Trans. R. ent. Soc. Lond.*, **84**, (Appendix 3), 530–2.

JACOBSON, E. (1909). Beobachtungen uber den polymorphismus von *Papilio memnon* L., *Tijdschr. Ent.*, **52**, 125–57.

JACOBSON, E. (1910). Corrigenda zu Beobachtungen über den Polymorphismus von *Pap. memnon*, *Tijdschr. voor Entomolog.*, **53**, 195.

JAIN, S. K. and BRADSHAW, A. D. (1966). Evolutionary divergence among adjacent plant populations, *Heredity*, **21**, 407–41.

JARVIS, F. V. L. (1963). The genetic relationship between *Aricia agestis* and its ssp. *artaxerxes*, *Proc. S. Lond. Ent. Nat. Hist. Soc.*, 106–22.

JOHNSON, F. M. and BURNS, J. M. (1966). Electrophoretic variation in esterases of *Colias eurytheme* (*Pieridae*), *J. Lepidopterists' Soc.*, **20**, 207–11.

JOHNSON, F. M., KANAPI, C. G., RICHARDSON, R. H., WHEELER, M. R., and STONE, W. S. (1966). An analysis of polymorphisms among isozyme loci in dark and light *Dros. ananassae* strains from American and Western Samoa, *Proc. Nat. Acad. Sci.*, **56**, 119–25.

JOHNSON, M. P., KEITH, A. D., and EHRLICH, P. R. (1968). The population biology of the butterfly *Euphydryas editha*. VII. Has *E. editha* evolved a serpentine race?, *Evolution*, **22**, 422–3.

JOHNSTON, R. F. and SELANDER, R. K. (1964). House Sparrows: Rapid evolution of races in North America, *Science*, **144**, 548–50.

JONES, D. A. (1962). Selective eating of the acyanogenic form of the plant *Lotus corniculatus* L. by various animals, *Nature*, **193**, 1109–10.

JONES, D. A. (1966). On the polymorphism of cyanogenesis in *Lotus corniculatus*. Selection by animals, *Can. J. Genet. Cytol.*, **8**, 556–67.

JONES, D. A. (1967). Polymorphism, plants and natural populations, *Sci. Prog. Oxford*, **55**, 379–400.

JONES, D. A., PARSONS, J., and ROTHSCHILD, THE HON. M. (1962). Release of hydrocyanic acid from crushed tissues of all stages in the life-cycle of species of the Zygaenidae (Lepidoptera), *Nature*, **193**, 52–3.

KALMUS, H. (1942). Differences in resistence to toxic substances shown by different body-colour mutants in *Drosophila*, *Proc. roy. Ent. Soc.*, **17**, 127–33.

KETTLEWELL, H. B. D. (1942). A survey of the insect *Panaxia* (*Callimorpha*) *dominula*, *Trans. S. Lond. ent. nat. Hist. Soc.* (1942–3), 1–49.

KETTLEWELL, H. B. D. (1944). Temperature effects on the pupae of *Panaxia dominula*, *Proc. S. Lond. ent. nat. Hist. Soc.* (1943–4), 79–81.

REFERENCES

KETTLEWELL, H. B. D. (1952). Use of Radioactive Tracer in the study of insect populations (Lepidoptera), *Nature*, **170**, 584.

KETTLEWELL, H. B. D. (1955*a*). Labelling locusts with radioactive isotopes, *Nature*, **175**, 821–2.

KETTLEWELL, H. B. D. (1955*b*). Recognition of appropriate backgrounds by the pale and black phase of Lepidoptera, *Nature*, **175**, 934.

KETTLEWELL, H. B. D. (1955*c*). Selection experiments on industrial melanism in the Lepidoptera, *Heredity*, **9**, 323–42.

KETTLEWELL, H. B. D. (1956). Further selection experiments on industrial melanism in the Lepidoptera, *Heredity*, **10**, 287–301.

KETTLEWELL, H. B. D. (1957*a*). Problems in industrial melanism, *Entomologist*, **90**, 98–105.

KETTLEWELL, H. B. D. (1957*b*). The contribution of industrial melanism in the Lepidoptera to our knowledge of evolution, *Adv. Sci.*, **52**, 245–52.

KETTLEWELL, H. B. D. (1957*c*). Industrial melanism in moths and its contribution to our knowledge of evolution, *Proc. roy. Instn, G.B.*, **36**, 1–14.

KETTLEWELL, H. B. D. (1958*a*). A survey of the frequencies of *Biston betularia* (L) (Lep.) and its melanic forms in Great Britain, *Heredity*, **12**, 51–72.

KETTLEWELL, H. B. D. (1958*b*). Industrial melanism in the Lepidoptera and its contribution to our knowledge of evolution, *Proc. 10th Int. Congr. Ent.* (1956), **2**, 831–41.

KETTLEWELL, H. B. D. (1959). New aspects of the genetic control of industrial melanism in the Lepidoptera, *Nature*, **183**, 918–21.

KETTLEWELL, H. B. D. (1961*a*). The phenomenon of industrial melanism in the Lepidoptera, *Ann. Rev. Ent.*, **6**, 245–62.

KETTLEWELL, H. B. D. (1961*b*). Geographic melanism in the Lepidoptera of Shetland, *Heredity*, **16**, 393–402.

KETTLEWELL, H. B. D. (1961*c*). Selection experiments on melanism in *Amathes glareosa* Esp. (Lepidoptera), *Heredity*, **16**, 415–34.

KETTLEWELL, H. B. D. (1965). Insect survival and selection for pattern, *Science*, **148**, 1290–6.

KETTLEWELL, H. B. D. and BERRY, R. J. (1961). The study of a cline, *Heredity*, **16**, 403–14.

KETTLEWELL, H. B. D. and BERRY, R. J. (1969*a*). Gene flow in a cline, *Heredity*, **24**, 1–14.

KETTLEWELL, H. B. D.; BERRY, R. J.; CADBURY, C. J. and PHILLIPS, G.

c. (1969b). Differences in behaviour, dominance and survival within a cline, *Heredity*, 24, 15–25.

KIMURA, M. (1968). Evolutionary Rate at the Molecular Level, *Nature*, 217, 624–6.

KING, J. L. and JUKES, T. H. (1969). Non-Darwinian Evolution, *Science*, 164, 788–98.

KIRBY, D. R. S., BILLINGTON, W. D., BRADBURY, S., and GOLDSTEIN, D. J. (1964). Antigen barrier of the mouse placenta, *Nature*, 204, 548–9.

KOEHN, R. K. (1969). Esterase Heterogeneity: Dynamics of a Polymorphism, *Science*, 163, 943–4.

KOEHN, R. K. and RASMUSSEN, D. I. (1967). Polymorphic and monomorphic serum esterase heterogeneity in Catostonid Fish populations, *Biochem. Genetics*, 1, 131–44.

KOJIMA, K. and TOBARI, Y. N. (1969). Selective modes associated with karyotypes in *Drosophila ananassae*. *Genetics*, 63, 639–51.

KOJIMA, K. and YARBOROUGH, K. M. (1967). Frequency dependent Selection at the Esterase-6 locus in *Drosophila melanogaster*, *Proc. U.S. Nat. Acad. Sci.*, 57, 645–9.

KOLLER, P. C. (1937). The Genetical and Mechanical properties of the Sex-chromosomes. III Man, *Proc. roy. Soc. Edinb.*, 57, 194–214.

KOMAI, TAKU (1956). Genetics of Ladybeetles, *Advanc. Genet.*, 8, 155–88.

KOMAI, T., CHINO, M., and HOSINO, Y. (1950). Contributions to the genetics of the Ladybeetle, *Harmonia*, I, *Genetics*, 35, 589–601.

LACK, D. (1954). *The Natural Regulation of Animal Numbers*, Oxford.

LACK, D. (1965). Evolutionary ecology, *J. Animal Ecol.*, 34, 223–31.

LAMBOURNE, W. A. (1912). *Hypolimnas (Euralia) anthedon*, Boisd., and *dubia*, Beauv., *Proc. ent. Soc. Lond.*, lxxvii-lxxviii.

LAMOTTE, M. (1951). Recherches sur la structure génétique des populations naturelles de *Cepaea nemoralis* L., *Bull. Biol. Suppl.*, 35, 1–239.

LAMOTTE, M. (1954). Sur le determinisme génétique du polymorphisme chez *Cepaea nemoralis* L., *C.R. Acad. Sci.*, 239, 365–7.

LAMOTTE, M. (1959). Polymorphism of natural populations of *Cepaea nemoralis*, *Cold Spr. Harb. Symp. quant. Biol.*, 24, 65–86.

LAMOTTE, M. (1966). Les facteurs de la diversité du polymorphisme dans les populations naturelles de *Cepaea nemoralis*, *Lavori della Società Malacologica Italiana*, 3, 33–73.

LANE, C. (1956). Preliminary notes on insects eaten and rejected by a tame Shama (*Kittacincla malabarica* Gm.), with the suggestion that in certain species of butterflies and moths females are less palatable than males, *Ent. mon. Mag.*, **93**, 172–9.

LANE, C. (1957). Notes on the brush organs and cervical glands of the Ruby Tiger (*Phragmatobia fuliginosa* L.), *Entomologist*, **90**, 148–51.

LANE, C. (1962). Notes on the Common Blue (*Polyommatus icarus*) egg-laying and feeding on the cyanogenic strains of the Bird's-foot Trefoil (*Lotus corniculatus*), *Entom. Gaz.*, **13**, 112–16.

LANE, C. and ROTHSCHILD, THE HON. M. (1959). A very toxic moth: the Five-spot Burnet (*Zygaena trifolii* Esp.), *Ent. mon. Mag.*, **95**, 93–4.

LANE, C. and ROTHSCHILD, THE HON. M. (1961). Observations on colonies of the Narrow-bordered Five-spot Burnet (*Zygaena lonicerae* von. Schev.) near Bicester, *Entomologist*, **94**, 79–81.

LEES, D. R. (1968). Genetic control of the melanic form *insularia* of the Peppered Moth *Biston betularia*, *Nature*, **220**, 1249–50.

LEES, D. R. (1970). The *medionigra* polymorphism of *Panaxia dominula* in 1969, *Heredity*, **25**, 470–5.

LEIGH, G. F. (1904). Synepigonic series of *Papilio cenea* (1902–3) and *Hypoliminas misippus* (1904), together with observations on the life-history of the former, *Trans. R. ent. Soc. Lond.*, 677–94.

LEIGH, G. F. (1906). Notes on *Euralia wahlbergi*, Wallgr., and *E. mima*, Trim., *Proc. ent. Soc. Lond.*, lii–lvii.

LEJUEZ, R. (1966). Comparison morphologique, biologique et génétique des quelques espèces genre *Sphaeroma* (Isopodes), *Arch. de Zool. Exp. et Gen.*, 407–668.

LERNER, I. M. (1954). *Genetic Homeostasis*, John Wiley, New York.

LESLIE, P. H., CHITTY, D. and H. (1953). An estimation of population parameters from data obtained by means of the capture-recapture method, *Biometrika*, **40**, 137–69.

LEVENE, H., PAVLOVSKY, O., and DOBZHANSKY, TH. (1954). Inter-action of the adaptive values in polymorphic experimental populations of *Drosophila pseudoobscura*, *Evolution*, **8**, 335–49.

LEWIS, D. (1954). Comparative incompatibility in Angiosperms and Fungi, *Advanc. Genet.*, **6**, 235–85.

LEWIS, H. and RAVEN, P. H. (1958). Rapid Evolution in *Clarkia*, *Evolution*, **12**, 319–36.

LEWIS, K. R. and JOHN, B. (1957). Studies on *Periplaneta americana*. II. Interchange heterozygosity in isolated populations, *Heredity*, **11**, 11–22.

LEWONTIN, R. C. and HUBBY, J. L. (1966). The amount of variation and

ECOLOGICAL GENETICS

degree of heterozygosity in natural populations of *Dros. psuedoobscura*, *Genetics*, **54**, 595–609.

LEWONTIN, R. C. and WHITE, M. J. D. (1960). Interaction between inversion polymorphisms of two chromosome pairs in the Grasshopper *Moraba scurra*, *Evolution*, **14**, 116–29.

LINCOLN, F. C. (1930). Calculating waterfowl abundance on the basis of banding returns, *Circ. U.S. Dep. Agric.*, No. 118.

LORKOVIC, Z. (1962). The genetics and reproductive isolating mechanisms of the *Pieris napi-bryoniae* Group, *J. Lep. Soc.*, **16**, 5–19, 105–27.

LORKOVIC, Z. and HERMAN, C. (1961). The solution of a long outstanding problem in the genetics of dimorphism in *Colias*, *J. Lep. Soc.*, **15**, 43–55.

LOWTHER, J. K. (1961). Polymorphism in the White-throated Sparrow, *Zonotrichia albicollis*, *Can. J. Zool.*, **39**, 281–92.

LUSIS, J. (1961). On the biological meaning of colour polymorphism of Lady-beetle *Adalia bipunctata*, *Latvijas Entomologs*, **4**, 3–29 (Russian with English summary).

MACLULICH, D. A. (1937). Fluctuations in the numbers of the Varying Hare (*Lepus americanus*), *Univ. Toronto Stud. biol.*, **43**, 1–136.

MCLAREN, I. A. (1967). Seals and Group Selection, *Evolution*, **48**, 104–10.

MCNEILLY, T. S. (1968). Evolution in closely adjacent plant populations, *Heredity*, **23**, 99–108.

MCNEILLY, T. S. and ANTONOVICS, J. (1968). Evolution in closely adjacent plant populations, *Heredity*, **23**, 205–18.

MCPHAIL, J. D. (1969). Predation and evolution of a Stickleback (*Gasterosteus*), *J. Fisheries Research Board, Canada*, **26**, 3183–3208.

MCWHIRTER, K. G. (1957). A further analysis of variability in *Maniola jurtina* L., *Heredity*, **11**, 359–71.

MCWHIRTER, K. G. (1967). Quantum genetics of human blood-groups and phoneme-preferences, *Heredity*, **22**, 162–3.

MCWHIRTER, K. G. (1969). Heritability of spot-number in Scillonian strains of the Meadow Brown Butterfly (*Maniola jurtina*), *Heredity*, **24**, 314–18.

MCWHIRTER, K. G. and SCALI, V. (1966). Ecological bacteriology of the Meadow Brown Butterfly, *Heredity*, **21**, 517–21.

MAGNUS, D. B. E. (1958). Experimental analysis of some 'over optimal' sign-stimuli in the mating-behaviour of the Fritillary butterfly *Argynnis paphia* L. (Lepidoptera. Nymphalidae), *Proc. 10th Int. Congr. Ent.*, **2**, 405–18.

386

REFERENCES

MAKINO, SAJIRO (1951). *An Atlas of the Chromosome Numbers in Animals*, Iowa State College Press.

MARRINER, T. F. (1926). A hybrid Coccinellid, *Ent. Rec.*, **38**, 81–3.

MARSDEN-JONES, E. B. and TURRILL, W. B. (1957). *The Bladder Campions* (pp. 378), Ray Society Publ., London.

MATHER, K. (1950). The genetical architecture of heterostyly in *Primula sinensis*, *Evolution*, **4**, 340–52.

MATHER, K. (1955). Polymorphism as an outcome of disruptive selection, *Evolution*, **9**, 52–61.

MATHER, K. and HARRISON, B. J. (1949a). The manifold effect of selection Part 1, *Heredity*, **3**, 1–52.

MATHER, K. and HARRISON, B. J. (1949b). The manifold effect of selection Part 2, *Heredity*, **3**, 131–62.

MATTHEWS, L. HARRISON (1952). *British Mammals*, New Naturalist Series, Collins, London.

MAYR, E. (1947, 3rd printing). *Systematics and the Origin of Species*, Columbia University Press, New York.

MAYR, E. (1954). Change of genetic environment and evolution, in *Evolution as a Process* (ed. Huxley, Hardy, and Ford), 157–80, Allen and Unwin, London.

MEDAWAR, P. (1960). *The Future of Man*, Methuen, London.

MEIJERE, J. C. H. DE (1910). Uber Jacobsons Züchtungsversuche bezüglich sekundärer Geschlechtsmerkmale, *Z. nid. Abst. u. Vererbunyslehere*, **3**, 161–81.

MILKMAN, R. D. (1967). Heterosis as a major cause of heterozygosity in nature, *Genetics*, **55**, 493–5.

MILNE H. and ROBERTSON, F. W. (1965). Polymorphism in egg albumen protein and behaviour in the Eider Duck, *Nature*, **205**, 367–9.

MOREAU, R. E. (1930). On the age of some races of birds, *Ibis* (Series **12**), **6**, 229–39.

MORGAN, T. H. (1927). *Experimental Embryology*, Columbia Univ. Press, N.Y.

MORGAN, T. H. (1929). The Variability of Eyeless, *Publ. Carneg. Inst.*, **399**, 139–68.

MORRELL, G. M. and TURNER, J. R. G. (1970). The response of wild birds to artificial prey, *Behaviour*, **36**, 116–30.

MORTON JONES, F. (1932). Insect coloration and the relative acceptability of insects to birds. *Trans. R. ent. Soc. Lond.*, **82**, 345–85.

MORTON JONES, F. (1934). Further experiments on coloration and relative acceptability of insects to birds, *Trans. R. ent. Soc. Lond.*, **82**, 443–53.

MOURANT, A. E. (1954). *The Distribution of the Human Blood Groups*, Blackwell Sci. Publications, Oxford.

MOURANT, A. E., KOPEĆ, A. C., and DOMANIEWSKA-SOBEZAK (1958). *The ABO Blood Groups: comprehensive tables and maps of world distribution*. Blackwell, Oxford.

MUHLMANN, H. (1934). Im Modellversuch künstlich erzeugte Mimikry und ihre Bedeutung für den Nachahmer, *Z. Morph. Okol. Tiere.*, **28** 259–96.

MÜLLER, F. (1878). [Notes on Brazilian entomology], *Trans. Ent. Soc. Lond.*, Pt. 3, 211–23.

MULLER, H. J. (1950). Our load of mutations, *Amer. J. hum. Genet.*, **2**, 111–76.

MULLER, H. J. (1958). The mutation theory re-examined, *Proc. 10th Int. Congr. Genet.*, 306–17.

MURRAY, J. (1962). Factors affecting gene-frequencies in some populations of *Cepaea*, *Thesis for the D.Phil. Degree*, deposited in the Bodleian Library, Oxford.

MURRAY, J. (1963). The inheritance of some characters in *Cepaea hortensis* and *Cepaea nemoralis* (Gastropoda). *Genetics*, **48**, 605–15.

MURRAY, J. and CLARKE, B. (1966). The inheritance of polymorphic shell characters in Partula (Gastropoda), *Genetics*, **54**, 1261–77.

NABOURS, R. K. (1929). The genetics of the Tettigidae (Grouse Locusts), *Bibliogr. genet.*, **5**, 27–104.

NABOURS, R. K., LARSON, I., and HARTWIG, N. (1933). Inheritance of color-patterns in the grouse locust *Acrydium arenosum* Burmeister (Tettigidae), *Genetics*, **18**, 159–71.

NEWMAN, E. (1869). *An Illustrated Natural History of British Moths*, pp. viii and 486; William Glaisher, London.

NEWTON, A. (1893). *A Dictionary of Birds*, Part 1, Black, London.

NICHOLSON, A. J. and BAILEY, V. A. (1935). The balance of Animal Populations, *Proc. Zool. Soc. Lond.*, 551–98.

OHNO, S., WEILER, C., POOLE, J., CHRISTIAN, L., and STENIUS, C. (1966). Autosomal polymorphism due to pericentric inversions in the Deer Mouse (*Peromyscus maniculatus*) . . ., *Chromosoma*, **18**, 177–87.

ONSLOW, THE HON. H. (1921). Inheritance of wing-colour in Lepidoptera, VI, *J. Genet.*, **11**, 277–92.

O'REILLY, R. A. and AGGELER, P. M. (1965). Coumarin anticoagulant drugs: hereditary resistance in Man, *Fed. Proc.*, **24**, 1266–73.

OWEN, D. F. (1961). Industrial melanism in North American moths, *Amer. Nat.*, **95**, 227–33.

OWEN, D. F. (1965). Density effects in polymorphic land snails, *Heredity*, **20**, 312–15.

OWEN, J. J. (1965). Karyotype studies on *Gallus domesticus*, *Chromosoma*, **16**, 601–8.

PARSONS, J. A. (1965). A digitalis-like toxin in the Monarch Butterfly, *Danaus plexippus*, *J. Physiol. Lond.*, **178**, 290–304.

PARSONS, P. A. (1958). Selection for increased recombination in *Drosophila melanogaster*, *Amer. Nat.*, **92**, 255–6.

PAVAN, C., DOBZHANSKY, TH., and DA CUNHA, A. B. (1957). Heterosis and elimination of weak homozygotes in natural populations of three related species of *Drosophila*, *Proc. nat. Acad. Sci.*, *Wash.*, **43**, 226–34.

PETERSEN, B. (1963). Breakdown of differentiation between *Pieris napi* and *Pieris bryoniae* and its causes, *Zool. bidrag Uppsala.*, **35**, 205–62.

PLATT, A. P. and BROWER, L. P. (1968). Mimetic versus disruptive colouration in integrating populations of *Limenitis arthemis* and *astyanax* butterflies, *Evolution*, **22**, 699–718.

PLATT, E. E. (1914). A large family of *Hypolimnas* (*Euralia*) *mima*, Trim., and *wahlbergi*, Wallgr., bred from known parents of the *wahlbergi* form at Durban, *Proc. R. ent. Soc. Lond.*, lxx–lxxv.

POST, R. H. (1962). Population differences in red and green color vision deficiency, *Eugen. Quart.*, **9**, 131–46.

POULTON, E. B. (1890). *The Colours of Animals, their meaning and use especially considered in the case of insects* (International Science Series Vol. 68), Kegan Paul, Trench, Trübner, London.

POULTON, E. B. (1924). The relation between the larvae of the Asilid genus *Hyperechia* (Laphriinae) and those of Xylocopid bees, *Trans. R. ent. Soc. Lond.*, 1924, 121–3.

PUNNETT, R. C. (1915). *Mimicry in Butterflies*, Cambridge, England.

PUNNETT, R. C. (1933). Inheritance of egg-colour in the 'parasitic' Cuckoo, *Nature*, **132**, 892–3.

RACE, R. R. and SANGER, R. (1958, 3rd Edn.). *Blood Groups in Man*, Blackwell, Oxford.

REICHSTEIN, T. (1967). Cardenolide (herzwirksame Glykoside) als Abwehrstoffe bei Insekten, *Naturwissenschaftliche Rundschau*, **20**, 499–511. (December). (The following co-authors were omitted from the title-page by the editor: ROTHSCHILD, THE HON. M., BROWER, L., FISHELSON, L., PARSONS, J. A., EUW, J. VON).

REICHSTEIN, T., EUW, J. VON, PARSONS, J. A., and ROTHSCHILD, THE HON. M. (1968). Heart poisons in the Monarch Butterfly, *Science*, **161**, 861–66 (August).

REMINGTON, C. L. (1954). The genetics of *Colias* (Lepidoptera), *Advanc. Genet.*, **6**, 403–50.

REMINGTON, C. L. (1958). Genetics of populations of Lepidoptera, *Proc. 10th Int. Congr. Ent.*, **2**, 787–805.

RENDEL, J. M. (1951). Mating of ebony, vestigal and wild type *Drosophila melanogaster* in light and dark, *Evolution*, **5**, 226–30.

RICHMOND, R. C. (1970). Non-Darwinian Evolution: A Critique, *Nature*, **225**, 1025–8.

RILEY, H. P. (1938). A character analysis of colonies of *Iris fulva* var. *giganticaerulea* and natural hybrids, *Amer. J. Bot.*, **25**, 727–38.

ROTHSCHILD, THE HON. M. (1961). Defensive odours and Müllerian mimicry among insects, *Trans. R. ent. Soc. Lond.*, **113**, 101–21.

ROTHSCHILD, THE HON. M. (1963). Is the Buff Ermine a mimic of the White Ermine?, *Proc. R. ent. Soc. Lond.*, Series A, **38**, 159–64.

ROTHSCHILD, THE HON. M. (1967). Mimicry, the deceptive way of life, *Natural History*, **76**, 44–51. (February).

ROTHSCHILD, THE HON. M. (1970). Les papillons qui se deguisent, *Science et Vie*, **117**, 88–158 (with intruded material).

ROTHSCHILD, THE HON. M. and CLAY, T. (1952). *Fleas, Flukes and Cuckoos*, Collins, London.

ROTHSCHILD, THE HON. M., EUW, J. VON, and REICHSTEIN, T. (1970a). Cardiac glycosides in the Oleander Aphid, *Aphis nerii*, *J. Insect Physiol.*, **16**, 1141–5.

ROTHSCHILD, THE HON. M., FORD, B., and HUGHES, M. (1969). Maturation of the male rabbit flea (*Spilopsyllus cuniculi*) and the oriental rat flea (*Xenopsylla cheopsis*): some effects of Mammalian hormones on development and impregnation, *Trans. zool. Soc. Lond.*, **32**, 105–88.

ROTHSCHILD, THE HON. M., REICHSTEIN, T., EUW, J. VON., and APLIN, R. (1970b). Toxic Lepidoptera, *Toxicon*, **8**, (2) (in press).

ROTHSCHILD, THE HON. M., REICHSTEIN, T., PARSONS, J. and APLIN, R. (1966). Poisons in aposematic Insects, *Descriptive Catalogue*, *Royal Society Conversazione*, May 1966, No. 19, p. 10.

RUITER, L. DE (1952). Some experiments on the camouflage of stick caterpillars, *Behaviour*, **4**, 222–32.

SCHULTE, A. (1952). Makrolepidopterologischr Sammeltage in Schwedisch-Lappland (Juni-Juli 1951), *Ent. Z.*, **61**, 169–74.

REFERENCES

SCHWERDTFEGER, F. (1941). Uber die Ursachen des Massenwechsels der Insekten, *Zeit. angew. Entom.*, **28**, 254–303.

SCHWERDTFEGER, F. (1950). *Grundriss der Forstpathologie*, Berlin.

SEDLMAIR, H. (1956). Verhaltense-, Resistenz- und Gehauseunterschiede bei den polymorphen Bänderschnecken *Cepaea hortensis* (Müll.) und *Cepaea nemoralis* (L.), *Biol.Zbl.*, **75**, 281–313.

SELANDER, R. K., HUNT, W. G., and YANG, S. Y. (1969). Protein polymorphism and genic heterozygosity in two European Subspecies of the House Mouse, *Evolution*, **23**, 379–90.

SEMEONOFF, R. and ROBERTSON, F. W. (1968). A biochemical and ecological study of plasma esterase polymorphism in natural populations of the Field Vole, *Microtus agrestis*, *Biochem. Genetics*, **1**, 205–27.

SHEPPARD, P. M. (1951*a*). A quantitative study of two populations of the moth *Panaxia dominula* (L.), *Heredity*, **5**, 349–78.

SHEPPARD, P. M. (1951*b*). Fluctuations in the selective value of certain phenotypes in the polymorphic land snail *Cepaea nemoralis* (L.), *Heredity*, **5**, 125–34.

SHEPPARD, P. M. (1952*a*). A note on non-random mating in the moth *Panaxia dominula* (L.), *Heredity*, **6**, 239–41.

SHEPPARD, P. M. (1952*b*). Natural selection in two colonies of the polymorphic land snail *Cepaea nemoralis*, *Heredity*, **6**, 233–8.

SHEPPARD, P. M. (1953). Polymorphism and population studies, *Symp. Soc. exp. Biol.*, **7**, 274–89.

SHEPPARD, P. M. (1954). Evolution in Bisexually reproducing organisms, in *Evolution as a Process*, pp. 201–18 (ed. Huxley, Hardy, and Ford), Allen and Unwin, London.

SHEPPARD, P. M. (1956). Ecology and its bearing on population genetics, *Proc. roy. Soc.*, B, **145**, 308–15.

SHEPPARD, P. M. (1958). *Natural Selection and Heredity*, Hutchinson, London.

SHEPPARD, P. M. (1959). The evolution of mimicry; a problem in ecology and genetics, *Cold Spr. Harb. Symp. quant. Biol.*, **24**, 131–40.

SHEPPARD, P. M. (1961*a*). Some contributions to population genetics resulting from the study of the Lepidoptera, *Advanc. Genet.*, **10**, 165–216.

SHEPPARD, P. M. (1961*b*). Recent genetical work on polymorphic mimetic *Papilios*, *Symposia R. ent. Soc.*, No. 1, 23–30.

SHEPPARD, P. M. (1963). Some genetic studies of Müllerian mimics in butterflies of the genus *Heliconius*, *Zoologica, N.Y.* **48**, 145–54.

SHEPPARD, P. M. (1969). Evolutionary genetics of animal populations: The study of natural populations, *Proc. XII int. Congr. Genetics*, **3**, 261–79.

SHEPPARD, P. M. and COOK, L. M. (1962). The manifold effects of the *medionigra* gene in the moth *Panaxia dominula* and the maintenance of a polymorphism, *Heredity*, **17**, 415–26.

SKALINSKA, M. (1950). Studies in cyto-ecology, geographic distribution and evolution of *Valeriana*, *Bull. Acad. Pol.*, *Series*, B, (**1**), 149–75.

SMITH, R. A. H. and BRADSHAW, A. D. (1970). Reclamation of toxic metalliferous wastes using tolerant populations of grass, *Nature*, **227**, 376–7.

SOBEY, W. R. (1969). Selection for resistance to myxomatosis in domestic rabbits, *J. Hygiene*, **67**, 743–54.

SOBEY, W. R., CONOLLY, D., HAYCOCK, P., and EDMONDS, J. W. (1970). Myxomatosis. The effect of age upon survival of wild and domestic rabbits with a genetic resistance and unselected domestic rabbits infected with myxoma virus, *J. Hygiene*, **68**, 137–49.

SOUTHERN, H. N. (1945). Polymorphism in *Poephila gouldiae*, *J. Genet.*, **47**, 51–7.

SOUTHERN, H. N. (1954). Mimicry in Cuckoo's Eggs, in *Evolution as a Process*, pp. 219–32 (edited by Huxley, Hardy, and Ford), Allen and Unwin, London.

SPIESS, E. B. (1957). Relation between frequencies and adaptive values of chromosomal arrangements in *Drosophila persimilis*, *Evolution*, **11**, 84–93.

SPIESS, E. B. (1958). Chromosomal adaptive polymorphism in *Drosophila persimilis*, II, *Evolution*, **12**, 234–45.

STAIGER, H. (1954). Der chromosomendimorphismus Bein Prosobranchier *Purpura lapillus* in Beziehung zur Ökologie der Art, *Chromosoma*, **6**, 419–78.

STALKER, H. D. and CARSON, H. L. (1947). Morphological variation in natural populations of *Drosophila robusta*, *Evolution*, **1**, 237–48.

STALKER, H. D. and CARSON, H. L. (1948). An altitudinal transect of *Drosophila robusta*, *Evolution*, **2**, 295–305.

STANDFUSS, M. (1896, 2nd. edn.). *Handbuch der paläarktischen Gross-Schmetterlinge*, G. Fischer, Jena.

STEHR, G. (1959). Hemolymph polymorphism in a moth and the nature of sex-controlled inheritance, *Evolution*, **13**, 537–60.

STRESEMANN, E. (1927–34). *Aves in Handbuch der Zoologie* (Kükenthal and Krumback, Berlin and Leipzig).

REFERENCES

STRUTHERS, D. (1951). ABO groups of infants and children dying in the West of Scotland (1949–51), *Brit. J. preventative soc. Med.*, **5**, 223–8.

STURTEVANT, A. H. and DOBZHANSKY, TH. (1936). Inversions in the third chromosome of wild races of *Drosophila pseudoobscura*, and their use in the study of the history of the species, *Proc. nat. Acad. Sci., Wash.*, **22**, 448–50.

SUFFERT, F. (1924). Bestimmungsfaktoren des Zeichnungsmusters beim Saison – Dimorphismus von *Araschnia levana-prorsa*, *Biol. zent.*, **44**, 173–88.

SUMNER, F. B. (1926). An analysis of geographic variation in mice of the *Peromyscus polionotus* group from Florida and Alabama, *J. Mammal.*, **7**, 149–84.

SUMNER, F. B. (1930). Genetic and distributional studies of three subspecies of *Peromyscus*, *J. Genet.*, **23**, 275–376.

SUOMALAINEN, E. (1938). Die erblichkeitsverhaltnisse des männlichen dimorphismus bei *Parasemia plantaginis*, *Hereditas*, **24**, 386–90.

SVED, J. A. (1968). Possible rates of gene substitution in evolution, *Am. Nat.*, **102**, 283–93.

TAN, C. C. (1946). Mosaic dominance in the inheritance of color-patterns in the lady-bird beetle *Harmonia axyridis*, *Genetics*, **31**, 195–210.

TAYLOR, J. W. (1907–14). *Monograph of the Land and Freshwater Mollusca of the British Isles*, Taylor, Leeds.

TEMPLADO, J. (1961). El mimetismo batesiano de *Paranthrene tabaniformis*, *Bol. R. Soc. esp. Hist. nat.* (B), **59**, 109–22.

THODAY, J. M. and BOAM, T. B. (1959). Effects of disruptive selection, 2. Polymorphism and divergence without isolation, *Heredity*, **13**, 205–18.

THOMSEN, M. and LEMCHE, H. (1933). Experimente zur Erzielung eines erblichen melanismus bei dem Spanner *Selenia bilunaria* Esp., *Biol. Zbl.*, **53**, 541–60.

THORNEYCROFT, H. B. (1966). Chromosomal polymorphism in the White-throated Sparrow, *Zonotrichia albicollis*, *Science*, **154**, 1571–2.

TIMOFEEFF-RESSOVSKY, N. W. (1940). Zur analyse des polymorphismus bei *Adalia bipunctata* L., *Biol. Zbl.*, **60**, 130–7.

TURNER, J. R. G. (1965). Evolution of complex polymorphism and mimicry in distasteful South American butterflies, *Int. Congr. Entom.*, **12**, 267.

TURNER, J. R. G. (1967a). Why does the genotype not congeal?, *Evolution*, **21**, 645–56.

TURNER, J. R. G. (1967b). The evolution of super-genes, *Am. Nat.*, **101**, 195–221.

TURNER, J. R. G. (1968a). The ecological genetics of *Acleris comariana* (Lepidoptera: Tortricidae), *J. Anim. Ecol.*, **37**, 489–520.

TURNER, J. R. G. (1968b). Some new *Heliconius* pupae: their taxonomic and evolutionary significance in relation to mimicry (Lepidoptera, Nymphalidae), *J. Zool., Lond.*, **155**, 311–25.

TURNER, J. R. G., CLARKE, C. A., and SHEPPARD, P. M. (1961). Genetics of a difference in the male genitalia of East and West African stocks of *Papilio dardanus* (Lep.), *Nature*, **191**, 935–6.

TURNER, J. R. G. and CRANE, J. (1962). The genetics of some polymorphic forms of the butterflies *Heliconius melpomene* and *H. erato*, *Zoologica, N.Y.*, **47**, 141–52.

TURNER, J. R. G. and WILLIAMSON, M. H. (1968). Population size, natural selection and the genetic load, *Nature*, **218**, 700.

VAUGHAN, H. E. N. and VAUGHAN, J. A. (1968). Some aspects of epizootiology of myxomatosis, *Symp. zool. Soc. Lond.*, **24**, 289–309.

VAURIE, C. (1951). Adaptive differences between two sympatric species of Nuthatches (*Sitta*), *Proc. Xth. Int. Ornith. Congr.*, Uppsala 1950, 163–6.

VERITY, R. (1950). Le Farfalle Diurne d'Italia, 4; Morzocco, Firenze.

WADDINGTON, C. H. (1953). Genetic assimilation of an acquired character, *Evolution*, **7**, 118–26.

WADDINGTON, C. H. (1957). *The Strategy of the Genes*, Allen and Unwin, London.

WAHRMAN, J., GOITEIN, R., and NEVO, E. (1969). Mole Rat *Spalax*: Evolutionary Significance of Chromosome Variation, *Science*, **164**, 82–4.

WALKER, M. F. (1966). Some observations on the behaviour and life history of . . . *Euplagia quadripunctaria* (Lepidoptera) in . . . Rhodes, *Entomologist*, **99**, 1–24.

WEATHERALL, D. J. (1969). The genetics of the Thalassaemias, *Brit. Med. Bull.*, **25**, (No. 1, New Aspects of Human Genetics), 24–9.

WELCH, D'A. A. (1938). Distribution and variation of *Achatinella mustelina* in the Waianae Mountains, Oahu, *Berenice P. Bishop Mus. Bull.*, **152**, 1–164.

WEST, D. A. (1964). Polymorphism in the Isopod *Sphaeroma rugicauda*, *Evolution*, **18**, 671–84.

WICKLER, W. (1968). *Mimicry in Plants and Animals* (translated by MARTIN, R. D.), World University Library.

WILLIAMS, H. B. (1946–7). *Angerona prunaria* L: its variation and genetics, *Proc. S. Lond. ent. nat. Hist. Soc.*, 123–39.

WILLIAMS, H. B. (1950). *Boarmia repandata* L. and ab. *conversaria* Hb., *Entom. Gaz.*, **1**, 36–9.

WILLIAMSON, D. L. and EHRMAN, L. (1967). Induction of hybrid sterility in nonhybrid males of *Drosophila paulistorum. Genetics*, **55**, 131–40.

WILLIAMSON, M. H. (1960). On the polymorphism of the moth *Panaxia dominula, Heredity*, **15**, 139–51.

WINGE, Ø. (1932). The nature of sex chromosomes, *Proc. 6th Int. Congr. Genet.*, 343–55.

WOLDA, H. (1969). Stability of a steep cline in morph frequencies of the snail *Cepaea nemoralis, J. anim. Ecol.*, **38**, 623–33.

WOOD, R. G. (1967). A comparative genetical study on D.D.T. resistance in adults and larvae of the mosquito *Aedes aegypti, Genet. Res., Camb.*, **10**, 219–28.

WOODELL, S. R. J. (1960). What pollinates Primulas?, *New Scientist*, **8**, 568–71.

WOOLF, B. (1954). On estimating the relation between blood groups and disease, *Ann. Hum. Genet.*, **19**, 251–3.

WORLD HEALTH ORGANIZATION (1963). Insecticide resistance and vector control, *Tech. Rept.*, 265, pp. 227.

WRIGHT, SEWALL (1931). Evolution in Mendelian populations, *Genetics*, **16**, 97–159.

WRIGHT, SEWALL (1932). The roles of mutation, inbreeding, cross-breeding, and selection in evolution, *Proc. 6th Int. Congr. Genet.*, 356–66.

WRIGHT, SEWALL (1940). The statistical consequences of Mendelian heredity in relation to speciation, in *The New Systematics* (ed. Julian Huxley), 161–83, Oxford.

WRIGHT, SEWALL (1948). On the roles of directed and random changes in gene frequency in the genetics of populations, *Evolution*, **2**, 279–94.

Index

397